Molecular Mechanisms Associated with Plant Tolerance upon Abiotic Stress

Molecular Mechanisms Associated with Plant Tolerance upon Abiotic Stress

Guest Editor

Emilia Apostolova

MDPI

Basel • Beijing • Wuhan • Barcelona • Belgrade • Novi Sad • Cluj • Manchester

Guest Editor
Emilia Apostolova
Institute of Biophysics and
Biomedical Engineering
Bulgarian Academy of Sciences
Sofia
Bulgaria

Editorial Office
MDPI AG
Grosspeteranlage 5
4052 Basel, Switzerland

This is a reprint of the Special Issue, published open access by the journal *Plants* (ISSN 2223-7747), freely accessible at: www.mdpi.com/journal/plants/special_issues/4Q8400B3N6.

For citation purposes, cite each article independently as indicated on the article page online and using the guide below:

Lastname, A.A.; Lastname, B.B. Article Title. *Journal Name* **Year**, *Volume Number*, Page Range.

ISBN 978-3-7258-3168-5 (Hbk)
ISBN 978-3-7258-3167-8 (PDF)
https://doi.org/10.3390/books978-3-7258-3167-8

Contents

About the Editor

Emilia Apostolova

Prof. Emilia Apostolova works at the Institute of Biophysics and Biomedical Engineering, Bulgarian Academy of Sciences. She studies the influence of abiotic stress factors on plants and the mechanisms of their adaptation to adverse environmental changes.

Preface

The important processes of plants are influenced by adverse environmental factors, which can have a negative impact on their growth and development. The last decade has seen an increase in the impact of abiotic stresses on plants due to climate changes. The impact of abiotic stress on plants and their defense mechanisms is presented in this Special Issue titled "Molecular Mechanisms Associated with Plant Tolerance upon Abiotic Stress". The studies within this Special Issue enhance our understanding of how biotic factors affect plants and their defense mechanisms. The important processes of plants are influenced by adverse environmental factors, which can have a negative impact on their growth and development. The past decade has seen an increase in the impact of abiotic stresses on plants due to climate changes.

Emilia Apostolova
Guest Editor

plants

Editorial

Molecular Mechanisms Associated with Plant Tolerance upon Abiotic Stress

Emilia L. Apostolova

Institute of Biophysics and Biomedical Engineering, Bulgarian Academy of Sciences, Acad. G. Bonchev Str., Bl. 21, 1113 Sofia, Bulgaria; emya@bio21.bas.bg or emilia.apostolova@gmail.com

Abstract: The important processes of plants are influenced by adverse environmental factors, which can have a negative impact on their growth and development. The last decade has seen an increase in the impact of abiotic stress on plants due to climate changes. The impact of abiotic stress on plants and their defense mechanisms is presented in the Special Issue "Molecular Mechanisms Associated with Plant Tolerance upon Abiotic Stress". The studies enhance our understanding of how abiotic factors affect plants and their defense mechanisms.

Citation: Apostolova, E.L. Molecular Mechanisms Associated with Plant Tolerance upon Abiotic Stress. *Plants* **2024**, *13*, 3532. https://doi.org/10.3390/plants13243532

Received: 29 November 2024
Revised: 6 December 2024
Accepted: 9 December 2024
Published: 18 December 2024

1. Introduction

The abiotic stress factors of the environment, such as drought, salinity, extreme temperatures, and others, are significantly increased by climate change [1,2]. In nature, plants are exposed to multiple stress conditions, and adaptation to these stressors involves integrative responses to individual stressors and the formation of a new type of response [3]. The effects of abiotic stress factors vary depending on the type of stress, its duration, and also the plant species [4–7]. Furthermore, it has been shown that there are variations in how stressors are tolerated between different varieties of a specific species [8,9]. These stresses greatly affect photosynthesis in plants. The reduction in yield is directly related to the decrease in the photosynthetic capacity of plants due to environmental stress [10]. For adaptation to multiple stresses, it is very important to have a transcription factor that can bind to the promoters of downstream genes [11]. Zinc finger-homeodomain (ZF-HD) proteins are specific transcription factors in plants that play an important role in their growth, development, and stress responses [12]. A recent genome-wide analysis of ZF-HD genes in *Phyllostachys edulis* has identified the key candidate genes related to abiotic stress in this plant [13].

2. Salt Stress

Salinity causes osmotic stress and ion toxicity in plants, which leads to oxidative stress. The overproduction of reactive oxygen species (ROS) under salt stress damage nucleic acids, proteins, lipids, and other macromolecules [14,15]. The activity of ROS triggers the membrane lipid peroxidation, chlorophyll degradation, and changes in the membrane integrity and its fluidity [16,17], which leads to changes in the structural arrangement of chloroplasts and modifications of the pigment-protein complexes of the thylakoid membranes [8,18]. Recent research revealed the important role of the rearrangement of mitochondrial metabolism for the adjustment of plant cells under salt stress. It has been shown that the regulation of glutamate metabolism through epigenetic mechanisms involves coordinated changes in the activities of 2-oxoglutarate-dehydrogenase, glutamate dehydrogenase, and glutamate decarboxylase [19].

The study of the adaptive mechanisms of *Poa pratensis* L. in saline environments revealed an increased expression of genes associated with chlorophyll biosynthesis, components of the complex of photosystem II (PSII), the Calvin cycle, and those involved in the biosynthesis of ROS-scavenging enzymes [20]. The authors also revealed a key role of the accumulation of proline and betaine as well as the upregulation of genes involved in

Na^+/K^+ transport (such as HKT1;5, HAK5, and SKOR) in improving the salt tolerance of *Poa pratensis* L. [20].

The influence of salt stress on different plant species was used to assess the reasons/mechanisms for the salt tolerance of plants. Previous studies revealed that salt-induced changes in the photosynthetic apparatus decrease in the open PSII reaction centers, influence the electron flow from Q_A^- to plastoquinone, and diminish the effective quantum yield of PSII, which were more strongly influenced in plant-sensitive species [6,7]. Moreover, the increase in energy losses after high salt treatment of more tolerant plant species is a result of enhancement mainly in regulated energy losses, while in the sensitive species, non-regulated energy losses occur [6,7]. The study of the impact of salinity on photosynthetic pigments, ions, osmolytes, oxidative stress markers, and antioxidant metabolites, as well as antioxidant enzyme activities in the *Imperata cylindrica*, *Phragmites australis*, and *Saccharum ravennae*, showed different sensitivities of these plant species [21]. The authors concluded that the tolerance mechanism to salinity is based on the inhibition of the Na^+ and Cl^- transport from roots to the aerial part of the plant and the accumulation of proline and soluble sugars in the photosynthetic tissues. It has been suggested that the activation of K^+ transport in leaves under salt stress contributes to an increase in salt tolerance [21].

Nitric oxide (NO) plays a significant role in the plant response to abiotic stress [22]. Nitric oxide acts as an antioxidant against ROS under stress and also can act as a signal molecule [23]. Previous research showed that NO alleviated the salt-induced changes in growth, development, and functions of the plants [24,25]. Recently, it has been shown that NO prevents changes in the fluidity of thylakoid membranes, energy transfer between the pigment protein complexes of the photosynthetic apparatus, and modification of the Mn clusters of the oxygen-evolving complex in the donor side of the PSII complex [17]. The impact of NO on the acceptor side of PSII was also registered, influencing Q_A^- reoxidation. In plants treated with SNP under salt stress, Q_A^- interaction with plastoquinone is more dominant than the recombination of electrons in $Q_AQ_B^-$ with the oxidized S_2 (S_3) state of the oxygen-evolving complex. The impact of the NO on the donor and acceptor side of the PSII under salt stress corresponded with better PSII efficiency and alleviation of the photosynthetic performance under salt stress [17].

3. Drought Stress

Global food production is negatively impacted by drought stress, a natural climatic phenomenon [26]. Plant development and crop production depend heavily on photosynthetic performance, which is significantly impacted by this stress [27]. The drought-induced changes in cell metabolism correspond with the overproduction of ROS, which disrupts proteins, lipids, and cell membranes [28]. Previous studies showed the changes in the composition, organization, and functions of the complexes of the photosynthetic apparatus depend on the drought tolerance of plant species [29,30]. A recent study with drought-susceptible and drought-tolerant wheat varieties revealed higher osmolyte accumulation and left water content in the tolerant variety in comparison to the variety that is sensitive under drought stress [31]. Moreover, it has been shown to increase antioxidant enzyme activity, which decreases membrane damage under drought. Proteomic analysis showed that drought induced the differential expression of proteins involved in biological processes in both wheat varieties (drought-tolerant and drought-susceptible) [31]. It has been shown that tolerant variety reprogramming protein synthesis in order to obtain the proteins with regulatory and/or protection roles against the effects of low water content. Moloi et al. [31] suggested that the observed downregulation of the proteins involved in photosynthesis in tolerant varieties might be a protective mechanism against oxidative stress in plants under drought. Increased levels of stress markers and non-enzymatic compounds as well as the activities of antioxidative and xenobiotic-detoxifying enzymes in triticale (*Triticum* spp.) under drought were shown [32].

The use of biostimulants in agriculture, as a key factor in the stimulation of plant growth and the enhancement of crop production under physiological and stress conditions, has increased [33,34]. In recent years, there has been increasing interest in the effects of melatonin on plants to prevent adverse environmental effects. Melatonin plays an important role in the integration of various environmental signals and the formation of the plant stress response [35]. It regulates the expression of the photosynthetic genes, preserves the chlorophyll molecules and photosynthetic functions, and interacts with NO and ROS to regulate the plant redox network [36,37]. Foliar spraying of oregano (*Origanum vulgare* L.) plants with 100 μM melatonin under moderate drought stress ameliorates PSII photochemistry [38]. It is suggested that under these conditions, melatonin triggered a non-photochemical mechanism to decrease $^1O_2$ production, which resulted in an increase in the open PSII centers and an increase in the electron transport rate. The effect of melatonin under drought depends on the plant species and light intensity, suggesting research with different plant species before the extensive use of melatonin in agriculture [38].

4. Cold Stress

Cold stress has harmful effects on plant growth, development, and their yield. Plants develop different mechanisms to copy this stress and induce a cold tolerance from the cellular level to the whole plant level [39]. Researchers demonstrate that NAC genes play an important role in controlling the cold tolerance in plants [40,41]. The study of Yang et al. [42] identified a total of 110 NAC genes in *Dendrobium officinale*, which were classified in 15 subfamilies. In addition, cold-responsive DoNACs were validated.

The study of the impacts of low temperature and low light on the wild type and mutant (tangerine) of *Solanum lycopersicum* L. demonstrates that the oxidative stress in both types of plants is higher after combined treatment with low temperature and low light in comparison to the low light only [43]. The author also revealed the protective role of the phenolic components, including anthocyanins, against cold-induced ROS.

Funding: This research received no external funding.

Conflicts of Interest: The authors declare no conflicts of interest.

References

1. Singh, J.; Thakur, J.K. Photosynthesis and abiotic stress in plants. In *Biotic and Abiotic Stress Tolerance in Plants*; Vats, S., Ed.; Springer: Singapore, 2018. [CrossRef]
2. Kumari, A.; Lakshmi, G.A.; Krishna, G.K.; Patni, B.; Prakash, S.; Bhattacharyya, M.; Singh, S.K.; Verma, K.K. Climate change and its impact on crops: A comprehensive investigation for sustainable agriculture. *Agronomy* **2022**, *12*, 3008. [CrossRef]
3. Zandalinas, S.I.; Fichman, Y.; Devireddy, A.R.; Sengupta, S.; Rajeev, K.; Azad, R.K.; Mittle, R. Systemic signaling during abiotic stress combination in plants. *Proc. Natl. Acad. Sci. USA* **2020**, *117*, 13810–13820. [CrossRef] [PubMed]
4. Rossi, L.; Borghi, M.; Francini, A.; Lin, X.; Xie, D.-Y.; Sebastiani, L. Salt stress induces differential regulation of the phenylpropanoid pathway in Olea europaea cultivars Frantoio (salt-tolerant) and Leccino (salt-sensitive). *J. Plant Physiol.* **2016**, *204*, 8–15. [CrossRef]
5. Acosta-Motos, J.R.; Ortuño, M.F.; Bernal-Vicente, A.; Diaz-Vivancos, P.; Sanchez-Blanco, M.J.; Hernandez, J.A. Plant responses to salt stress: Adaptive mechanisms. *Agronomy* **2017**, *7*, 18. [CrossRef]
6. Stefanov, M.A.; Rashkov, G.D.; Yotsova, E.K.; Borisova, P.B.; Dobrikova, A.G.; Apostolova, E.L. Different sensitivity levels of the photosynthetic apparatus in *Zea mays* L. and *Sorghum bicolor* L. under salt stress. *Plants* **2021**, *10*, 1469. [CrossRef]
7. Stefanov, M.A.; Rashkov, G.D.; Apostolova, E.L. Assessment of the photosynthetic apparatus functions by chlorophyll fluorescence and P700 absorbance in C3 and C4 slants under physiological conditions and under salt stress. *Int. J. Mol. Sci.* **2022**, *23*, 3768. [CrossRef]
8. Zeeshan, M.; Lu, M.; Sehar, S.; Holford, P.; Wu, F. Comparison of biochemical, anatomical, morphological, and physiological responses to salinity stress in wheat and barley genotypes deferring in salinity tolerance. *Agronomy* **2020**, *10*, 127. [CrossRef]
9. Stefanov, M.A.; Rashkov, G.D.; Yotsova, E.K.; Borisova, P.B.; Dobrikova, A.G.; Apostolova, E.L. Protective effects of sodium nitroprusside on photosynthetic performance of *Sorghum bicolor* L. under salt stress. *Plants* **2023**, *12*, 832. [CrossRef] [PubMed]
10. Long, S.P.; Zhu, X.G.; Naidu, S.L.; Ort, D.R. Can improvement in photosynthesis increase crop yields? *Plant Cell Environ.* **2006**, *29*, 315–330. [CrossRef]
11. Amorim, L.L.B.; da Fonseca Dos Santos, R.; Neto, J.P.B.; Guida-Santos, M.; Crovella, S.; Benko-Iseppon, A.M. Transcription factors involved in plant resistance to pathogens. *Curr. Protein Pept. Sci.* **2017**, *18*, 335–351. [CrossRef] [PubMed]
12. Tan, Q.K.; Irish, V.F. The *Arabidopsis* zinc finger-homeodomain genes encode proteins with unique biochemical properties that are coordinately expressed during floral development. *Plant Physiol.* **2006**, *140*, 1095–1108. [CrossRef]

13. Huang, F.; Wang, J.; Tang, C. Genome-wide identification and analysis of ZF-HD gene family in moso bamboo (*Phyllostachys edulis*). *Plants* **2023**, *12*, 4064. [CrossRef] [PubMed]
14. Azeem, M.; Pirjan, K.; Qasim, M.; Mahmood, A.; Javed, T.; Muhammad, H.; Yang, S.; Dong, R.; Ali, B.; Rahimi, M. Salinity stress improves antioxidant potential by modulating physio-biochemical responses in *Moringa oleifera* Lam. *Sci. Rep.* **2023**, *13*, 2895. [CrossRef]
15. Kesawat, M.S.; Satheesh, N.; Kherawat, B.S.; Kumar, A.; Kim, H.-U.; Chung, S.-M.; Kumar, M. Regulation of reactive oxygen species during salt stress in plants and their crosstalk with other signaling molecules—Current perspectives and future directions. *Plants* **2023**, *12*, 864. [CrossRef] [PubMed]
16. Taïbi, K.; Taïbi, F.; Ait Abderrahim, L.; Ennajah, A.; Belkhodja, M.; Mulet, J.M. Effect of salt stress on growth, chlorophyll content, lipid peroxidation and antioxidant defence systems in *Phaseolus vulgaris* L. *S. Afr. J. Bot.* **2016**, *105*, 306–312. [CrossRef]
17. Rashkov, G.D.; Stefanov, M.A.; Yotsova, E.K.; Borisova, P.B.; Dobrikova, A.G.; Apostolova, E.L. Exploring nitric oxide as a regulator in salt tolerance: Insights into photosynthetic efficiency in maize. *Plants* **2024**, *13*, 1312. [CrossRef]
18. Liu, Z.; Zou, L.; Chen, C.; Zhao, H.; Yan, Y.; Wang, C.; Liu, X. iTRAQ-based quantitative proteomic analysis of salt stress in *Spica Prunellae*. *Sci. Rep.* **2019**, *9*, 9590. [CrossRef]
19. Eprintsev, A.T.; Anokhina, G.B.; Selivanova, P.S.; Moskvina, P.P.; Igamberdiev, A.U. Biochemical and epigenetic regulation of glutamate metabolism in maize (*Zea mays* L.) leaves under salt stress. *Plants* **2024**, *13*, 2651. [CrossRef]
20. Wang, R.; Yan, S.-J.; Liu, C.; Guo, H.; Cui, Y.-N. Comparative physiological and gene expression analyses reveal mechanisms involved in maintaining photosynthesis capacity, alleviating ion toxicity and oxidative stress of kentucky bluegrass under NaCl treatment. *Plants* **2024**, *13*, 2107. [CrossRef] [PubMed]
21. Sapiña-Solano, A.; Boscaiu, M.; Collado, F.; Vicente, O.; Ruiz-González, M.X. Effects of high salinity and water stress on wetland grasses from the Spanish Mediterranean coast. *Plants* **2024**, *13*, 1939. [CrossRef]
22. Nabi, R.B.S.; Tayade, R.; Hussain, A.; Kulkarni, K.P.; Imran, Q.M.; Mun, B.-G.; Yun, B.-W. Nitric oxide regulates plant responses to drought, salinity, and heavy metal stress. *Environ. Exp. Bot.* **2019**, *161*, 120–133. [CrossRef]
23. Ferreira, L.C.; Cataneo, A.C. Nitric oxide in plants: A brief discussion on this multifunctional molecule. *Sci. Agric.* **2010**, *67*, 236–243. [CrossRef]
24. Misra, A.N.; Vladkova, R.; Singh, R.; Misra, M.; Dobrikova, A.G.; Apostolova, E.L. Action and target sites of nitric oxide in chloroplasts. *Nitric Oxide—Biol. Chem.* **2014**, *39*, 35–45. [CrossRef] [PubMed]
25. Hajihashemi, S.; Skalicky, M.; Brestic, M.; Pavla, V. Cross-talk between nitric oxide, hydrogen peroxide and calcium in salt-stressed Chenopodium quinoa Willd. At seed germination stage. *Plant Physiol. Biochem.* **2020**, *154*, 657–664. [CrossRef]
26. Hussain, M.; Farooq, S.; Hasan, W.; Ul-Allah, S.; Tanveer, M.; Farooq, M.; Nawaz, A. Drought stress in sunflower: Physiological effects and its management through breeding and agronomic alternatives. *Agric. Water Manag.* **2018**, *201*, 152–166. [CrossRef]
27. Zhang, X.; Liu, W.; Lv, Y.; Li, T.; Tang, J.; Yang, X.; Bai, J.; Jin, X.; Zhou, H. Effects of drought stress during critical periods on the photosynthetic characteristics and production performance of naked oat (*Avena nuda* L.). *Sci. Rep.* **2022**, *12*, 11199. [CrossRef] [PubMed]
28. Huseynova, I.M.; Rustamova, S.M.; Suleymanov, S.Y.; Aliyeva, D.R.; Mammadov, A.C.; Aliyev, J.A. Drought-induced changes in photosynthetic apparatus and antioxidant components of wheat (*Triticum durum* Desf.) varieties. *Photosynth. Res.* **2016**, *130*, 215–223. [CrossRef] [PubMed]
29. Salehi-Lisar, S.Y.; Bakhshayeshan-Agdam, H. Drought stress in plants: Causes, Consequences, and tolerance. In *Drought Stress Tolerance in Plants*; Springer International Publishing: Cham, Switzerland, 2016; Volume 1, pp. 1–16.
30. Stefanov, M.; Rashkov, G.; Borisova, P.; Apostolova, E. Sensitivity of the photosynthetic apparatus in maize and sorghum under different drought levels. *Plants* **2023**, *12*, 1863. [CrossRef]
31. Moloi, S.J.; Alqarni, A.O.; Brown, A.P.; Goche, T.; Shargie, N.G.; Moloi, M.J.; Gokul, A.; Chivasa, S.; Ngara, R. Comparative physiological, biochemical, and leaf proteome responses of contrasting wheat varieties to drought stress. *Plants* **2024**, *13*, 2797. [CrossRef]
32. Katerova, Z.; Todorova, D.; Shopova, E.; Brankova, L.; Dimitrova, L.; Petrakova, M.; Sergiev, I. Biochemical alterations in triticale seedlings pretreated with selective herbicide and subjected to drought or waterlogging stress. *Plants* **2023**, *12*, 2803. [CrossRef] [PubMed]
33. Nephali, L.; Piater, L.A.; Dubery, I.A.; Patterson, V.; Huyser, J.; Burgess, K.; Tugizimana, F. Biostimulants for plant growth and mitigation of abiotic stresses: A metabolomics perspective. *Metabolites* **2020**, *10*, 505. [CrossRef] [PubMed]
34. Ma, Y.; Freitas, H.; Dias, M.C. Strategies and prospects for biostimulants to alleviate abiotic stress in plants. *Front. Plant Sci.* **2022**, *13*, 1024243. [CrossRef]
35. Khan, M.S.S.; Ahmed, S.; Ikram, A.U.; Hannan, F.; Yasin, M.U.; Wang, J.; Zhao, B.; Islam, F.; Chen, J. Phytomelatonin: A key regulator of redox and phytohormones signaling against biotic/abiotic stresses. *Redox Biol.* **2023**, *64*, 102805. [CrossRef]
36. Arnao, M.; Hernández-Ruiz, J. Melatonin and reactive oxygen and nitrogen species: A model for the plant redox network. *Melatonin Res.* **2019**, *2*, 152–168. [CrossRef]
37. Ahmad, I.; Song, X.; Hussein Ibrahim, M.E.; Jamal, Y.; Younas, M.U.; Zhu, G.; Zhou, G.; Adam Ali, A.Y. The role of melatonin in plant growth and metabolism, and its interplay with nitric oxide and auxin in plants under different types of abiotic stress. *Front. Plant Sci.* **2023**, *14*, 1108507. [CrossRef]

38. Moustaka, J.; Sperdouli, I.; İşgören, S.; Şaş, B.; Moustakas, M. Deciphering the mechanism of melatonin-induced enhancement of photosystem II function in moderate drought-stressed oregano Plants. *Plants* **2024**, *13*, 2590. [CrossRef]
39. Wu, J.; Nadeem, M.; Galagedara, L.; Thomas, R.; Cheema, M. Recent insights into cell responses to cold stress in plants: Signaling, defence, and potential functions of phosphatidic acid. *Environ. Exp. Bot.* **2022**, *203*, 105068. [CrossRef]
40. Puranik, S.; Sahu, P.P.; Srivastava, P.S.; Prasad, M. NAC proteins: Regulation and role in stress tolerance. *Trends Plant Sci.* **2012**, *17*, 369–381. [CrossRef] [PubMed]
41. Mao, X.; Zhang, H.; Qian, X.; Li, A.; Zhao, G.; Jing, R. TaNAC2, a NAC-type wheat transcription factor conferring enhanced multiple abiotic stress tolerances in Arabidopsis. *J. Exp. Bot.* **2012**, *63*, 2933–2946. [CrossRef]
42. Yang, Q.; Li, Z.; Wang, X.; Jiang, C.; Liu, F.; Nian, Y.; Fu, X.; Zhou, G.; Liu, L.; Wang, H. Genome-wide identification and characterization of the NAC gene family and its involvement in cold response in *Dendrobium officinale*. *Plants* **2023**, *12*, 3626. [CrossRef]
43. Popova, A.V.; Stefanov, M.; Mihailova, G.; Borisova, P.; Georgieva, K. Response of tomato plants, ailsa craig and carotenoid mutant tangerine, to simultaneous treatment by low light and low temperature. *Plants* **2024**, *13*, 1929. [CrossRef] [PubMed]

plants

MDPI

Article

Comparative Physiological, Biochemical, and Leaf Proteome Responses of Contrasting Wheat Varieties to Drought Stress

Sellwane J. Moloi [1], Ali O. Alqarni [2], Adrian P. Brown [2], Tatenda Goche [2,3], Nemera G. Shargie [4], Makoena J. Moloi [5], Arun Gokul [1], Stephen Chivasa [2] and Rudo Ngara [1,*]

1 Department of Plant Sciences, University of the Free State, Qwaqwa Campus, P. Bag X13, Phuthaditjhaba 9866, South Africa; moloisj@ufs.ac.za (S.J.M.); gokula@ufs.ac.za (A.G.)

2 Department of Biosciences, Durham University, South Road, Durham DH1 3LE, UK; ali.alqarni@durham.ac.uk (A.O.A.); a.p.brown@durham.ac.uk (A.P.B.); tatenda.goche@durham.ac.uk (T.G.); stephen.chivasa@durham.ac.uk (S.C.)

3 Department of Crop Science, Bindura University of Science Education, P. Bag 1020, Bindura, Zimbabwe

4 Agricultural Research Council-Grain Crops, P. Bag X1251, Potchefstroom 2520, South Africa; shargien@arc.agric.za

5 Department of Plant Sciences-Botany Division, University of the Free State, 205 Nelson Mandela Drive, Bloemfontein 9301, South Africa; moloimj@ufs.ac.za

* Correspondence: ngarar@ufs.ac.za; Tel.: +27-(0)58-718-5332

Abstract: Drought stress severely affects crop productivity and threatens food security. As current trends of global warming are predicted to exacerbate droughts, developing drought-resilient crops becomes urgent. Here, we used the drought-tolerant (BW35695) and drought-sensitive (BW4074) wheat varieties to investigate the physiological, biochemical, and leaf proteome responses underpinning drought tolerance. In response to drought, the tolerant variety had higher osmolyte accumulation and maintained higher leaf water content than the sensitive variety. BW35695 also had an enhanced antioxidant enzyme capacity and reduced reactive oxygen species (ROS), resulting in diminished membrane lipid damage, as reflected by malondialdehyde content. Proteomic analysis revealed that drought-induced differential expression of proteins involved in diverse biological processes in both wheat varieties, including primary and secondary metabolism, protein synthesis/folding/degradation, defense/ROS detoxification, energy, transcription, and cell structure. Notably, photosynthesis emerged as the most enriched biochemical process targeted for suppression in the drought-tolerant BW35695 wheat, but not in drought-sensitive BW4074, possibly as a survival strategy for averting cell damage inflicted by photosynthesis-derived ROS. Additionally, protein synthesis-related proteins were highly upregulated in BW35695, presumably to drive cell-wide stress-adaptive responses. The protein network identified here will be useful in further studies to understand the molecular basis for divergent drought response phenotypes in crops.

Keywords: *Triticum aestivum*; drought; proline; oxidative stress; lipid peroxidation; antioxidant enzymes; proteomics; iTRAQ; photosynthesis; gene expression

check for updates

Citation: Moloi, S.J.; Alqarni, A.O.; Brown, A.P.; Goche, T.; Shargie, N.G.; Moloi, M.J.; Gokul, A.; Chivasa, S.; Ngara, R. Comparative Physiological, Biochemical, and Leaf Proteome Responses of Contrasting Wheat Varieties to Drought Stress. *Plants* **2024**, *13*, 2797. https://doi.org/10.3390/plants13192797

Academic Editor: Emilia Apostolova

Received: 8 August 2024
Revised: 26 September 2024
Accepted: 1 October 2024
Published: 5 October 2024

1. Introduction

Drought stress threatens global crop production, exposing vulnerable populations to food insecurity [1]. In Sub-Saharan Africa, global warming is projected to intensify the prevalence of hot and dry spells, which are currently causing devastating yield losses, across the region [2,3]. Consequently, more drought-tolerant crops are required to meet the global food demand under the changing climate [4]. This has led to growing research interest in understanding plant responses to water deficit stress and identifying heritable traits for improved drought resilience [5,6].

Numerous reviews have highlighted the detrimental effects of drought stress on plant growth and development, as well as the complex stress-adaptive responses [7–10]. For example, inadequate water supply negatively affects various morpho-physiological and biochemical processes, including plant growth, reproduction, photosynthesis, respiration, and nutrient uptake and assimilation [11]. The disruption of normal cell metabolism may result in the overproduction and accumulation of reactive oxygen species (ROS) leading to oxidative stress. Oxidative stress, in turn, disrupts the structure and function of lipids, proteins, nucleic acids, and cell membranes, further compromising cellular homeostasis [12,13].

Plants utilize diverse mechanisms to mitigate the detrimental effects of water deficits on cell structure and function. Some plant genotypes avoid drought by enhancing their water-foraging capacity through extensive root systems or by reducing transpiration-dependent water loss via leaf rolling, thick cuticles, and increased stomatal control of transpiration [11]. Conversely, drought tolerance mechanisms enable plants to maintain normal physiological and metabolic functions under water deficit stress [11,14]. Such adaptive responses result from complex processes of stress perception, signal transduction, and alterations in gene expression aimed at restoring cellular homeostasis for survival [15,16]. Some drought-induced physiological and metabolic responses are modulated by the stress hormone abscisic acid (ABA), which accumulates in leaves and roots upon exposure to drought [17].

ABA-dependent and ABA-independent response pathways have been extensively described [18,19] and contribute towards drought tolerance by promoting the production of proteins and metabolites with regulatory and protective functions [15]. For instance, drought-induced metabolites, including sugars, proline, and glycine betaine, participate in osmotic adjustment and protective roles against the osmotic and oxidative effects of water scarcity [20,21]. Likewise, increased enzymatic and non-enzymatic antioxidant activities in plants under drought stress alleviate oxidative cell damage [13,22]. Nevertheless, the efficiency of these adaptive responses may vary depending on the plant species, genotype or developmental stage, and the duration and severity of the prevailing stress [23].

Wheat (*Triticum aestivum*) is one of the most widely grown and consumed cereals globally [24,25]. However, its production yields are negatively affected by drought stress [26,27]. In addition, the wheat germplasm is genetically diverse [28,29] and is grown under equally diverse climatic conditions, worldwide [30,31]. Therefore, comparative studies of wheat varieties with contrasting drought phenotypes are required for deeper insights into the crop's response to drought stress. Such studies could assist in identifying target genes for developing more drought-resistant wheat genotypes for improved food security. Comparative analyses of physiological and biochemical [32–37], transcriptomic [34,38,39], proteomic [37,40,41], and metabolomic [39,41,42] responses of wheat plants to drought stress have been reported. Results of such studies suggest that drought-tolerant wheat genotypes possess greater capacity to maintain cell structure and function, by increasing membrane stability, osmoregulation, and cell redox homeostasis under water-scarce conditions than their drought-sensitive counterparts [32,33,35–38,40–42]. Furthermore, systems biology studies [7,43,44] that integrate plant physiological analyses with molecular cell biochemistry using "omics" technologies are pivotal in unraveling the complex networks of adaptive responses of cells, organelles, and whole plants to drought. Some of the identified drought-responsive genes have been functionally validated using transgenic plants [45,46] and serve as promising candidates for improving the drought resilience of food crops [47].

Therefore, this study investigated the growth, physiological, biochemical, and leaf proteome responses of two contrasting wheat cultivars to drought stress. We hypothesized that contrasting wheat varieties utilize networks of divergent plant responses to water deficit stress, which ultimately determine their drought phenotypes.

2. Results

2.1. Growth and Physiological Responses of BW4074 and BW35695 Wheat Varieties towards Drought Stress

When BW4074 (drought-susceptible) and BW35695 (drought-tolerant) wheat seedlings were two weeks old, watering was stopped for 28 days to induce drought stress. During this period, the extent of soil drying in the water-deprived pots was estimated gravimetrically and presented as a percentage relative to the well-watered controls. The results showed that withholding water from the drought-treatment pots progressively dried the soil with time (Figure 1a). At 14 and 28 days after watering was stopped, the water-deprived soil contained between 76 and 77 and 53 and 59% of moisture relative to the well-watered controls, respectively (Figure 1a). However, there was no significant difference in soil moisture content between the two wheat varieties at each time point, suggesting comparable levels of water deficit stress.

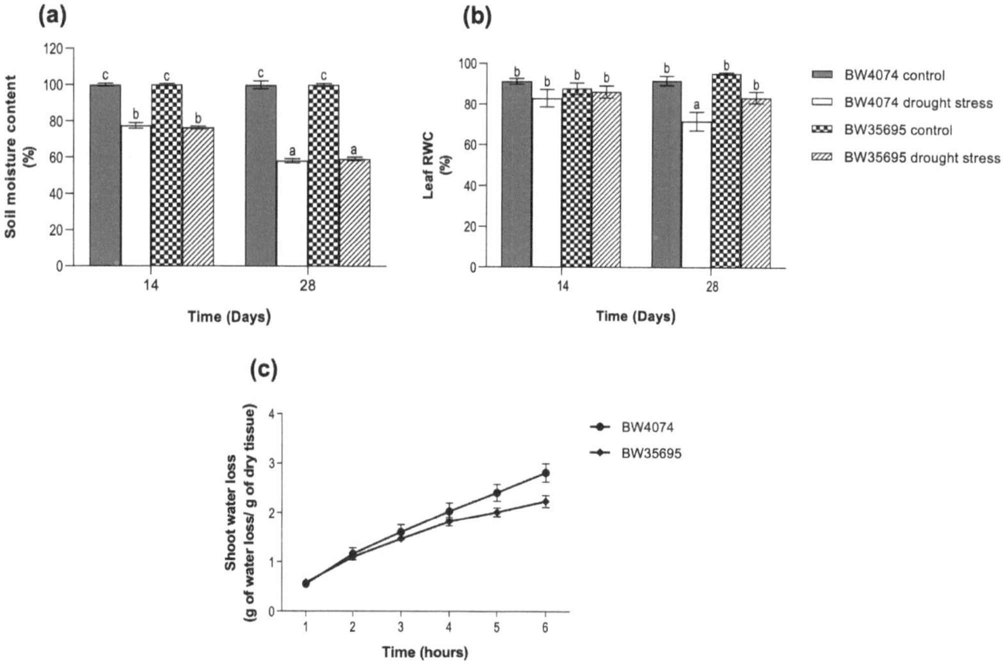

Figure 1. Soil moisture content, relative water content, and shoot water loss in potted wheat plants exposed to drought stress treatment. (**a**) Soil moisture content (n = 3); (**b**) leaf relative water content (RWC; n = 4); (**c**) relative shoot water loss (n = 5). Two-week old plants of BW4074 (drought-susceptible) and BW35695 (drought-tolerant) wheat varieties were exposed to drought stress by withholding water for 28 days and samples were harvested at the indicated time points for the respective measurements. Data presented as mean ± SE. Different letters indicate significant differences between means at ($p \leq 0.05$) according to ANOVA and Tukey–Kramer test.

We observed a decline in the leaf relative water content (RWC) of both wheat varieties only after 28 days of no watering (Figure 1b). However, the decrease was only statistically significant in the drought-susceptible variety BW4074 (Figure 1b). We also assessed the rate of water loss from detached shoots of both wheat varieties as a proxy for the plants' capacity to retain water under dehydration stress. The results indicated that the rate of water loss was not statistically different between the two varieties; however, BW4074 was more likely to lose water faster than BW35695 (Figure 1c). Plant growth measurements taken on day 28 of the drought stress treatment showed no significant differences in all parameters measured compared to the controls, except for the decline in shoot fresh weight observed in BW35695 (Figure S1). Taken together, our results suggest that upon exposure to similar levels of soil moisture deficit stress, the drought-tolerant wheat variety BW35695 has a greater water retention capacity compared to the drought-susceptible BW4074.

2.2. Drought Stress Increases the Levels of Photosynthetic Pigments Mainly in the Drought-Tolerant Wheat Variety

Chlorophyll A and B and carotenoid contents were evaluated in both wheat varieties during the 28 days of water deprivation. As mentioned in our methods description, chlorophyll and carotenoid content measurements were performed on days 0, 14, and 28 to investigate the effects of different levels of drought stress on the parameters and inform us on the possible harvest times for subsequent experiments on reactive oxygen species (ROS) and antioxidant enzymatic assays, and leaf proteome and gene expression analyses. When watering was stopped on day 0, both wheat varieties had similar amounts of photosynthetic pigments in leaves, and these levels remained unchanged between treatments and varieties on day 14 (Figure 2). However, as the duration of stress progressed to day 28, chlorophyll A and B and carotenoid levels increased by 92, 91, and 50%, respectively, in the water-deprived BW35695 plants relative to the well-watered controls (Figure 2a–c). In the drought-susceptible BW4074 variety, chlorophyll A and carotenoid contents remained unchanged following 28 days of drought stress (Figure 2a,c), while that of chlorophyll B marginally increased by 49% relative to the control (Figure 2b). These results suggest that BW35695 increased the biosynthesis and/or accumulation of photosynthetic pigments under drought conditions.

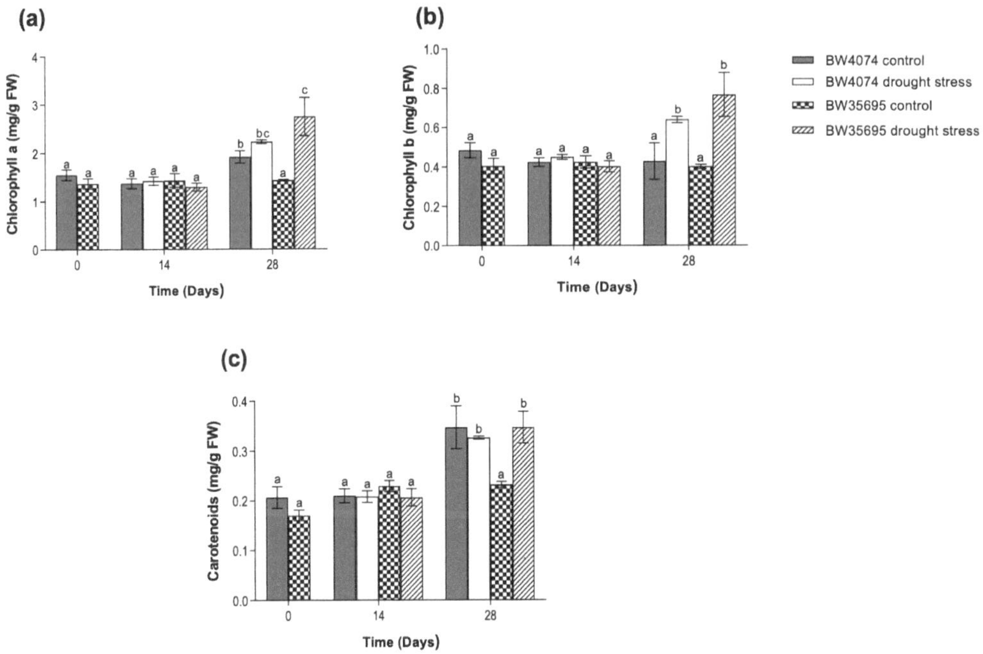

Figure 2. Effects of drought stress on chlorophyll and carotenoid content of wheat. (**a**) Chlorophyll A; (**b**) chlorophyll B; (**c**) carotenoid content. Two-week old plants of BW4074 (drought-susceptible) and BW35695 (drought-tolerant) wheat varieties were exposed to drought stress by withholding water for 28 days and samples were harvested at the indicated time points for the respective measurements. Data presented as mean ± SE (n = 3). Different letters indicate significant differences between means at ($p \leq 0.05$) according to ANOVA and Tukey–Kramer test.

2.3. Drought Stress Induces Oxidative Stress in Wheat Tissues

Since oxidative stress is a secondary stress of many environmental factors, including drought, we evaluated lipid peroxidation levels and ROS contents in leaves and roots of both wheat varieties following 28 days of drought stress (Figure 3). In this study, malondialdehyde (MDA) content was used as a proxy for lipid peroxidation, while ROS contents were used to assess the extent of oxidative stress. Compared to their respective well-watered controls, the drought-stressed BW4074 plants exhibited a 118 and 172% increase in leaf and root MDA content, respectively (Figure 3a,b). For the drought-tolerant BW35695, the MDA content increased by a modest 64% and 60% in leaves and roots following drought stress, respectively, and this increase was only statistically significant in the root tissue (Figure 3a,b).

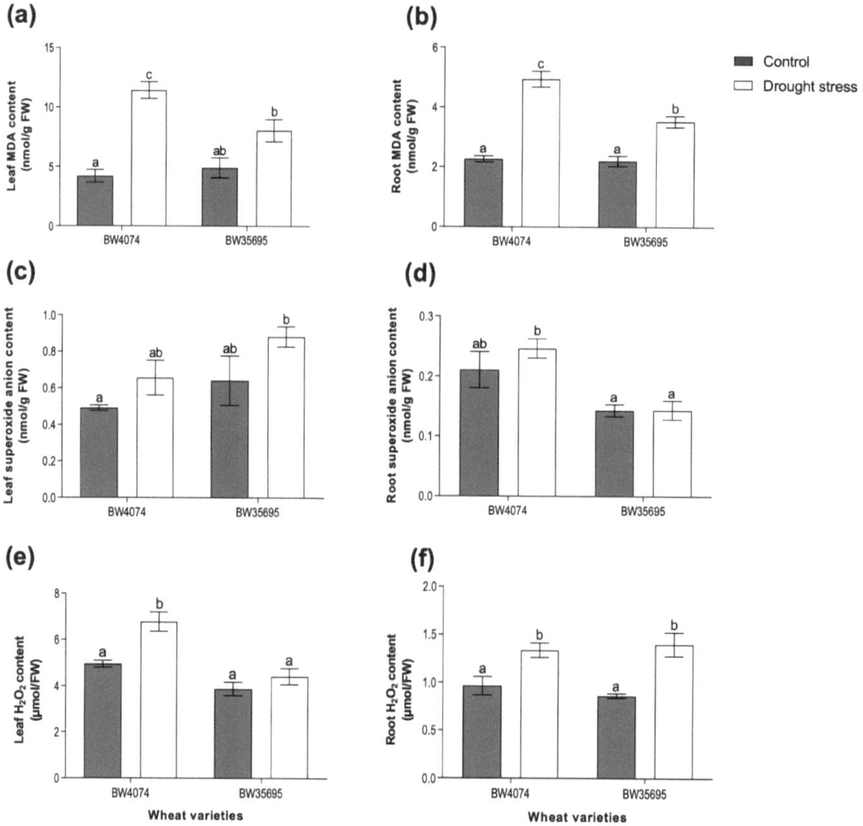

Figure 3. Effects of drought stress on lipid peroxidation and ROS accumulation in wheat plants. Malondialdehyde (MDA) content was used as a proxy for lipid peroxidation, while H_2O_2 and superoxide (O_2^-) content analyses were used to assess the extent of oxidative stress. Levels of (**a**) leaf MDA; (**b**) root MDA; (**c**) leaf O_2^-; (**d**) root O_2^-; (**e**) leaf H_2O_2; (**f**) root H_2O_2. Two-week old plants of BW4074 (drought-susceptible) and BW35695 (drought-tolerant) wheat varieties were exposed to drought stress by withholding water for 28 days for the various measurements. Data presented as mean ± SE (n = 3). Different letters indicate significant differences between means at ($p \leq 0.05$) according to ANOVA and Tukey–Kramer test.

Although an upward trend in leaf superoxide anion (O_2^-) content was observed in both wheat varieties following 28 days of drought stress, the increase was not significantly different from the respective controls (Figure 3c). Similarly, the root superoxide content of both varieties remained unchanged following drought stress (Figure 3d). Hydrogen peroxide (H_2O_2) content increased by 38% and 62% in the roots of BW4074 and BW35695, respectively, following 28 days of drought stress (Figure 3f). In the leaves, however, only BW4074 drought-stressed plants exhibited a 37% significant increase in H_2O_2, while that of BW35695 remained unchanged (Figure 3e). Taken together, results from the lipid peroxidation and ROS content analyses suggest that the 28 days of drought stress induced oxidative stress in both wheat varieties; however, the extent of oxidative damage was more pronounced in the drought-susceptible variety BW4074 compared to BW35695.

2.4. Drought-Induced Accumulation of Osmoprotectants in Wheat Leaf and Root Tissues

Leaf and root tissue samples were harvested on days 0, 14, and 28 of drought stress treatment to analyze the differential accumulation of the osmoprotectants, proline and glycine betaine, in both wheat varieties. On days 0 and 14 of the stress treatment, no significant changes in the levels of both osmolytes were observed between treatments within and across the wheat varieties (Figure 4a–d). However, on day 28 of the drought stress treatment, leaf proline levels massively increased only in water-deprived BW35695 plants relative to the control but remained unaltered in the drought-susceptible BW4074 (Figure 4a). In contrast, both proline and glycine betaine accumulated in the leaves and roots of the two wheat varieties on day 28 of drought stress (Figure 4b–d). However, no significant differences were observed in the levels of each osmolyte between similar tissues of the two wheat varieties. We also observed that both osmoprotectants were more abundant in the leaves than the roots (Figure 4). Collectively, our results suggest tissue-specific and varietal differences in the accumulation of proline and glycine betaine under the experimental conditions of this study.

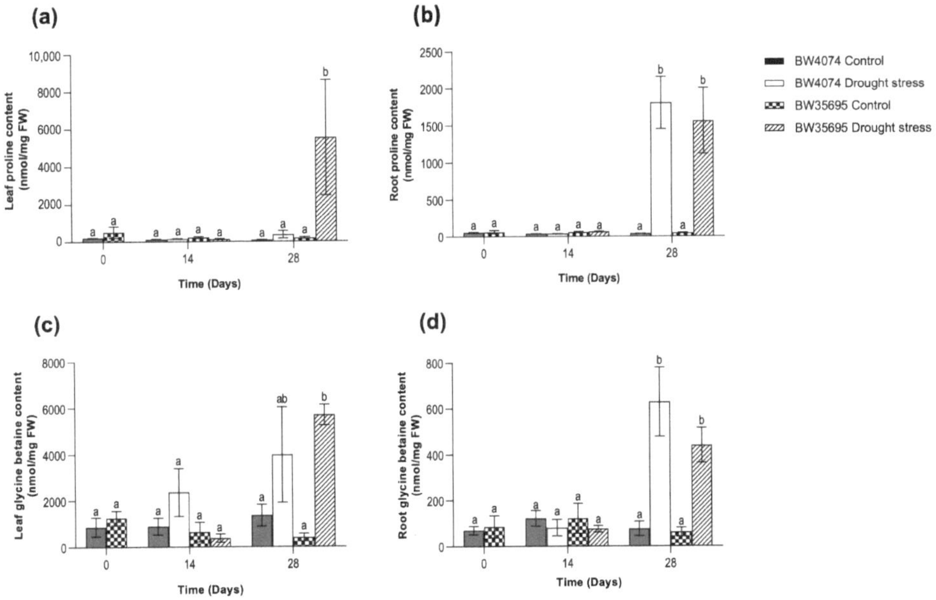

Figure 4. Effects of drought stress on proline and glycine betaine content of wheat plants. Levels of (**a**) leaf proline; (**b**) root proline; (**c**) leaf glycine betaine content; (**d**) root glycine betaine. Two-week old plants of BW4074 (drought-susceptible) and BW35695 (drought-tolerant) wheat varieties were exposed to drought stress by withholding water for 28 days and samples were harvested at the indicated time points for the respective measurements. Data presented as mean ± SE (n = 3). Different letters indicate significant differences between means at ($p \leq 0.05$) according to ANOVA and Tukey–Kramer test.

2.5. The Drought-Tolerant Wheat Variety BW35695 Possesses Greater Antioxidant Enzyme Activities Compared to Drought-Susceptible BW4074 under Drought Stress

Following the observed drought-induced oxidative stress and oxidative damage in wheat leaves and roots (Figure 3), we evaluated the antioxidant capacity of the plants to scavenge for ROS. This was achieved by measuring the activities of selected antioxidant enzymes in the leaf and root tissues of both wheat varieties following 28 days of drought stress (Figure 5). A 191% increase in the activity of superoxide dismutase (SOD) was observed in BW35695 drought-stressed leaves relative to the controls (Figure 5a), while that of the roots remained unchanged (Figure 5b). Conversely, a downward trend in SOD activity was observed in the BW4074 drought-stressed plant leaves and roots relative to the controls, with a 42% significant decrease observed only in the roots (Figure 5a,b).

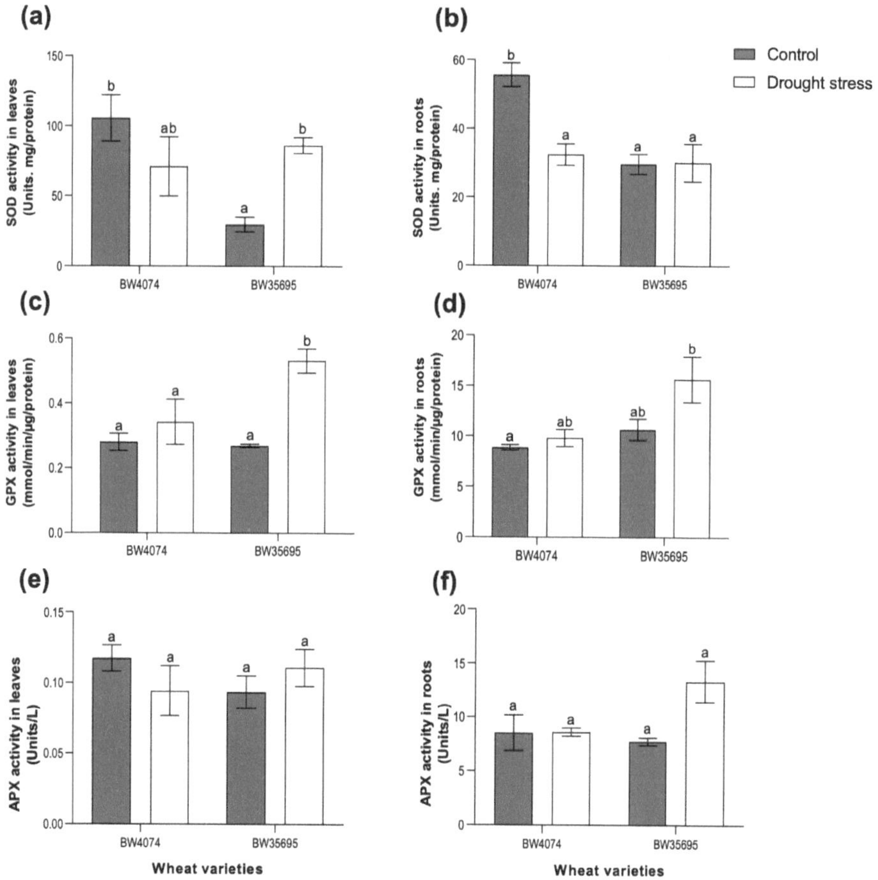

Figure 5. Effects of drought stress on antioxidant enzyme activities of wheat plants. Enzymatic activities of (**a**) leaf superoxide dismutase (SOD); (**b**) root SOD; (**c**) leaf guaiacol peroxidase (GPX); (**d**) root (GPX); (**e**) leaf ascorbate peroxidase (APX); and (**f**) root APX. Two-week old plants of BW4074 (drought-susceptible) and BW35695 (drought-tolerant) wheat varieties were exposed to drought stress by withholding water for 28 days for the various measurements. Data presented as mean ± SE (n = 3). Different letters indicate significant differences between means at ($p \leq 0.05$) according to ANOVA and Tukey–Kramer test.

Guaiacol peroxidase (GPX) activity increased by 98% in the leaf tissue of BW35695 following 28 days of drought stress, while that of BW4074 remained unaltered (Figure 5c).

Similarly, the root GPX activity of drought-stressed BW4074 plants remained unchanged relative to the controls, while a non-significant increase was observed for the drought-tolerant BW35695 (Figure 5d). Ascorbate peroxidase (APX) activities of the leaves and roots remained largely unchanged following 28 days of drought stress treatment in both wheat varieties, except for a 72% non-significant increase in BW35695 (Figure 5e–f). Our results suggest that the drought-tolerant wheat cultivar BW35695 possesses greater enzymatic antioxidant capacity to scavenge ROS and limit oxidative stress damage under conditions of water scarcity compared to the drought-susceptible BW4074.

2.6. The Drought-Responsive Leaf Proteome of Two Contrasting Wheat Varieties

2.6.1. Drought Stress Modulates the Accumulation of Total Soluble Leaf Proteins of Wheat

We complemented the drought-induced physiological and biochemical analyses of the two contrasting wheat varieties with a gel-free leaf proteomic investigation under the same drought conditions. For this experiment, we used the isobaric tags for relative and absolute quantification (iTRAQ) method in combination with liquid chromatography-tandem mass spectrometry (LC-MS/MS) to identify the drought-responsive wheat leaf proteins following 28 days of no watering. In this study, proteins with at least one sequenced peptide were regarded as positively identified, resulting in 1062 and 882 positively identified leaf proteins of BW4074 and BW35695, respectively. The full peptide data are given in Tables S2 and S3. A Student's t-test at $p \leq 0.05$ was used to analyze the iTRAQ data for differential protein expression patterns between the control and drought-stressed treatment groups of each wheat variety.

For the drought-susceptible BW4074 variety, 69 of the 1062 proteins were differentially expressed in response to drought stress. Of these BW4074 leaf proteins, 41 (59%) were upregulated while 28 (41%) were downregulated. For the drought-tolerant BW35695 variety, 110 leaf proteins were drought responsive, of which 58 (52%) were upregulated while 52 (48%) were downregulated. A summary of this proteome data is given in Table 1, while the full iTRAQ quantitation data are given in Tables S4 and S5 for BW4074 and BW35695, respectively. We also found nine drought-responsive proteins that were common between the two wheat varieties, while 60 and 101 proteins were unique to BW4074 and BW35695, respectively (Figure 6a).

Table 1. Summary list of wheat leaf proteome data obtained using iTRAQ and LC-MS/MS.

Wheat Variety	Positively Identified Proteins	Drought-Responsive Proteins	Protein Regulation	
			Up	Down
BW4074	1062	69	41	28
BW35695	882	110	58	52

We then retrieved Gene Ontology (GO) data and protein family names of the drought-responsive proteins from the UniProt and InterPro databases and used this information to functionally group the proteins. The GO terms, protein family names, and the putative functional categories of the drought-responsive proteins are summarized in Tables S6 and S7 for BW4074 and BW35695, respectively. Due to the extensive sizes of Tables S6 and S7, we generated Tables 2 and 3 as shortened protein lists for illustrative purposes. The drought-responsive proteins in Tables 2 and 3 are those with a minimum fold change of 1.5; however, all results presentations and discussions in this study are based on the full lists of drought-responsive proteins in Tables S6 and S7.

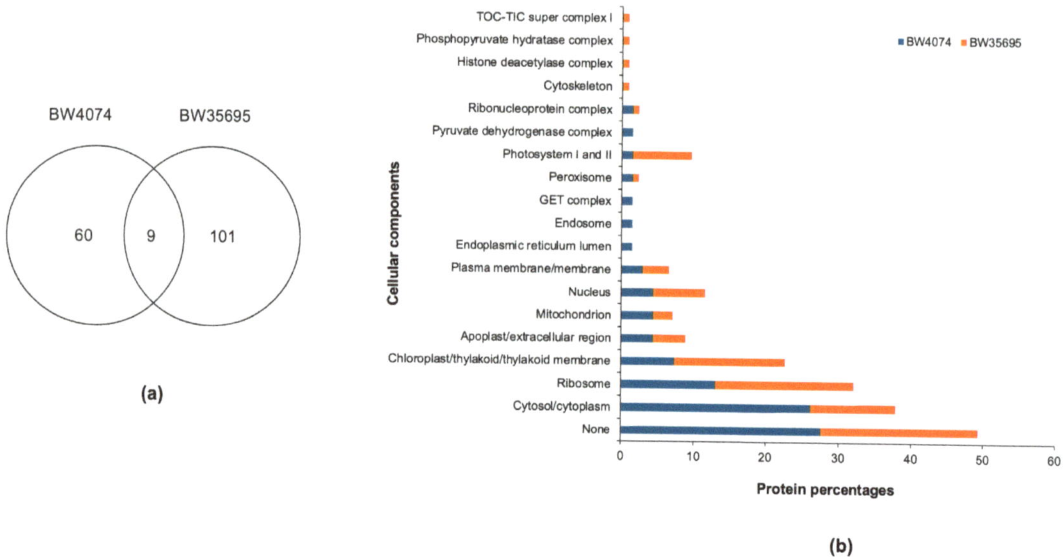

Figure 6. Total number of drought-responsive leaf proteins of two wheat varieties and their putative cellular locations. (**a**) A Venn diagram of the drought-responsive leaf proteins of two wheat varieties. (**b**) Gene ontology terms for cellular locations retrieved from the UniProt database. BW4074 is drought susceptible, while BW35695 is drought tolerant.

Table 2. List of drought-responsive leaf proteins of the drought-susceptible wheat variety BW4074 with a minimum fold change of 1.5.

Accession [a]	Protein Name [b]	Ratio [c]	SD [d]	*p*-Value [e]	Family Name [f]
		Primary Metabolism			
A0A3B6QKI8	GDSL esterase/lipase	−1.86	0.22	1.29×10^{-2}	GDSL lipase/esterase-like, plant
A0A1D5URN5	Fibronectin type III-LIKE domain-containing protein	−1.55	0.08	2.99×10^{-2}	Beta-D-xylosidase
A0A077RPJ4	Tryptophan synthase	−1.67	0.19	4.51×10^{-2}	Tryptophan synthase, alpha chain
Q45NB2	Glutamine synthetase	−1.68	0.26	3.75×10^{-2}	Glutamine synthetase
A0A3B6FGE7	Enoyl reductase (ER) domain-containing protein	1.89	0.41	2.32×10^{-2}	Medium-chain dehydrogenase/reductase
A0A3B6DHI0	Glutamate dehydrogenase	−2.08	0.19	4.52×10^{-2}	Glutamate dehydrogenase
W5ACM8	Delta-1-pyrroline-5-carboxylate synthase	1.52	0.24	3.17×10^{-2}	Delta-l-pyrroline-5-carboxylate synthetase
A0A3B6MSL2	Aminocyclopropanecarboxylate oxidase	1.58	0.36	2.09×10^{-2}	Plant 2-oxoglutarate-dependent oxidoreductases
		Protein synthesis/folding/degradation			
W5I1R7	30S ribosomal protein S3, chloroplastic	1.63	0.18	8.04×10^{-3}	Small ribosomal subunit protein uS3
A0A3B6TJK6	Small ribosomal subunit protein uS10 domain-containing protein	1.64	0.40	2.86×10^{-2}	Small ribosomal subunit protein uS10
W5H9B7	Peptidyl-prolyl cis-trans isomerase	−1.57	0.21	3.52×10^{-2}	Cyclophilin-type peptidyl-prolyl cis-trans isomerase/CLD
A0A2 × 0SLI3	40S ribosomal protein S6	1.94	0.33	4.09×10^{-2}	Small ribosomal subunit protein eS6-like
A0A3B6QHV4	Anion-transporting ATPase-like domain-containing protein	−1.68	0.25	2.63×10^{-2}	Arsenical pump ATPase, ArsA/GET3

Table 2. *Cont.*

Accession [a]	Protein Name [b]	Ratio [c]	SD [d]	*p*-Value [e]	Family Name [f]
		Defense/ROS detoxification			
A0A3B6HMK6	Glutathione reductase	1.94	0.66	4.69×10^{-2}	Glutathione reductase
S6AWC2	Cold induced 16	1.68	0.42	4.43×10^{-2}	Nodulin-related protein1/2
D8L9B5	Putative PDI-like protein	−2.57	0.11	1.11×10^{-2}	Protein disulfide isomerase A6
		Energy			
A0A3B5Z298	Phosphoglycerate kinase	−1.51	0.22	4.29×10^{-2}	Phosphoglycerate kinase family
A0A3B6FUA0	Fructose-bisphosphate aldolase	1.71	0.27	2.64×10^{-3}	Fructose-bisphosphate aldolase, class I
A0A3B6TBB9	Ferredoxin--NADP reductase, chloroplastic	−2.02	0.31	1.77×10^{-2}	Ferredoxin-NADP reductase
A0A3B6KSW8	Glyceraldehyde-3-phosphate dehydrogenase	−1.68	0.18	4.84×10^{-2}	Glyceraldehyde-3-phosphate dehydrogenase, type 1
A0A3B6BY66	Dihydrolipoyllysine-residue succinyltransferase	−1.61	0.18	3.67×10^{-2}	Dihydrolipoamide succinyltransferase
A0A3B6BZB5	Transketolase	−1.88	0.04	1.89×10^{-2}	Transketolase, bacterial-like
		Transcription			
A0A3B6LV22	H15 domain-containing protein	1.50	0.19	4.14×10^{-3}	Linker histone H1/H5
A0A3B6A3G3	Hyaluronan/mRNA-binding protein domain-containing protein	1.99	0.59	2.54×10^{-2}	RNA binding protein HABP4/SERBP1-like
		Secondary metabolism			
A0A3B6C785	Zeta-carotene desaturase	−1.50	0.18	4.20×10^{-2}	Zeta-carotene desaturase
A0A3B6DPV0	AB hydrolase-1 domain-containing protein	−1.78	0.07	3.66×10^{-2}	Epoxide hydrolase-like
		Cell structure			
A0A3B6SEK0	Cyanobacterial aminoacyl-tRNA synthetase CAAD domain-containing protein	−1.87	0.19	2.29×10^{-2}	Protein curvature thylakoid I
		Unclear classification			
A0A3B6FKP0	PH domain-containing protein	1.71	0.34	7.96×10^{-3}	Ricin B-like lectin EULS3-like
A0A3B5XZW3	Remorin C-terminal domain-containing protein	−1.65	0.02	1.98×10^{-2}	None predicted

[a] Protein accessions obtained from UniProt (https://www.uniprot.org/) database searches against sequences of *Triticum aestivum* only, accessed on 6 June 2024. [b] Protein names retrieved from the UniProt database on 6 June 2024. [c] Fold change in the stress-responsive proteins calculated relative to the control samples, each with four biological replicates. A positive value indicates upregulation, while a negative value indicates downregulation of the respective protein. [d] Standard deviation of the fold changes ($n = 4$). [e] Probability value ($p \leq 0.05$) obtained from a Student's *t*-test comparing the fold changes between the drought stress treatment and the control ($n = 4$). [f] Protein family names retrieved from the InterPro (https://www.ebi.ac.uk/interpro/) database, accessed on 6 June 2024.

The GO cell localization data show that the drought stress treatment affected the accumulation of leaf proteins present in various cellular locations, irrespective of the wheat variety (Figure 6b). Some of the highly represented cellular locations common to both wheat proteomes include the cytoplasm/cytosol, ribosome, chloroplast, Photosystem I and II, mitochondrion, and the nucleus (Figure 6b). Likewise, the drought-responsive proteins are potentially involved in diverse biological processes (Figure S2) and molecular functions (Figure S3). Biological process terms of 'photosynthesis' and 'translation' were highly represented in the proteome of the drought-tolerant variety BW35695 while 'translation' dominated the BW4074 proteome. Consistent with the biological process terms, molecular functional terms of 'chlorophyll binding' and 'structural constituent of ribosome' were highly represented in BW35695, while 'structural constituent of ribosome' emerged as the most represented molecular functional term for the drought-sensitive BW4074. Overall, our GO analyses data demonstrate that drought stress modulates the accumulation of proteins in various cellular locations and is involved in diverse biological processes. Both wheat varieties seemed to have largely reprogrammed translation-related processes, while the drought-tolerant variety BW35695 also reconfigured various biochemical aspects of photosynthesis (Figures S2 and S3).

Table 3. List of drought-responsive leaf proteins of the drought-tolerant wheat variety BW35695 with a minimum fold change of 1.5.

Accession [a]	Protein Name [b]	Ratio [c]	SD [d]	p-Value [e]	Family Name [f]
		Primary Metabolism			
A0A3B6UC94	Acid phosphatase	2.05	0.28	5.05×10^{-4}	Acid phosphatase, plant
A0A3B6FKL0	Nucleoside phosphorylase domain-containing protein	1.66	0.27	3.78×10^{-3}	Phosphorylase superfamily
A0A3B6B442	Aspartate/glutamate/uridylate kinase domain-containing protein	−1.92	0.34	4.08×10^{-2}	Glutamate/acetylglutamate kinase
A0A3B6TUD9	Thiamine thiazole synthase, chloroplastic	−1.74	0.13	3.42×10^{-2}	Thiamine thiazole synthase
A0A3B6NHD8	O-methyltransferase ZRP4	1.56	0.41	3.89×10^{-2}	O-methyltransferase COMT-type
A0A3B5ZXG4	Glycosyltransferase	−1.50	0.12	2.23×10^{-2}	UDP-glucuronosyl/UDP-glucosyltransferase
A0A3B6MS26	Glucose-6-phosphate 1-epimerase	−2.37	0.15	1.86×10^{-3}	Glucose-6-phosphate 1-epimerase
A0A3B6KPK9	Beta-glucosidase	2.27	0.88	4.53×10^{-2}	Cellulase degradation glycosyl hydrolase 3
		Energy			
P24065	Photosystem II CP47 reaction center protein	−2.02	0.12	1.01×10^{-2}	Photosystem II CP47 reaction centre protein
A0A3B6AYY2	23 kDa subunit of oxygen evolving system of photosystem II	1.56	0.35	4.27×10^{-2}	PsbP
A0A3B6N1I7	Photosystem I reaction center subunit II, chloroplastic	−1.57	0.19	1.60×10^{-2}	Photosystem I PsaD
A0A3B6LQN1	Chlorophyll A-B binding protein, chloroplastic	−1.63	0.11	2.36×10^{-2}	Chlorophyll A-B binding protein
A0A3B6JMT4	Glyceraldehyde-3-phosphate dehydrogenase	−2.20	0.42	4.92×10^{-2}	Glyceraldehyde-3-phosphate dehydrogenase, type 1
W5C4P1	Uncharacterized protein	−1.92	0.08	1.06×10^{-2}	Oxygen-evolving enhancer protein 3, plants
W5AY52	Chlorophyll A-B binding protein, chloroplastic	−1.67	0.08	1.53×10^{-3}	Chlorophyll A-B binding protein
W5D4R0	Photosystem II 22 kDa protein, chloroplastic	−1.63	0.13	5.52×10^{-3}	Chlorophyll A-B binding protein
A0A3B6QKY1	Aconitate hydratase	2.05	0.47	5.72×10^{-3}	Aconitase/Iron-responsive element-binding protein 2
P58386	Photosystem I P700 chlorophyll A apoprotein A2	−2.39	0.09	2.93×10^{-3}	Photosystem I PsaB
A0A3B6QDB2	Photosystem II protein D1	−1.57	0.14	2.89×10^{-2}	Photosynthetic reaction centre, L/M
P58311	Photosystem I P700 chlorophyll A apoprotein A1	−2.39	0.04	9.95×10^{-3}	Photosystem I PsaA
A0A3B6HUR4	Chlorophyll A-B binding protein, chloroplastic	−1.91	0.09	4.14×10^{-3}	Chlorophyll A-B binding protein
A0A3B6AWZ1	Chlorophyll A-B binding protein, chloroplastic	−1.58	0.05	3.46×10^{-3}	Chlorophyll A-B binding protein
W5F8Z5	Chlorophyll A-B binding protein, chloroplastic	−1.69	0.07	1.88×10^{-4}	Chlorophyll A-B binding protein
A0A3B5Z4J5	ATP synthase subunit b, chloroplastic OS = *Triticum aestivum*	−1.55	0.09	2.09×10^{-2}	ATPase, FO complex, subunit b/b′
P69415	Photosystem I iron-sulfur center	−1.65	0.06	3.97×10^{-2}	Photosystem I protein PsaC
A0A3B6PUD8	Photosystem II 10 kDa polypeptide, chloroplastic	−1.64	0.08	5.55×10^{-3}	Photosystem II PsbR
F6K744	Chlorophyll A-B binding protein, chloroplastic	−2.27	0.06	1.71×10^{-3}	Chlorophyll A-B binding protein
A0A3B6U9Q7	Cytochrome b/b6 C-terminal region profile domain-containing protein	−3.44	0.21	1.90×10^{-2}	Cytochrome b6/f complex, subunit IV

Table 3. *Cont.*

Accession [a]	Protein Name [b]	Ratio [c]	SD [d]	p-Value [e]	Family Name [f]
Protein synthesis/folding/degradation					
W5ASA4	Uncharacterized protein	1.61	0.41	4.51×10^{-2}	Small ribosomal subunit protein uS19
A0A3B6JIR3	Heat shock cognate 70kDa protein	1.55	0.34	4.59×10^{-2}	Heat shock protein 70 family
A0A3B6B204	Large ribosomal subunit protein uL23 N-terminal domain-containing protein	1.59	0.16	5.55×10^{-4}	Large ribosomal subunit protein uL23
W5D739	KOW domain-containing protein	1.64	0.24	4.49×10^{-3}	Large ribosomal subunit protein uL24
A0A3B6MTE3	Peptidylprolyl isomerase	1.64	0.38	3.20×10^{-2}	Peptidyl-prolyl cis-trans isomerase FKBP18-like
A0A3B6N0K3	50S ribosomal protein L17, chloroplastic	1.51	0.21	3.22×10^{-3}	Large ribosomal subunit protein bL17
A0A3B6A2B9	60S ribosomal protein L37a	2.54	0.42	9.27×10^{-3}	Large ribosomal subunit protein eL43
A0A3B6QGX5	60S ribosomal protein L6	1.65	0.29	4.00×10^{-2}	Large ribosomal subunit protein eL6
A0A3B6UD00	50S ribosomal protein L20	1.83	0.45	1.28×10^{-2}	Large ribosomal subunit protein bL20
Transporters					
A0A3B6GKQ2	Non-specific lipid-transfer protein	1.71	0.40	1.31×10^{-2}	Plant non-specific lipid-transfer protein/Par allergen
A0A3B5YX09	Chloroplast inner envelope protein	−1.50	0.10	2.53×10^{-2}	Protein TIC110, chloroplastic
A0A3B6I0D3	STI1/HOP DP domain-containing protein	1.74	0.39	1.84×10^{-2}	None predicted
Transcription					
A0A3B6LSN3	MBD domain-containing protein	1.52	0.18	7.18×10^{-3}	Methyl-CpG-binding domain-containing protein 10/11
Q8LRU5	HMG-I/Y protein HMGa	1.60	0.40	3.88×10^{-2}	High-mobility group protein HMGA
A0A3B6MXZ6	H15 domain-containing protein	1.84	0.44	4.79×10^{-2}	Linker histone H1/H5
A0A3B6GNG0	Histone H2B	1.53	0.24	5.14×10^{-3}	Histone H2B
Defense/ROS detoxification					
A0A3B6EFA0	Uncharacterized protein	2.66	0.73	4.24×10^{-3}	Nodulin-related protein $\frac{1}{2}$ family
Q8W428	Chitinase	−1.56	0.17	9.40×10^{-3}	Glycoside hydrolase, family 19
A0A172WCB1	Cold-responsive LEA/RAB-related COR protein	3.23	1.19	1.01×10^{-2}	None predicted
A0A3B6TZ07	GH18 domain-containing protein	1.84	0.28	1.98×10^{-3}	Glycoside hydrolase 18 family chitinases
A0A3B6MJX1	Pathogen-related protein	−2.00	0.13	1.68×10^{-2}	Pathogen-related defense protein
Secondary metabolism					
A0A3B6TV37	Amine oxidase domain-containing protein	1.67	0.33	2.10×10^{-2}	Flavin monoamine oxidase and related enzymes
A0A3B6QDX1	Delta-aminolevulinic acid dehydratase	−1.59	0.17	8.35×10^{-3}	Delta-aminolevulinic acid dehydratase family
A0A3B5Y2F9	Dienelactone hydrolase domain-containing protein	1.60	0.22	2.30×10^{-3}	Dienelactone hydrolase family
Cell structure					
W5FAY5	Actin	−1.50	0.15	3.53×10^{-2}	Actin family
Unclear classification					
A0A3B6GTL4	DJ-1/PfpI domain-containing protein	−1.79	0.11	5.38×10^{-4}	Protein/nucleic acid deglycase DJ-1
A0A3B6TRL4	Thylakoid membrane protein slr0575	−1.78	0.19	1.58×10^{-2}	Protein of unknown function DUF2854
A0A3B6AVR1	Uncharacterized protein	−1.63	0,06	4.05×10^{-2}	RidA family
A0A3B6N353	Pentacotripeptide-repeat region of PRORP domain-containing protein	2.10	0.41	3.42×10^{-3}	Tetratricopeptide-like helical domain superfamily

Table 3. *Cont.*

Accession [a]	Protein Name [b]	Ratio [c]	SD [d]	*p*-Value [e]	Family Name [f]
	Unclear classification				
W5CRR3	DUF538 domain-containing protein	−2.38	0.13	1.91×10^{-3}	Protein of unknown function DUF538
A0A3B5ZXF0	Protein kinase domain-containing protein	1.92	0.44	9.42×10^{-3}	None predicted

[a] Protein accessions obtained from UniProt (https://www.uniprot.org/) database against sequences of *Triticum aestivum* only, accessed on 6 June 2024. [b] Protein names retrieved from the UniProt database on 6 June 2024. [c] Fold change in the stress-responsive proteins calculated relative to the control samples, each with four biological replicates. A positive value indicates upregulation, while a negative value indicates downregulation of the respective protein. [d] Standard deviation of the fold changes (n = 4). [e] Probability value ($p \leq 0.05$) obtained from a Student's t-test comparing the fold changes between the drought stress treatment and the control (n = 4). [f] Protein family names retrieved from the InterPro (https://www.ebi.ac.uk/interpro/) database, accessed on 6 June 2024.

2.6.2. Drought-Responsive Wheat Leaf Proteins Are Implicated in Diverse Functional Roles

To understand the functional roles of the identified differentially expressed leaf proteins in drought response, we used the GO data and protein family names (Tables S6 and S7) to group proteins into theoretical functional categories (Tables S6 and S7; Figure 7). For this task, we mainly followed the classification scheme suggested by Bevan et al. [48]. The 110 drought-responsive leaf proteins of BW35695 had putative functions in energy (28%), protein synthesis/folding/degradation (25%), primary metabolism (14%), defense/ROS detoxification (10%), transcription (7%), transporters (4%), secondary metabolism (3%), and cell structure (1%). Likewise, the 69 drought-responsive leaf proteins of BW4074 had putative functions in primary metabolism (23%), energy (23%), protein synthesis/folding/degradation (20%), defense/ROS detoxification (13%), transcription (4%), secondary metabolism (6%), and cell structure (2%). However, 9% and 8% of the differentially expressed proteins for BW4074 and BW35695 had limited GO data to facilitate their functional grouping and thus remained unclassified (Tables S6 and S7; Figure 7). The transporters' functional group with four (4%) proteins was only present in the proteome of BW35695 (Figure 7).

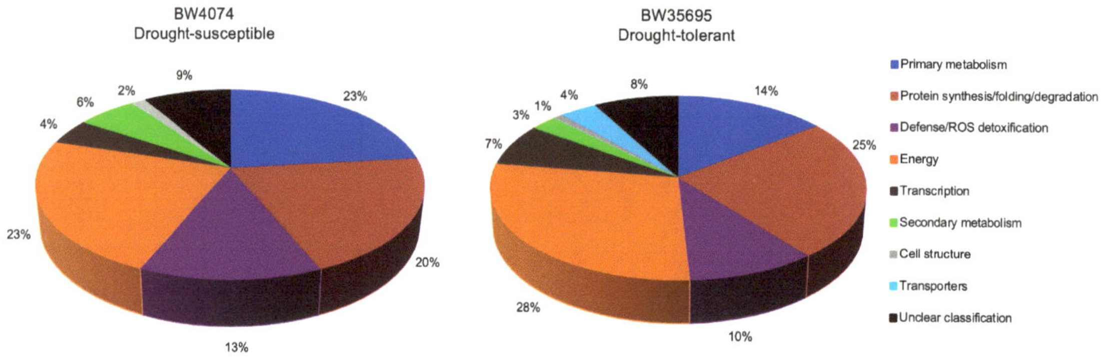

Figure 7. Putative functional groups of the drought stress-responsive leaf proteins of wheat.

The distribution patterns of up and downregulated proteins of each functional group per wheat variety are shown in Figure 8. Although the absolute counts of up- and downregulated proteins varied between functional group across the two wheat varieties, we noted comparable trends in some functional catagories. For example, (i) there were generally more upregulated proteins in the defense/ROS detoxification, protein synthesis/folding/degradation, and transcription functional groups of both wheat varieties; (ii) most energy-related proteins were downregulated in both varieties; and (iii) diverse transporters were only identified in BW35695 and most were upregulated (Figure 8). These results suggest that wheat plants under drought stress conditions reprogram various physiological and biochemical processes, and some of these changes may be similar, dissimilar, or even unique between varieties with contrasting drought phenotypes. Brief results presentations of the main trends in selected functional groups are outlined below.

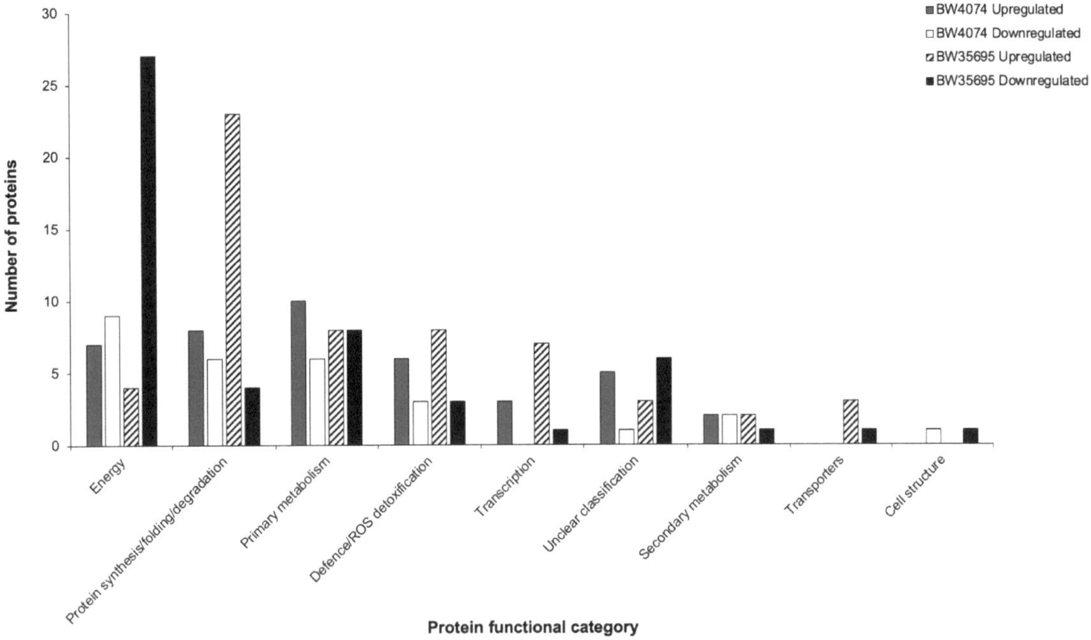

Figure 8. Number of up- and downregulated leaf proteins per functional group for each wheat variety under drought stress. BW4074 is drought susceptible, while BW35695 is drought tolerant.

Drought Stress Largely Downregulates Energy-Related Proteins, Particularly Those Associated with Photosynthesis

The energy group constituted the largest functional category in BW35695 with 31 (28%) drought-responsive proteins (Table S7, Figure 7). Of these proteins, 27 (87%) were downregulated and mostly involved in photosynthesis and ATP synthesis (Table S7; Figure 8). For example, three ATP synthase subunits involved in the production of ATP, and 18 chlorophyll-binding proteins of Photosystems (PS) I and II, a large subunit of ribulose biphosphate carboxylase (Rubisco), and an oxygen-evolving enhancer protein, all with core functions in photosynthesis, were downregulated (Table S7). Conversely, aconitate hydratase with accession A0A3B6QKY1 was the most upregulated protein with a 2.05 fold change, while a cytochrome b6/f complex protein with accession A0A3B6U9Q7 was the most downregulated by a fold change of −3.44 (Table S7). For BW4074, 9 of the 16 (56%) energy-related proteins were downregulated; however, no prominent trends were observed in this protein list. Nevertheless, an ATP synthase subunit and photosynthesis-related proteins such as the ferrodoxin-NADP reductase and Photosystem I iron-sulfur center protein were downregulated. Collectively, our results suggest that wheat plants, in particular the drought-tolerant BW35695 variety, respond to dehydration stress by downregulating proteins involved in photosynthesis and ATP production.

Drought Stress Upregulates Proteins Involved in Defense Response, Protein Homeostasis, and Transcription

The transcription functional group consisted of three (4%) drought-responsive proteins in BW4074 and all were upregulated (Table S6, Figures 7 and 8). Similarly, the drought-tolerant BW35695 variety had eight (7%) transcription-related proteins and seven were upregulated (Table S7, Figures 7 and 8). Collectively, the upregulated proteins in both wheat varieties were histones of types H2A and H2B, linker histones H1/H5, and various regulators of transcription within the STI1 domain-containing protein, RNA binding protein HABP4/SERBP1-like, methyl-CpG binding domain-containing protein 10/11, and the high-mobility group protein HMGA families (Tables S6 and S7). The only downregulated protein in BW35695 was another histone H2A isoform named protein H2A.7 with accession Q43312 (Table S7).

The protein synthesis/folding/degradation functional category consisted of 14 (20%) drought-responsive proteins in BW4074 and 27 (25%) for BW35695 (Figure 7). Of these proteins, 8 and 23 were upregulated in BW4074 and BW35695, respectively, while the rest were downregulated (Tables S6 and S7; Figure 8). Most of the small and large ribosomal subunit proteins involved in translation, amounting to five in BW4074 and 18 in BW35695, were upregulated. Examples of downregulated proteins include those with protein folding/refolding functions such as peptidyl-prolyl cis-trans isomerase (accession Q93XQ6) and a trigger factor ribosomal binding protein (accession W5D1B3) in BW4074, and chaperonin CP60-2 with accession A0A3B6JPZ3 in BW35695. Only two proteolysis-related proteins were identified in this study and both were found in the BW35695 leaf proteome (Table S7). Of these two proteins, the ATP-dependent zinc metalloprotease (accession A0A3B6TR29) was downregulated, while a dipeptiylpeptidase IV N-terminal domain-containing protein (accession A0A3B6HYY4) with serine-type peptidase activity was upregulated (Table S7).

The defense/ROS detoxification functional group consitituted 13% of the drought-responsive proteins of BW4074 with nine proteins (Table S6; Figure 7). The same group made up 10% of the drought-responsive proteome of BW35695 with 11 proteins (Table S7; Figure 7). Of these proteins, six and eight were upregulated in BW4074 and BW35695, respectively (Figure 8). Antioxidant enzymes such as SOD, gluthathione reductase (GR), and monodehydroascorbate reductase (MDAR) were upregulated in BW4074, while a secretory peroxidase was downregulated (Table S6). Other upregulated proteins in the drought-susceptible variety include a cold-induced 16 protein which regulates ABA biosynthesis and a late embrogenesis abundant (LEA) protein, while a putative protein disulphide isomerase-like (accession D8L9B5) which responds to ER stress was the most downregulated protein with a −2.57 fold change (Table S6). Similar trends were observed in the drought-tolerant BW35695 where a cold-responsive LEA-RAB-related COR protein with accession A0A172WCB1 and an uncharacterized nodulin-related protein with accession A0A3B6EFA0 were the two most upregulated proteins with fold changes of 3.23 and 2.66, respectively (Table S7). Taken together, these results suggest that wheat plants under drought stress reprogram transcriptional activities, resulting in increased protein synthesis, and the accumulation of various proteins involved in defense response, regulation of ABA biosynthesis, ROS detoxification, and other responses to water deficit stress.

Other Functional Categories of Drought-Responsive Wheat Leaf Proteins

Other drought-responsive proteins identified in this study were categorized into theoretical functional groups of primary metabolism, secondary metabolism, cell structure, and transporters (Tables S6 and S7; Figure 7). Although the primary metabolism group consitituted a large proportion of the drought-responsive proteomes of both wheat varieties (Figure 7), there were no striking trends emerging from these data, except that various proteins involved in the metabolism of fatty acids, amino acids, phosphates, polysaccharides, and nucleosides were modulated (Tables S6 and S7). In the drought-sensitive BW4074 variety, a delta-1-pyrroline-5-carboxylate synthase (accession W5ACM8) involved in proline biosynthesis was upregulated with a fold change of 1.52 (Table S6). In BW35695, a beta-glucosidase (accession A0A3B6KPK9) belonging to the cellulose degradation glycosyl hydrolase 3 family was highly upregulated by a factor of 2.27 (Table S7). In the secondary metabolism group, a zeta-carotene desaturase (accession A0A3B6C785) was involved in the biosynthesis of carotenoids, and an AB hydrolase-1 domain-containing protein (accession A0A3B6DPV0), belonging to the epoxide hydrolase-like protein family were both downregulated in BW4074 (Table S6).

Furthermore, in BW35695, a delta-aminolevulinic acid dehydratase (accession A0A3B6QDX1) involved in chlorophyll biosynthesis was also downregulated. An actin protein (accession W5FAY5) identified in BW35695 and a protein curvature thylakoid I-related protein (accession A0A3B6SEK0) identified in BW4074 were both downregulated, further illustrating the negative effects of drought on the strutural intergrity of the cytoskeleton and the chloroplast thylakoid membrane (Tables S6 and S7). The BW35695 drought-responsive leaf proteome also contained transporter proteins such as a porin-related uncharacterized protein, a lipid-transporting non-specific lipid transfer protein, a chloroplast inner envelope protein, and an STI1/HOP DP-containing protein both with putative roles in protein transport or targeting into the chloroplast (Table S7). All these proteins, except the chloroplast inner envelope protein, were upregulated, indicating that the drought-tolerant wheat variety also reconfigures various intracellular transport systems in response to drought. Although no particular transporter was identified in the BW4074, the anion-transporting ATPase-like domain-containing protein grouped under protein synthesis due to its asssociated role with translation has a putative function in the post-translational protein targeting to the ER membrane (Table S6). Together with the unclassified proteins listed in Tables S6 and S7, our wheat leaf proteome results illustrate a vast array of cellular processes that are modulated by dehydration stress.

Common Drought-Responsive Proteins between BW4074 and BW35695

Following our observation that nine of the differentially expressed proteins of this study are common between the two wheat varieties (Figure 6a), we statistically analyzed the fold changes of each protein between the two varieties (Table 4). The data show that five of these common proteins with accessions A0A3B6TJK6, A0A3B6MJZ2, A0A3B6MTE3, A0A3B6QKY1, and A0A3B6MQA1 had comparable upward or downward fold changes in response to the drought stress treatment. However, a fibronectin type III-like domain-containing protein with accession A0A1D5URN5, belonging to the beta-D-xylosidase protein family, was more downregulated in the drought-sensitive BW4074 than BW35695. Conversely, a Photosystem I iron-sulfur center protein (accession P69415) which functions in photosynthesis was more downregulated in the drought-tolerant wheat variety than in BW4074. Contrasting expression patterns were also observed for the 30S ribosomal protein S20 (accession W5D1D3), which was downregulated in BW4074 but upregulated in BW35695. Likewise, an unclassified RRM domain-containing protein (accession A0A3B6MQA1) with eukaryotic RNA-binding properties was upregulated in BW4074 but downregulated in BW35695. Overall, the observed similarities and differences in expression trends of these common stress-responsive proteins illustrate the complex dynamics in stress responses to drought between varieties of a single species with contrasting phenotypes. Some of these common changes may represent fundamental stress responses required by plants during exposure to unfavorable conditions, while contrasting expression patterns of the same protein isoform in the two wheat varieties possibly highlight key biological processes that may determine the probability of plant survival under stress conditions.

Table 4. List of drought stress-responsive proteins common to both BW4074 and BW35695 wheat varieties.

Accession	Protein Name	Ratio [a]		BW4074 vs. BW35698 Ratio p-Value [b]
		BW4074	BW35698	
	Primary Metabolism			
A0A3B6MJZ2	5-methyltetrahydropteroyltriglutamate-homocysteine S-methyltransferase	1.47	1.37	3.39×10^{-1}
A0A1D5URN5	Fibronectin type III-like domain-containing protein	−1.55	−1.15	3.30×10^{-3} *

Table 4. *Cont.*

Accession	Protein Name	Ratio [a]		BW4074 vs. BW35698 Ratio *p*-Value [b]
		BW4074	BW35698	
Protein synthesis/folding/degradation				
A0A3B6TJK6	Small ribosomal subunit protein uS10 domain-containing protein	1.64	1.44	3.98×10^{-1}
A0A3B6MTE3	Peptidylprolyl isomerase	1.28	1.64	1.28×10^{-1}
W5D1D3	30S ribosomal protein S20, chloroplastic	−1.27	1.36	2.49×10^{-3} *
Energy				
A0A3B6QKY1	Aconitate hydratase	1.26	2.05	1.83×10^{-2} *
P69415	Photosystem I iron-sulfur center	−1.46	−1.65	4.52×10^{-1}
Transcription				
A0A3B6LSN3	MBD domain-containing protein	1.36	1.52	2.86×10^{-1}
Unclear classification				
A0A3B6MQA1	RRM domain-containing protein	1.20	−1.24	9.37×10^{-4} *

[a] Fold change in each common protein as illustrated in Tables S6 and S7 for BW4074 and BW35695 wheat variety, respectively. [b] The *p*-value from a Student t-test ($p \leq 0.05$) comparing the fold change in each protein for the two wheat varieties. These data were extracted from Tables S6 and S7 for BW4074 and BW35695, respectively. * Proteins with significant differences in fold changes between the two wheat varieties according to a Student *t*-test ($p \leq 0.05$). The four biological replicate fold changes used in this analysis are those used to generate data shown in Tables S4 and S5 for BW4074 and BW35695, respectively.

2.7. Pathway Enrichment Analysis and Protein–Protein Interactions

2.7.1. KEGG Pathway Enrichment Analysis

To analyze the functions of the drought-responsive leaf proteins of the two wheat varieties, a Kyoto Encyclopedia of Genes and Genomes (KEGG) pathway enrichment analysis was conducted (Figure 9). Ten of the identified pathways were present in both varieties. These include citrate cycle (TCA cycle), carbon fixation in photosynthetic organisms, photosynthesis, glyoxylate and dicarboxylate metabolism, carbon metabolism, biosynthesis of amino acids, ribosome, and metabolic pathways. Pathways exclusive to BW4074 were alanine aspartate and glutamate metabolism, arginine biosynthesis, nitrogen metabolism, peroxisome, pyruvate metabolism, cysteine and methionine metabolism, glycolysis-gluconeogenesis, biosynthesis of secondary metabolites, and biosynthesis of various plant secondary metabolites. Conversely, pathways unique to BW35695 were photosynthesis-antenna proteins, 2-Oxocarboxylic acid metabolism, RNA degradation, and amino sugar and nucleotide sugar metabolism. The top two most significantly enriched pathways in BW4074 were associated with alanine aspartate and glutamate metabolism and arginine biosynthesis whilst those in BW35695 were related to photosynthesis-antennae and photosynthesis (Figure 9). Results of our KEGG pathway enrichment analysis further complement the putative functional groupings data (Figure 7), which suggest the drought stress modulates various biological processes in wheat leaves, but mainly primary metabolism in BW4074 and photosynthesis in BW35695 (Tables S6 and S7).

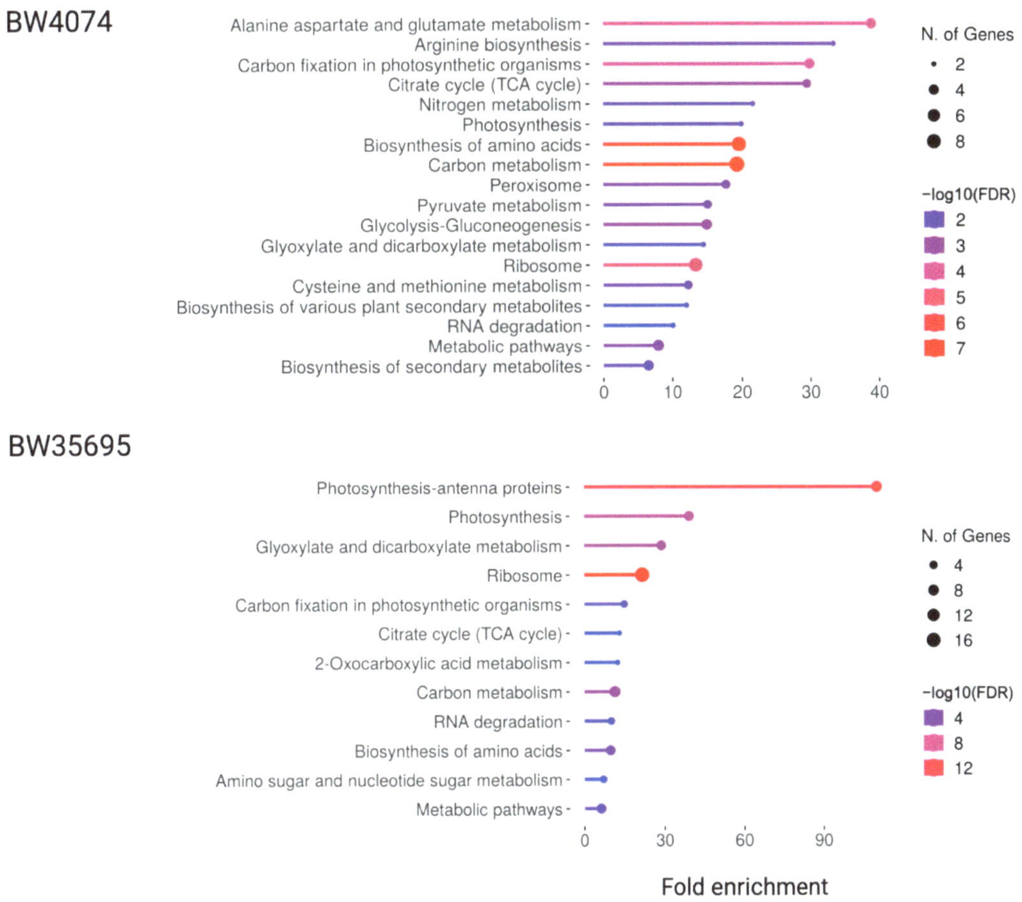

Figure 9. KEGG pathway enrichment analysis of drought-responsive leaf proteins of wheat. BW4074 is drought susceptible, while BW35695 is drought tolerant. The significance of the KEGG path is based on the Student's *t*-test ($p \leq 0.05$). The sizes of dots represent the number of genes in each pathway at a scale indicated by the legend within the graphics. To correct for multiple testing, False Discovery Rate (FDR) was calculated using the Benjamini–Hochberg method, and the measure of FDR is indicated by the colors at the logarithmic scale provided in the legend within the graphics.

2.7.2. Protein–Protein Interaction Analysis

To identify the interactions between drought-responsive leaf proteins of BW4074 and BW35695 wheat varieties, a protein–protein interaction analysis was conducted using the STRING (Search Tool for the Retrieval of Interacting Genes/Proteins) database. Primary metabolism and protein synthesis were the two main interacting biological processes for BW4074, whilst photosynthesis and protein synthesis emerged as the two main interacting functional groups for BW35695 (Figure 10). The abundance of the leaf proteins involved in the photosynthesis functional group decreased following drought stress treatment as indicated by blue nodes. In the protein synthesis category, the majority of the proteins were upregulated in response to drought stress as represented by red nodes, and only two ribosomal proteins, A0A341U912 and W5D067, were downregulated. Overall, the protein–protein interaction maps suggest wheat plants under drought stress modulate various biological processes by altering the expression levels of proteins. These stress-responsive proteins associate with each other to varying degrees to coordinate the desired response mechanism; such protein interactions may also differ between wheat plants of contrasting drought phenotypes.

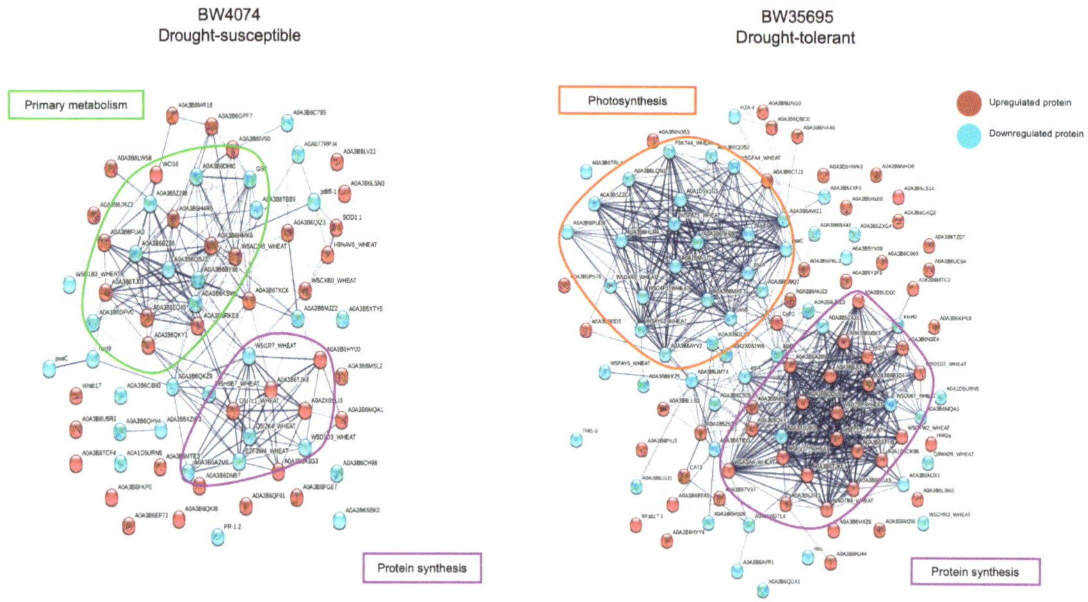

Figure 10. Protein–protein interactions analysis of BW4074 and BW35695 drought-responsive leaf protein using the STRING database. In this network, nodes are proteins and lines represent functional associations between proteins. Nodes in red are upregulated proteins while nodes in blue are downregulated proteins. Thicker lines represent stronger associations between the interacting proteins.

2.8. Gene Expression Analysis Using qRT-PCR

We selected some targets for analysis at the gene expression level in leaf tissues on both wheat varieties following 28 days of drought stress (Table S1). The six target genes were randomly selected from the lists of differentially expressed proteins in the two wheat varieties (Tables S6 and S7). Our results show that qRT-PCR data validated the protein data for some targets. For example, gene expression results show that the *delta-1-pyrroline-5-carboxylate synthase* (*W5ACM8*) gene was upregulated by a factor of two in BW4074 following drought stress treatment (Figure 11), which corresponded with the iTRAQ result showing that its protein with accession W5ACM8 was upregulated by a factor of 1.52 in the same wheat variety (Table S6).

For the rest of the targets (Figure 11), the qRT-PCR data revealed that differential expression is either controlled post-transcriptionally or that the time points where transcript abundance aligns with protein abundance are different. For instance, the iTRAQ data showed that the Photosystem I P700 chlorophyll A apoprotein A1 with accession P58311 and the CYTB_CTER domain-containing protein with accession A0A3B6U9Q7 were highly downregulated in BW35695 (Table S7), but not listed among the drought-responsive proteins of BW4074 (Table S6). Nonetheless, gene expression data revealed that both genes, *P58311* and *A0A3B6U9Q7*, were downregulated in both wheat varieties but were only statistically significant in the drought-sensitive BW4074 (Figure 11). Similarly, the *GH18 domain-containing protein* (*A0A3B6TZ07*) and the *Uncharacterized protein* (*A0A3B6EFA0*) genes did not show any significant change in expression following the drought treatment in both wheat varieties (Figure 11), yet their corresponding proteins were highly upregulated in BW35695 during drought stress (Table S7). Overall, our qRT-PCR analysis suggests that the drought-induced upregulation of the delta-1-pyrroline-5-carboxylate synthase

(W5ACM8) protein observed in BW4074 (Table S6) is transcriptionally regulated under the experimental conditions of this study. The other five protein targets used in gene expression (Figure 11) could either be post-transcriptionally regulated or the abundances of the proteins and transcripts differentially accumulate at different time points.

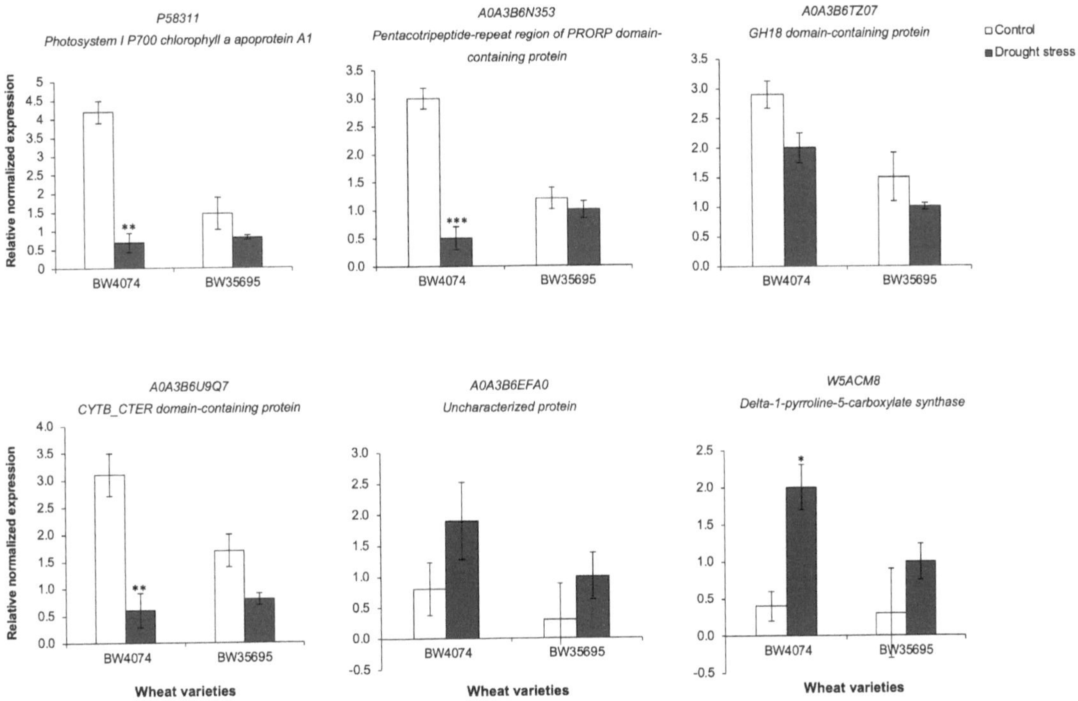

Figure 11. Gene expression analysis in wheat leaf tissue following drought stress treatment. Drought-susceptible BW4074 and drought-tolerant BW35695 wheat plants were exposed to drought stress by withholding water for 28 days and gene expression analysis was performed using qRT-PCR. Data presented as the mean ± SE (n = 3). ** and *** represent statistical significance between the control and drought-stressed samples at $p \leq 0.05$, 0.01, and 0.001, respectively, using a Student's t-test on the CFX Maestro software version 4.1.2433.1219.

3. Discussion

We used the drought-susceptible BW4074 and drought-tolerant BW35695 wheat varieties to comparatively investigate the differential effects of drought stress on the growth, physiology, biochemistry, and leaf proteome expression of wheat plants. Our results showed that withholding water supply to potted wheat plants for 28 days resulted in progressively drying soils (Figure 1a), equivalent to mild and moderate drought stress conditions [49,50], on days 14 and 28 of the stress treatment, respectively. The intensity of water deficit stress at each time point was not significantly different between the two wheat varieties (Figure 1a), suggesting that our experimental system could be used to compare the differential responses of the wheat varieties to drought. Consequently, any differences in physiological and/or molecular responses between the two wheat varieties could be attributed to the genetic constitution and drought phenotypes of the plants, rather than variations in the levels of water deficit stress imposed.

The leaf relative water content (RWC) and shoot water loss results (Figure 1b,c) suggest that the drought-tolerant variety BW35695 possesses greater water retention capacity under dehydration stress than the drought-susceptible BW4074. Similar findings have been reported in drought-stressed wheat (*Triticum aestivum*) [37,51], sorghum (*Sorghum bicolor*) [52], and soybean (*Glycine max*) [53] plants under drought conditions, where the drought-tolerant genotypes remained more hydrated than their drought-sensitive counterparts. A higher plant tissue water content under drought conditions contributes towards optimal cellular hydration, and the preservation of cell structure and function [14]. However, more studies are required to evaluate the contributions of stomatal and/or non-stomatal responses [32,54] of BW4074 and BW35695 in water retention processes under drought stress.

Although the 28 days of moderate drought stress treatment did not cause elaborate changes in the measured growth parameters in both wheat varieties (Figure S1), the chlorophyll A and b and carotenoid levels increased in BW35695 (Figure 2). Conversely, only chlorophyll B marginally increased in BW4074 relative to the controls (Figure 2). Chlorophyll A and B and carotenoids are essential light-harvesting molecules of photosynthesis [55,56] and their increased accumulation under drought conditions may enhance the photosynthetic potential of plants for increased growth and survival. In contrast to our results, another comparative study between drought-tolerant and drought-susceptible wheat varieties revealed a decline in chlorophyll A and B content of both varieties following drought stress [51]. However, it is unclear whether the increased levels of chlorophyll A and B in the BW35695 wheat variety (Figure 2) are due to increased biosynthesis and/or reduced drought-induced degradation of the pigments. Nevertheless, we suggest more physiological measurements of other parameters including the chlorophyll index and photosynthesis rate [32,41,54] to understand the differential effects of drought on pigment content, photosynthesis, and yield components of wheat.

Dehydration stress disrupts normal cell metabolism which may result in increased production of reactive oxygen species (ROS). Any imbalances between ROS hyperaccumulation and their detoxification may lead to oxidative stress damage of cell components [13,57]. In our study, the 28 days of drought stress caused increased lipid peroxidation in BW4074 plant tissues (Figure 3a,b), possibly due to increased ROS accumulation (Figure 3c–f), coupled with a reduced enzymatic antioxidant capacity (Figure 5). Conversely, the drought-tolerant BW35695 had lower ROS tissue content, better enzymatic antioxidant activities, and thus reduced membrane damage (Figures 3 and 5). These results further underscore the superior drought-tolerant nature of BW35695 compared to BW4074.

Some drought-induced metabolites possess various regulatory and protective functions in plant cells [15]. For example, apart from their light-harvesting roles during photosynthesis, carotenoids may also function as non-enzymatic antioxidants [57]. Proline and glycine betaine act as osmolytes and osmoprotectants under water-limiting conditions [58–60]. Therefore, the observed increase in the level of carotenoids (Figure 2c) and proline (Figure 4a) in the leaves of BW35695 may signify greater potential for osmotic adjustment and ROS scavenging capacity, which could have contributed towards higher RWC (Figure 1b) and lower lipid peroxidation (Figure 3a). The significant increase in the levels of root proline, as well as root and leaf glycine betaine in both wheat varieties following 28 days of stress (Figure 4), also signifies the critical roles of these metabolites in drought response. As reported in other studies of wheat [51], maize (*Zea mays*) [61], and rice (*Oryza sativa*) [62], under drought conditions, we observed higher levels of these osmoprotectants in leaves than roots (Figure 4). However, the reasons behind these tissue-specific differences in metabolite content are unclear from the results of our current study.

We complemented the physiological and biochemical analyses with a leaf proteome study of the two wheat varieties following 28 days of drought stress (Table 1). Our proteome data suggest that the imposed drought stress modulated the levels of proteins involved in various plant biological processes including primary and secondary metabolism, photosynthesis, ATP synthesis, transcription, protein synthesis and homeostasis, ROS detoxification, and other defense-related systems (Tables S6 and S7; Figure 7). We have chosen to focus our discussion on the key common and unique trends observed in the two wheat varieties, highlighting their potential significance in drought response. Understandably so, our proteomics discussion is largely general and speculative in nature, until functional validation studies are performed for the identified drought-responsive proteins and their isoforms.

Our results suggest that both wheat varieties modulated the accumulation profiles of transcription-related proteins, with a marked increase in histones, linker histones, and other regulators of transcription particularly in the drought-tolerant BW35695 variety (Tables S6 and S7; Figures 7 and 8). Histones are involved in chromosomal DNA packing [63], and their upregulation during stress response facilitates transcriptional regulation of stress-responsive genes [64]. Similar results have been reported in other comparative proteome studies of wheat leaves [65], sorghum roots [52], as well as transcriptome analyses of wheat leaves and roots [34,38] under drought stress. It is also well established that changes in transcriptional activities under stress conditions allow plants to reprogram their gene expression patterns and ultimately cell metabolism to produce regulatory and protective proteins and metabolites for stress survival [15]. Therefore, the observed upregulation of transcription-related proteins especially in BW35695 potentially highlights its superior ability to quickly respond to the prevailing stress by modulating gene expression patterns to maintain cell homeostasis for survival.

Our proteome data also suggest that the transcription-related protein changes resulted in increased synthesis and accumulation of drought-responsive proteins. This viewpoint is partly supported by the upregulation of numerous ribosomal proteins involved in translation, ROS-detoxifying enzymes, and other defense proteins identified in both wheat varieties (Tables S6 and S7; Figure 8). However, more ribosomal proteins were upregulated in BW35695 than in BW4074 (Figure 8). Furthermore, two peptidyl-prolyl cis-trans isomerases (Q93XQ6 and A0A3B6MTE3) were upregulated in BW35695. In BW4074, A0A3B6MTE3 was also upregulated, while W5H9B7 was downregulated (Table S6). Peptidyl-prolyl cis-trans isomerases function in protein folding activities [66] and have been implicated in drought responses of wheat [40] and maize [67] leaves. The observed contrasting accumulation patterns of peptidyl-prolyl cis-trans isomerases in the current study possibly points towards differential capacities to correctly fold proteins in BW4074 versus BW35695 under drought stress conditions, which may influence the overall structure and functions of drought-responsive proteins.

Contrary to our antioxidant enzymatic activity results, which implied that the drought-tolerant BW35695 had greater ROS scavenging capacity than BW4074 (Figure 5), we identified drought-responsive antioxidant enzymes in both BW4074 and BW35695 (Tables S6 and S7), albeit from different protein families and with varying fold changes. In BW4074, two superoxide dismutase (SOD) isoforms, a glutathione reductase (GR) and a monodehydroascorbate reductase (MDAR), were upregulated (Table S6), while catalase, a thioredoxin domain-containing protein, germin-like protein, and a secretory peroxidase were upregulated in BW35695 (Table S7). It is plausible that the differences in protein families of the identified antioxidant enzymes between the two wheat genotypes reflect genotypic differences in the wheat responses to oxidative stress. Nevertheless, our proteomics data suggest the existence of diverse drought-responsive enzymatic antioxidants in wheat leaves. Furthermore, the two wheat varieties may utilize different types of ROS scavengers to maintain cellular redox homeostasis and these proteins inevitably require different activity assays to study their involvement in stress response. Therefore, future enzymatic activities assays should be performed for a wider pool of enzymes and substrates.

Some transcriptomics studies of soybean [53], maize [68], and sorghum ([69–71] under drought stress have suggested that the constitutive expression of transcripts in leaves and/or roots may vary between contrasting genotypes under normal growth conditions. As such, the overall drought responses of such genotypes may be influenced by the basal levels of specific proteins and metabolites of such transcripts. In addition, tissue-specific proteome responses may also be influenced by multi-gene families where different protein isoforms may have different functional roles depending on their cellular location [72]. Therefore, we recommend meta-analyses of "omics" data of various wheat genotypes and related species under control and drought stress conditions to better understand the implication of constitutive gene/protein/metabolite expression in drought-adaptive responses.

The massive downregulation of proteins involved in photosynthesis and ATP production is another noteworthy trend of this study, particularly for the drought-tolerant BW35695 variety (Tables S6 and S7; Figure 8). It is well established that some plants under drought reduce their rate of photosynthesis [47], possibly as a measure to avoid the overproduction and accumulation of ROS [73]. Furthermore, stress tolerant genotypes can enhance the levels of energy metabolism enzymes, thereby increasing ATP production in response to drought stress [73]. However, in our study, both photosynthesis-related proteins and those involved in ATP synthesis were generally downregulated (Table S7). Similar results have also been reported in proteomic studies of wheat leaves [40] and cassava (*Manihot aeculenta*) chloroplasts [74], where the drought-tolerant genotypes reduced the accumulation of proteins related to photosynthesis and ATP production. However, it is unclear why the biochemical contents of chlorophyll A and B increased in BW35695 more than in BW4074 (Figure 2), yet the accumulation patterns of a delta-aminolaevulinic acid dehydratase involved in chlorophyll biosynthesis and that of light-harvesting chlorophyll A-B binding proteins were downregulated in the proteome of this wheat variety (Table S7).

The divergent KEGG pathway enrichment results (Figure 9) and equally distinct protein–protein interaction maps (Figure 10) further demonstrate the extent to which the two genotypes differ in response to drought stress. The BW35695 drought-responsive leaf proteome is highly enriched with photosynthesis-antennae and photosynthesis-related pathways as opposed to primary and secondary metabolism-related pathways in BW4074 (Figure 9). In addition, our protein–protein interactomes highlight the overly complex but interconnected networks of proteins and biological processes in wheat leaves under water deficits (Figure 10). As discussed earlier, these protein interaction networks illustrate that the drought-tolerant BW35695 reprograms its cellular metabolic activities by increasing protein synthesis-related proteins possibly to increase the biosynthesis of other stress-responsive proteins.

4. Materials and Methods

4.1. Plant Materials, Growth Conditions, and Drought Stress Treatments

Two pure wheat (*Triticum aestivum*) lines, BW4074 (an improved cultivar that is drought susceptible) and BW35695 (a drought-tolerant breeder's line), obtained from the International Maize and Wheat and Improvement Center (CIMMYT) in Mexico were used in this study. The seeds were germinated on moist paper towels in a growth chamber (Model: GC-539DH, Already Enterprise Inc., Tapei, Taiwan) at 18 °C under dark conditions. Twelve six-day-old seedlings of each wheat variety were transplanted into plastic pots with dimensions of 10 cm top diameter × 6.5 cm bottom diameter × 8.5 cm height containing 150 g potting soil (Culterra, Muldersdrift, South Africa), previously saturated with Nitrosol nutrient solution (Envirogreen (Pty) Ltd., Braamfontein, South Africa). About 32 pots were prepared for each wheat variety. The plants were grown for a further 8 days in the growth chamber at 18/15 °C day/night temperatures under a 16 h photoperiod. All plants were well watered during this growth period.

Fourteen-day-old plants were thinned to ten plants per pot, and the pots were split into two groups for each wheat variety. The first group was for the drought stress treatment, where watering was withheld for 28 days. The second group consisted of control plants that were well watered throughout the experiment. On days 14 and 28 of the stress treatment, we collected three biological replicate pots from each treatment group per wheat variety to measure the soil moisture content as a proxy for the level of drought stress. The soil moisture content was estimated using the gravimetric method [75] with minor modifications. For these measurements, plants were harvested and the fresh weights of soil samples were immediately taken, before soil oven-drying at 105 °C for 48 h for dry weight readings. Three biological replicate soil samples were used for each treatment group per wheat variety.

The effect of water deprivation on wheat plants was assessed by measuring various growth, physiological, and biochemical parameters at days 0, 14, and 28 of treatment. In this study, day 0 denotes the time when watering of the two-week old seedlings was stopped, and the treatment group plants received no further watering thereafter. Leaf and root tissues harvested at these time points were stored at −80 °C for use in biochemical, leaf proteomic, and gene expression analyses. Growth, physiological, and biochemical analyses that required freshly harvested tissues were performed immediately after sampling. All experimental procedures performed in this study had a minimum of three biological replicates as described below.

The first sets of measurements such as leaf relative water content and chlorophyll, carotenoid, proline, and glycine betaine content analyses were conducted using plant tissue samples harvested at 0, 14, and 28 days of drought stress. Our results revealed that both wheat varieties mostly showed some changes in these parameters following 28 days of drought stress and not at day 14; hence, subsequent ROS measurements, antioxidant enzymatic activity assays, and protein and gene expression analyses were conducted on tissue samples harvested on day 28 of the drought stress treatment.

4.2. *Plant Growth Parameters, Relative Water Content, and Relative Shoot Water Loss Measurements*

Plant growth parameters, leaf relative water content (RWC), and relative shoot water loss were determined following 28 days of drought stress. The RWC was estimated on freshly harvested whole leaf samples as described previously [76] with minor modifications [52]. Four biological replicates of the third oldest leaf samples were used per treatment per wheat variety.

The shoot water loss experiment was performed following a previously described method [77] and used as a proxy for the plants' water retention capacity. Briefly, the above-ground shoots were cut off from five biological replicate plants per treatment per variety, and their fresh weights were immediately taken. The shoots were left on a benchtop at room temperature (~25 °C) and re-weighed at 1 h intervals for 6 h. The relative shoot water loss was calculated using the formula described by Rahman et al. [77].

The length and fresh and dry weight measurements of the roots and shoots were determined for the control and drought-stressed plants of each wheat variety as described previously [52]. Five biological replicates were used per treatment for each measurement.

4.3. *Biochemical Assays*

4.3.1. Chlorophyll and Carotenoid Content

Chlorophyll A and B and carotenoids were extracted using a dimethyl sulfoxide (DMSO) extraction method [78]. Three biological replicates were used per treatment per wheat variety for plants harvested on day 0, 14, and 28 following drought stress. Briefly, 100 mg of each frozen leaf tissue sample was mixed with DMSO and incubated in a water bath at 65 °C for 3 h. The samples were pulse vortexed, and 1 mL aliquots of the supernatant were used for absorbance measurements at 480, 649, and 665 nm against a DMSO blank solution. Chlorophyll A and B and carotenoid contents were estimated using a formula described by Wellburn [79].

4.3.2. Lipid Peroxidation, Hydrogen Peroxide, and Superoxide Content

The levels of malondialdehyde (MDA), hydrogen peroxide (H_2O_2), and superoxide anion (O_2^-) were estimated in leaf and root samples harvested on day 28 of the drought stress treatment. Three biological replicates were used for each tissue sample per treatment and wheat variety. Superoxide anion and H_2O_2 content analyses were conducted to assess the levels of oxidative stress-causing molecules in the drought-stressed plants, while the MDA content was used as a proxy for oxidative stress damage by lipid peroxidation.

Trichloroacetic acid extracts of each tissue sample were prepared for use in the MDA and H_2O_2 content analyses. Briefly, 100 mg of frozen leaf and root tissues were separately ground and homogenized in 5 mL of 6% (*w*/*v*) trichloroacetic acid (TCA), vortexed and centrifuged at 15,000× g for 10 min. The resultant supernatant was used as the TCA extract for the respective assays. MDA formation was assessed using the thiobarbituric acid (TBA) method as previously described [80]. The H_2O_2 content was estimated as described previously [80] with minor modifications. The reaction mixture containing 93.8 µL of the TCA extract, 187.5 µL 20 mM K_2HPO_4 (pH 5), and 375 µL 500 mM potassium iodide in a final volume of 750 µL was incubated for 20 min at 25 °C. Hydrogen peroxide standards were also prepared and processed as described previously [81]. Thereafter, absorbance readings of all samples were measured at 390 nm. The superoxide anion content was determined following a method described by Gokul et al. [80]. The supernatant was analyzed at 600 nm and superoxide content was calculated using the nitroblue tetrazolium (NBT) extinction coefficient of 12.8 $mM^{-1}cm^{-1}$.

4.3.3. Proline and Glycine Betaine Content

Proline and glycine betaine content was determined for leaf and root samples using the hydrophilic interaction chromatography in tandem with liquid chromatography-mass spectrometry (HILIC LC-MS) as described previously [52,82]. The assays were performed on a QTRAP 6500 MS (Applied Biosystems Sciex, Foster City, CA, United States of America (USA)) following procedures described in our previous study [52]. Three biological replicates of tissue samples harvested on days 0, 14, and 28 of drought stress treatment were used. Leaf discs of 7 mm in diameter and 50 mg of ground frozen root samples were used for metabolite extraction and analyses.

4.3.4. Enzymatic Antioxidant Activity Assays

Antioxidant enzymatic activity assays were determined for leaf and root samples harvested on day 28 of drought stress. All assays used three biological replicates of each tissue sample, per treatment and wheat variety. Plant tissues were separately ground to a fine powder using ice-cold mortar and pestles. The ground tissue of 100 mg was homogenized in 500 µL extraction buffer containing 40 mM phosphate buffer (pH 7.4), 1 mM ethylenediaminetetraacetic acid (EDTA), and 5% polyvinylpyrrolidone, vortexed and centrifuged at 9000× g for 5 min. The resultant supernatant was used as the enzyme extract, and protein concentrations were determined using a modified Bradford assay [83].

Superoxide dismutase (SOD) activity was determined as described previously [80]. Briefly, 10 µL of the enzyme extract was mixed with 300 µL 20 mM phosphate buffer (pH 7.8), 100 µL 0.1 mM EDTA, 100 µL 10 mM methionine, 100 µL 0.005 mM riboflavin, 100 µL 0.1 mM NBT, and 90 µL distilled water in a final volume of 800 µL. The reaction mixture was incubated at 25 °C for 20 min under light exposure and absorbance readings were taken at 560 nm. One unit of SOD activity was calculated based on the amount of enzyme required to cause a 50% reduction of NBT.

Ascorbate peroxidase (APX) activity was determined as described by Moloi and van der Merwe [84]. The reaction mixture contained 530 µL 50 mM potassium phosphate buffer (pH 7.0), 150 µL 0.5 mM ascorbate, 50 µL 0.1 mM EDTA, 200 µL 0.1 mM H_2O_2, 70 µL for leaf enzyme extract in a final volume of 1 mL. For the root samples, a 50 µL enzyme extract was used in a similar reaction mixture described above. The blank solution contained all reagents except for the enzyme extracts. Absorbance readings were measured at 290 nm, and the APX activity was calculated using an extinction coefficient of 2.8 $mM^{-1}cm^{-1}$.

Guaiacol peroxidase (GPX) activity was determined as described previously [84]. The reaction mixture of 1 mL consisted of 505 µL 80 mM phosphate buffer (pH 5.5), 100 µL 50 mM guaiacol, 50 µL 20 mM H_2O_2, 5 µL leaf enzyme extract, and 340 µL distilled water. For root samples, a 2 µL root enzyme extract was used in a 1 mL reaction mixture. The blank solution contained all reagents except for the enzyme extracts. The GPX activity was measured by observing an increase in absorbance at 470 nm over 3 min at 30 °C using a UV—visible spectrophotometer (Cary 100 Bio, Varian, Australia), using an extinction coefficient of 26.6 $mM^{-1}cm^{-1}$.

4.4. Leaf Protein Extraction, iTRAQ Labeling, LC-MS/MS, and Identification

4.4.1. Leaf Protein Extraction, iTRAQ Labeling, and Tandem Mass Spectrometry

Proteomic analysis was conducted on leaf samples of both wheat varieties harvested on day 28 of the drought stress treatment. Four biological replicates were used for each treatment group per wheat variety. Total soluble proteins were extracted from 1 g of frozen leaf samples and quantified following previously described methods [83].

Full details of all protocols used for iTRAQ labeling, trypsin digestion, and protein identification by liquid-chromatography tandem mass spectrometry (LC-MS/MS) are described in our previous publication [52]. Briefly, 10 µg of each protein sample was prepared for labeling using an iTRAQ Reagent-Multiplex Buffer Kit (Applied Biosystems Sciex, Foster City, CA, USA) following the manufacturer's protocol. Protein samples were subsequently digested with trypsin overnight at 37 °C and labeled with an 8-plex ITRAQ Reagent Kit (Applied Biosystems Sciex) following the manufacturer's protocol with minor modifications [52]. Peptides of the four control samples of each wheat variety were separately labeled with iTRAQ tags 113, 114, 115, and 116, while those of the drought-stressed samples were labeled with tags 117, 118, 119, and 121. The four control and four drought-stressed labeled leaf samples of each wheat variety were pooled into one composite sample prior to sample clean-up and LC-MS/MS. Tandem mass spectrometry was performed on a TripleTOF 6600 spectrometer (Applied Biosystems Sciex) linked to an Eksigent 425 LC system via a Duospray source (Applied Biosystems Sciex, Foster City, CA, USA). Mass spectrometry data were acquired using the Applied Biosystems Sciex Analyst TF 1.7.1 instrument control and data processing software.

4.4.2. Protein Identification and Quantification

Mass spectrometry protein identification was performed against the TrEMBL database for *T. aestivum* only with sequences downloaded in July 2022. The MS data analysis settings are as fully described by Goche et al. [52]. Peptide and protein tables were exported from the ProteinPilot 5.01 version 4895 software with the Paragon Algorithm 5.0.1.0.4874 and manually processed and filtered. In this study, all positively identified proteins were identified based on at least one peptide. This gave rise to 1062 and 882 positively identified proteins in the leaf extracts of BW4074 and BW35695, respectively. The relative expression of the drought-responsive leaf proteins is presented as fold change, calculated as a ratio to the 113-tagged control sample of each wheat variety. Upregulated proteins are given as positive values above 1, while downregulated proteins are denoted with negative fold change. The drought-responsive proteins were statistically analyzed using a Student's *t*-test at $p \leq 0.05$.

4.4.3. Bioinformatics Analyses of Differentially Expressed Proteins

The drought-responsive leaf proteins were ascribed with Gene Ontology (GO) terms and protein family names using the UniProt [85] and InterPro [86] databases, respectively. The differentially expressed proteins were assigned functional groups using the classification scheme suggested by Bevan et al. [48]. We analyzed the functions of the drought-responsive proteins using a pathway enrichment analysis on the Kyoto Encyclopedia of Genes and Genomes (KEGG) database [87] using gene IDs as input data. Protein–protein interactions were predicted on the STRING (Search Tool for the retrieval of Interacting Genes/Proteins) version 12.0 database [88] using protein accessions as input data. The pathway enrichment analysis and protein–protein interactions were conducted using default settings of each bioinformatic tool. The UniProt and InterPro databases were accessed on 6 June 2024, while the KEGG and STRING databases were accessed on 20 June 2024.

4.5. Gene Expression Analysis

Total RNA was extracted from leaf samples of the control and drought-stressed plants of both wheat varieties following 28 days of stress treatment for use in gene expression analysis. Day 28 leaf samples were used for gene expression to correspond with the harvest time used for the proteome analysis. Three biological replicates were used for each treatment and wheat variety. Total RNA extraction with a DNase digestion step were performed using the Spectrum™ Plant Total RNA kit (Sigma Aldrich, St. Louis, MI, USA) and the On-column DNase 1 Digestion Set (Sigma), respectively. The GoScript Reverse Transcriptase System (Promega, Southampton, UK) was used for complementary DNA (cDNA) synthesis on a 1 μg total RNA template.

Quantitative real-time PCR (qRT-PCR) was performed using the SsoAdvanced Universal SYBR Green Supermix kit (Biorad, Hercules, CA, USA) following the manufacturer's protocol. Reaction mixtures were prepared and run on a CFX Connect Real-Time System (Biorad) using the reaction component volumes and thermal cycling conditions described by Ngwenya et al. [89]. Two *T. aestivum* reference control genes, a wheat α-tubulin [90] and elongation factor-1 alpha [91], were used for data analysis performed on the CFX Maestro software version 4.1.2433.1219 (Biorad). Primer sequences (Table S1) were designed using the Primer-BLAST tool [92], synthesized and supplied by Inqaba Biotechnical Industries (Menlo Park, South Africa). Target genes for qRT-PCR analysis were randomly selected from the drought-responsive leaf proteins identified in the current study (Tables S6 and S7).

4.6. Statistical Analysis

A two-way analysis of variance (ANOVA) test was used to statistically analyze the data unless stated otherwise. In cases where three factors of stress type (control and drought stress), duration of stress treatment (across the different sampling times), and wheat genotypes (drought-susceptible and drought-tolerant) were analyzed, three-way ANOVA was performed. Means were compared according to the Tukey multiple comparison test at $p \leq 0.05$. The ANOVA and Tukey multiple comparison statistical tests were performed using GraphPad Prism analysis software version 8.0.2.263.

5. Conclusions

We used two wheat varieties, the drought-tolerant BW35695 and the drought-sensitive BW4074, to gain insights into the physiological and molecular basis for drought tolerance in wheat. Our results suggest that when the plants were deprived of water for 28 days, the drought-tolerant variety maintains a higher leaf water content, possibly by accumulating higher levels of osmolytes such as proline, than the drought-sensitive variety. In addition, BW35695 protected its cell components from oxidative stress damage by limiting ROS accumulation and enhancing the enzymatic antioxidant capacity in plant tissues.

Our leaf proteomic results further suggest that drought stress modulates various cellular processes in both wheat varieties. Some of the identified drought-responsive proteins have putative functions in primary and secondary metabolism, protein synthesis/folding/degradation, defense/ROS detoxification, energy production, transcription, and cell structure. However, the number of the up- and downregulated proteins varied between functional groups within and between the two wheat varieties.

Nevertheless, we observed two noteworthy trends in the proteome data of the drought-tolerant variety BW35695: proteins involved in protein synthesis were highly upregulated, while those related to photosynthesis were downregulated. Although further functional validation of these proteomic results is required, it is plausible that the BW35695 responds to drought stress by reprogramming protein synthesis to generate proteins with regulatory and/or protective roles against the primary and secondary effects of drought. On the other hand, photosynthesis is pivotal in generating carbohydrates and energy for the growth and development of plants. However, during drought stress, photosynthesis may generate excess ROS, which exacerbates drought-induced oxidative stress and its adverse effects on plant cell structure and function. Therefore, the observed downregulation of proteins involved in photosynthesis in BW35695 might be a protective mechanism of this variety against oxidative stress in plants under drought. The effects of diminished photosynthetic capacity on yield losses of wheat plants under drought ought to be further investigated in field studies.

Apart from the future functional validation of these proteomic results using a broader range of enzyme assays and transgenic plant biology of a few target genes from these data, we propose more time-course gene expression analysis studies to investigate the transcriptional regulation in the two wheat varieties under drought conditions. In addition, future studies could analyze the drought responses of the root transcriptome, proteome and metabolome of these two wheat varieties to gain insights into the below-ground stress responses. Meta-analysis of datasets across different crops could also help to rank drought-adaptive proteins in different tissues and prioritize targets with predicted essential roles for reverse genetic analysis. Our results make an important contribution towards such an endeavor and the functional validation of candidate genes would ascertain the roles of selected proteins in plant stress response.

Supplementary Materials: The following supporting information can be downloaded at: https://www.mdpi.com/article/10.3390/plants13192797/s1, Figure S1: Effect of drought stress on the growth of wheat plants. (a) Shoot fresh weight; (b) shoot dry weight; (c) root fresh weight; (d) root dry weight; (e) shoot length; (f) root length. Two-week old plants of BW4074 (drought-susceptible) and BW35695 (drought-tolerant) wheat varieties were exposed to drought stress by withholding water for 28 days for the respective measurements. Data presented as mean ± SE (n = 3). Different letters indicate significant differences between means at ($p \leq 0.05$) according to ANOVA and Tukey–Kramer test; Figure S2: Putative biological processes of the drought stress-responsive leaf proteins of wheat; Figure S3: Putative molecular functions of the drought stress-responsive leaf proteins of wheat; Table S1: List of leaf wheat primer sequences used in gene expression analysis; Table S2: List of leaf proteins identified in BW4074 wheat variety; Table S3: List of leaf proteins identified in BW35695 wheat variety; Table S4: List of differentially expressed leaf proteins in BW4074 in response to drought stress; Table S5: List of differentially expressed leaf proteins in BW35695 in response to drought stress; Table S6: List of drought-responsive leaf proteins of the drought-susceptible wheat variety BW4074 identified using iTRAQ and LC-MS/MS; Table S7: List of drought-responsive leaf proteins of the drought-tolerant wheat variety BW35695 identified using iTRAQ and LC-MS/MS.

Author Contributions: Conceptualization, R.N. and S.C.; methodology, S.J.M., A.O.A., A.P.B., N.G.S., T.G., M.J.M., A.G., S.C. and R.N.; software, A.P.B., S.C. and R.N.; formal analysis, S.J.M. and T.G.; investigation, S.J.M.; resources, N.G.S., M.J.M., A.G., S.C. and R.N.; data curation, S.J.M., N.G.S., S.C. and R.N.; writing—original draft preparation, S.J.M. and R.N.; writing—review and editing, S.J.M., A.O.A., N.G.S., T.G., M.J.M., A.G., S.C. and R.N.; supervision, R.N. and S.C.; project administration, R.N.; funding acquisition, R.N. and S.C. All authors have read and agreed to the published version of the manuscript.

Funding: This research was funded by the National Research Foundation of South Africa-Thuthuka Grant [number 113966] awarded to R.N. and the Royal Society-Newton Advanced Fellowship Grant [number NA160140] jointly awarded to R.N. and S.C.

Data Availability Statement: The datasets generated and/or analyzed during this study are available from the corresponding author on request.

Acknowledgments: This work emanates from the doctoral research study of S.J.M. S.J.M. was supported by the National Research Foundation (NRF) postgraduate student scholarship. A.O.A and T.G. were financially supported by the Najran University scholarship (Resolution No: 443-16-590) and The Royal Society's Newton International Fellowship (NIF\R1\221653), respectively.

Conflicts of Interest: The authors declare no conflicts of interest. The funders had no role in the design of this study; in the collection, analyses, or interpretation of data; in the writing of the manuscript; or in the decision to publish the results.

References

1. Mirzabaev, A.; Kerr, R.B.; Hasegawa, T.; Pradhan, P.; Wreford, A.; Pahlen, M.C.T.V.; Gurney-Smith, H. Severe climate change risks to food security and nutrition. *Clim. Risk Manag.* **2023**, *39*, 100473. [CrossRef]
2. Gan, T.Y.; Ito, M.; Hulsmann, S.; Qin, X.; Lu, X.X.; Liong, S.Y.; Rutschman, P.; Disse, M.; Koivusalo, H. Possible climate change/variability and human impacts, vulnerability of drought-prone regions, water resources and capacity building for Africa. *Hydrol. Sci. J.* **2016**, *61*, 1209–1226. [CrossRef]

3. Lottering, S.; Mafongoya, P.; Lottering, R. Drought and its impacts on small-scale farmers in sub-Saharan Africa: A review. *S. Afr. Geogr. J.* **2021**, *103*, 319–341. [CrossRef]
4. Mittler, R.; Blumwald, E. Genetic engineering for modern agriculture: Challenges and perspectives. *Annu. Rev. Plant Biol.* **2010**, *61*, 443–462. [CrossRef] [PubMed]
5. Passioura, J.B. Phenotyping for drought tolerance in grain crops: When is it useful to breeders? *Funct. Plant Biol.* **2012**, *39*, 851–859. [CrossRef]
6. Zheng, C.; Bochmann, H.; Liu, Z.; Kant, J.; Schrey, S.D.; Wojciechowski, T.; Postma, J.A. Plant root plasticity during drought and recovery: What do we know and where to go? *Front. Plant Sci.* **2023**, *14*, 1084355. [CrossRef]
7. Cramer, G.R.; Urano, K.; Delrot, S.; Pezzotti, M.; Shinozaki, K. Effects of abiotic stress on plants: A systems biology perspective. *BMC Plant Biol.* **2011**, *11*, 163. [CrossRef]
8. Ghatak, A.; Chaturvedi, P.; Weckwerth, W. Cereal crop proteomics: Systemic analysis of crop drought stress responses towards marker-assisted selection breeding. *Front. Plant Sci.* **2017**, *8*, 757. [CrossRef]
9. Ngara, R.; Goche, T.; Swanevelder, D.Z.H.; Chivasa, S. Sorghum's whole-plant transcriptome and proteome responses to drought stress: A review. *Life* **2021**, *11*, 704. [CrossRef]
10. Thanmalagan, R.R.; Jayaprakash, A.; Roy, A.; Arunachalam, A.; Lakshmi, P.T.V. A review on applications of plant network biology to understand the drought stress response in economically important cereal crops. *Plant Gene* **2022**, *29*, 100345. [CrossRef]
11. Farooq, M.; Wahid, A.; Kobayashi, N.; Fujita, D.; Basra, S.M.A. Plant drought stress: Effects, mechanisms and management. *Agron. Sustain. Dev.* **2009**, *29*, 185–212. [CrossRef]
12. Qamer, Z.; Chaudhary, M.T.; Du, X.; Hinze, L.; Azhar, M.T. Review of oxidative stress and antioxidative defense mechanisms in *Gossypium hirsutum* L. in response to extreme abiotic conditions. *J. Cotton Res.* **2021**, *4*, 9. [CrossRef]
13. Sharma, P.; Jha, A.B.; Bubey, R.S.; Pessarakli, M. Reactive oxygen species, oxidative damage, and antioxidative defense mechanism in plants under stressful conditions. *J. Bot.* **2012**, *2012*, 217037. [CrossRef]
14. Fang, Y.J.; Xiong, L.Z. General mechanisms of drought response and their application in drought resistance improvement in plants. *Cell. Mol. Life Sci.* **2015**, *72*, 673–689. [CrossRef]
15. Shinozaki, K.; Yamaguchi-Shinozaki, K. Gene networks involved in drought stress response and tolerance. *J. Exp. Bot.* **2007**, *58*, 221–227. [CrossRef]
16. Takahashi, F.; Kuromori, T.; Sato, H.; Shinozaki, K. Regulatory gene networks in drought stress responses and resistance in plants. *Adv. Exp. Med. Biol.* **2018**, *1081*, 189–214.
17. Rai, G.K.; Khanday, D.M.; Choudhary, S.M.; Kumar, P.; Kumari, S.; Martínez-Andújar, C.; Martinez-Melgarejo, P.A.; Rai, P.K.; Pérez-Alfocea, F. Unlocking nature's stress buster: Abscisic acid's crucial role in defending plants against abiotic stress. *Plant Stress* **2024**, *11*, 100359. [CrossRef]
18. Yoshida, T.; Mogami, J.; Yamaguchi-Shinozaki, K. ABA-dependent and ABA-independent signaling in response to osmotic stress in plants. *Curr. Opin. Plant Biol.* **2014**, *21*, 133–139. [CrossRef]
19. Liu, S.W.; Lv, Z.Y.; Liu, Y.H.; Li, L.; Zhang, L.D. Network analysis of ABA-dependent and ABA-independent drought responsive genes in *Arabidopsis thaliana*. *Genet. Mol. Biol.* **2018**, *41*, 624–637. [CrossRef]
20. Singh, M.; Kumar, J.; Singh, S.; Singh, V.P.; Prasad, S.M. Roles of osmoprotectants in improving salinity and drought tolerance in plants: A review. *Rev. Environ. Sci. Biotechnol.* **2015**, *14*, 407–426. [CrossRef]
21. Turner, N.C. Turgor maintenance by osmotic adjustment: 40 years of progress. *J. Exp. Bot.* **2018**, *69*, 3223–3233. [CrossRef] [PubMed]
22. Dumanovic, J.; Nepovimova, E.; Natic, M.; Kuca, K.; Jacevic, V. The significance of reactive oxygen species and antioxidant defense system in plants: A concise overview. *Front. Plant Sci.* **2021**, *11*, 552969. [CrossRef] [PubMed]
23. Tardieu, F.; Simonneau, T.; Muller, B. The physiological basis of drought tolerance in crop plants: A scenario-dependent probabilistic approach. *Annu. Rev. Plant Biol.* **2018**, *69*, 733–759. [CrossRef] [PubMed]
24. Grote, U.; Fasse, A.; Nguyen, T.T.; Erenstein, O. Food security and the dynamics of wheat and maize value chains in Africa and Asia. *Front. Sustain. Food Syst.* **2021**, *4*, 617009. [CrossRef]
25. Sendhil, R.; Kiran Kumara, T.M.; Kandpal, A.; Kumari, B.; Mohapatra, S. Wheat production, trade, consumption, and stocks: Global trends and prospects. In *Wheat Science: Nutritional and Anti-Nutritional Properties, Processing, Storage, Bioactivity, and Product Development*, 1st ed.; Gupta, O.P., Kumar, S., Pandey, A., Khan, M.K., Singh, S.K., Singh, G.P., Eds.; CRC Press: Boca Raton, USA, 2023; p. 23.
26. Ahmad, Z.; Waraich, E.A.; Akhtar, S.; Anjum, S.; Ahmad, T.; Mahboob, W.; Hafeez, O.B.; Tapera, T.; Labuschagne, M.; Rizwan, M. Physiological responses of wheat to drought stress and its mitigation approaches. *Acta Physiol. Plant.* **2018**, *40*, 80. [CrossRef]
27. Farooq, M.; Hussain, M.; Siddique, K.H.M. Drought stress in wheat during flowering and grain-filling periods. *Crit. Rev. Plant Sci.* **2014**, *33*, 331–349. [CrossRef]
28. Lopes, M.S.; El-Basyoni, I.; Baenziger, P.S.; Singh, S.; Royo, C.; Ozbek, K.; Aktas, H.; Ozer, E.; Ozdemir, F.; Manickavelu, A.; et al. Exploiting genetic diversity from landraces in wheat breeding for adaptation to climate change. *J. Exp. Bot.* **2015**, *66*, 3477–3486. [CrossRef]
29. Sansaloni, C.; Franco, J.; Santos, B.; Percival-Alwyn, L.; Singh, S.; Petroli, C.; Campos, J.; Dreher, K.; Payne, T.; Marshall, D.; et al. Diversity analysis of 80,000 wheat accessions reveals consequences and opportunities of selection footprints. *Nat. Commun.* **2020**, *11*, 4572. [CrossRef]

30. Tadesse, W.; Bishaw, Z.; Assefa, S. Wheat production and breeding in Sub-Saharan Africa: Challenges and opportunities in the face of climate change. *Int. J. Clim. Chang. Strateg. Manag.* **2019**, *11*, 696–715. [CrossRef]
31. De Sousa, T.; Ribeiro, M.; Sabença, C.; Igrejas, G. The 10,000-Year Success Story of Wheat! *Foods* **2021**, *10*, 2124. [CrossRef]
32. Abid, M.; Ali, S.; Qi, L.K.; Zahoor, R.; Tian, Z.W.; Jiang, D.; Snider, J.L.; Dai, T.B. Physiological and biochemical changes during drought and recovery periods at tillering and jointing stages in wheat (*Triticum aestivum* L.). *Sci Rep.* **2018**, *8*, 4615. [CrossRef] [PubMed]
33. Wang, J.R.; Zhang, X.Y.; Han, Z.D.; Feng, H.X.; Wang, Y.Y.; Kang, J.; Han, X.J.; Wang, L.F.; Wang, C.Y.; Li, H.; et al. Analysis of physiological indicators associated with drought tolerance in wheat under drought and re-watering conditions. *Antioxidants* **2022**, *11*, 2266. [CrossRef]
34. Amoah, J.N.; Ko, C.S.; Yoon, J.S.; Weon, S.Y. Effect of drought acclimation on oxidative stress and transcript expression in wheat (*Triticum aestivum* L.). *J. Plant Interact.* **2019**, *14*, 492–505. [CrossRef]
35. Kirova, E.; Moskova, I.; Geneva, M.; Kocheva, K. Antioxidant potential of tolerant and susceptible wheat varieties under drought and recovery. *Cereal Res. Commun.* **2022**, *50*, 841–849. [CrossRef]
36. Popova, A.V.; Mihailova, G.; Geneva, M.; Peeva, V.; Kirova, E.; Sichanova, M.; Dobrikova, A.; Georgieva, K. Different responses to water deficit of two common winter wheat varieties: Physiological and biochemical characteristics. *Plants* **2023**, *12*, 2239. [CrossRef]
37. Faghani, E.; Gharechahi, J.; Komatsu, S.; Mirzaei, M.; Khavarinejad, R.A.; Najafi, F.; Farsad, L.K.; Salekdeh, G.H. Comparative physiology and proteomic analysis of two wheat genotypes contrasting in drought tolerance. *J. Proteom.* **2015**, *114*, 1–15. [CrossRef]
38. Chaichi, M.; Sanjarian, F.; Razavi, K.; Gonzalez-Hernandez, J.L. Analysis of transcriptional responses in root tissue of bread wheat landrace (*Triticum aestivum* L.) reveals drought avoidance mechanisms under water scarcity. *PLoS ONE* **2019**, *14*, e0212671. [CrossRef]
39. Lv, L.J.; Chen, X.Y.; Li, H.; Huang, J.A.; Liu, Y.P.; Zhao, A.J. Different adaptive patterns of wheat with different drought tolerance under drought stresses and rehydration revealed by integrated metabolomic and transcriptomic analysis. *Front. Plant Sci.* **2022**, *13*, 1008624. [CrossRef]
40. Ford, K.L.; Cassin, A.; Bacic, A. Quantitative proteomic analysis of wheat cultivars with differing drought stress tolerance. *Front. Plant Sci.* **2011**, *2*, 44. [CrossRef]
41. Michaletti, A.; Naghavi, M.R.; Toorchi, M.; Zolla, L.; Rinalducci, S. Metabolomics and proteomics reveal drought-stress responses of leaf tissues from spring-wheat. *Sci. Rep.* **2018**, *8*, 5710. [CrossRef]
42. Kang, Z.Y.; Babar, M.A.; Khan, N.; Guo, J.; Khan, J.; Islam, S.; Shrestha, S.; Shahi, D. Comparative metabolomic profiling in the roots and leaves in contrasting genotypes reveals complex mechanisms involved in post-anthesis drought tolerance in wheat. *PLoS ONE* **2019**, *14*, e0213502. [CrossRef] [PubMed]
43. Yuan, J.S.; Galbraith, D.W.; Dai, S.Y.; Griffin, P.; Stewart, C.N. Plant systems biology comes of age. *Trends Plant Sci.* **2008**, *13*, 165–171. [CrossRef] [PubMed]
44. Fukushima, A.; Kusano, M.; Redestig, H.; Arita, M.; Saito, K. Integrated omics approaches in plant systems biology. *Curr. Opin. Chem. Biol.* **2009**, *13*, 532–538. [CrossRef] [PubMed]
45. Gomez-Sanchez, A.; Gonzalez-Melendi, P.; Santamaria, M.E.; Arbona, V.; Lopez-Gonzalvez, A.; Garcia, A.; Hensel, G.; Kumlehn, J.; Martinez, M.; Diaz, I. Repression of drought-induced cysteine-protease genes alters barley leaf structure and responses to abiotic and biotic stresses. *J. Exp. Bot.* **2019**, *70*, 2143–2155. [CrossRef]
46. Wang, C.L.; Gao, B.; Chen, N.N.; Jiao, P.; Jiang, Z.Z.; Zhao, C.L.; Ma, Y.Y.; Guan, S.Y.; Liu, S.Y. A novel senescence-specific gene (*ZmSAG39*) negatively regulates darkness and drought responses in maize. *Int. J. Mol. Sci.* **2022**, *23*, 15984. [CrossRef]
47. Shao, H.B.; Chu, L.Y.; Jaleel, C.A.; Manivannan, P.; Panneerselvam, R.; Shao, M.A. Understanding water deficit stress-induced changes in the basic metabolism of higher plants—Biotechnologically and sustainably improving agriculture and the ecoenvironment in arid regions of the globe. *Crit. Rev. Biotechnol.* **2009**, *29*, 131–151. [CrossRef]
48. Bevan, M.; Bancroft, I.; Bent, E.; Love, K.; Goodman, H.; Dean, C.; Bergkamp, R.; Dirkse, W.; Van Staveren, M.; Stiekema, W.; et al. Analysis of 1.9 Mb of contiguous sequence from chromosome 4 of *Arabidopsis thaliana*. *Nature* **1998**, *391*, 485–488. [PubMed]
49. Samarah, N.H. Effects of drought stress on growth and yield of barley. *Agron. Sustain. Dev.* **2005**, *25*, 145–149. [CrossRef]
50. Bai, L.-P.; Sui, F.-G.; Ge, T.-D.; Sun, Z.-H.; Lu, Y.-Y.; Zhou, G.-S. Effect of soil drought stress on leaf water status, membrane permeability and enzymatic antioxidant system of maize. *Pedosphere* **2006**, *16*, 326–332. [CrossRef]
51. Akter, N.; Brishty, T.A.; Karim, M.A.; Ahmed, M.J.U.; Islam, M.R. Leaf water status and biochemical adjustments as a mechanism of drought tolerance in two contrasting wheat (*Triticum aestivum* L.) varieties. *Acta Physiol. Plant.* **2023**, *45*, 50. [CrossRef]
52. Goche, T.; Shargie, N.G.; Cummins, I.; Brown, A.P.; Chivasa, S.; Ngara, R. Comparative physiological and root proteome analyses of two sorghum varieties responding to water limitation. *Sci. Rep.* **2020**, *10*, 11835. [CrossRef] [PubMed]
53. Xu, Y.W.; Song, D.; Qi, X.L.; Asad, M.; Wang, S.; Tong, X.H.; Jiang, Y.; Wang, S.D. Physiological responses and transcriptome analysis of soybean under gradual water deficit. *Front. Plant Sci.* **2023**, *14*, 1269884. [CrossRef] [PubMed]
54. Cui, G.B.; Zhao, Y.F.; Zhang, J.L.; Chao, M.N.; Xie, K.L.; Zhang, C.; Sun, F.L.; Liu, S.D.; Xi, Y.J. Proteomic analysis of the similarities and differences of soil drought and polyethylene glycol stress responses in wheat (*Triticum aestivum* L.). *Plant Mol. Biol.* **2019**, *100*, 391–410. [CrossRef]

55. Melkozernov, A.N.; Blankenship, R.E. Photosynthetic Functions of Chlorophylls. In *Chlorophylls and Bacteriochlorophylls: Biochemistry, Biophysics, Functions and Applications*; Grimm, B., Porra, R.J., Rudiger, W., Scheer, H., Eds.; Springer: Dordrecht, The Netherlands, 2006; Volume 25, pp. 397–412.
56. Taiz, L.; Zeiger, E. *Plant Physiology*, 4th ed.; Sinauer Associates Inc.: Sunderland, MA, USA, 2010.
57. Gill, S.S.; Tuteja, N. Reactive oxygen species and antioxidant machinery in abiotic stress tolerance in crop plants. *Plant Physiol. Biochem.* **2010**, *48*, 909–930. [CrossRef]
58. Ashraf, M.; Foolad, M.R. Roles of glycine betaine and proline in improving plant abiotic stress resistance. *Environ. Exp. Bot.* **2007**, *59*, 206–216. [CrossRef]
59. Hayat, S.; Hayat, Q.; Alyemeni, M.N.; Wani, A.S.; Pichtel, J.; Ahmad, A. Role of proline under changing environments: A review. *Plant Signal. Behav.* **2012**, *7*, 1456–1466. [CrossRef]
60. Szabados, L.; Savouré, A. Proline: A multifunctional amino acid. *Trends Plant Sci.* **2010**, *15*, 89–97. [CrossRef]
61. Zeng, W.J.; Peng, Y.L.; Zhao, X.Q.; Wu, B.Y.; Chen, F.Q.; Ren, B.; Zhuang, Z.L.; Gao, Q.H.; Ding, Y.F. Comparative proteomics analysis of the seedling root response of drought-sensitive and drought-tolerant maize varieties to drought stress. *Int. J. Mol. Sci.* **2019**, *20*, 2793. [CrossRef]
62. Pyngrope, S.; Bhoomika, K.; Dubey, R.S. Reactive oxygen species, ascorbate-glutathione pool, and enzymes of their metabolism in drought-sensitive and tolerant indica rice (*Oryza sativa* L.) seedlings subjected to progressing levels of water deficit. *Protoplasma* **2013**, *250*, 585–600. [CrossRef]
63. Smith, K.T.; Workman, J.L. Chromatin proteins: Key responders to stress. *PLoS Biol.* **2012**, *10*, e1001371. [CrossRef]
64. Yuan, L.Y.; Liu, X.C.; Luo, M.; Yang, S.G.; Wu, K.Q. Involvement of histone modifications in plant abiotic stress responses. *J. Integr. Plant Biol.* **2013**, *55*, 892–901. [CrossRef] [PubMed]
65. Li, N.; Zhang, S.; Liang, Y.J.; Qi, Y.H.; Chen, J.; Zhu, W.N.; Zhang, L.S. Label-free quantitative proteomic analysis of drought stress-responsive late embryogenesis abundant proteins in the seedling leaves of two wheat (*Triticum aestivum* L.) genotypes. *J. Proteom.* **2018**, *172*, 122–142. [CrossRef] [PubMed]
66. Singh, H.; Kaur, K.; Singh, M.; Kaur, G.; Singh, P. Plant cyclophilins: Multifaceted proteins with versatile roles. *Front. Plant Sci.* **2020**, *11*, 585212. [CrossRef]
67. Li, Y.H.; Cui, J.Y.; Zhao, Q.; Yang, Y.Z.; Wei, L.; Yang, M.D.; Liang, F.; Ding, S.T.; Wang, T.C. Physiology and proteomics of two maize genotypes with different drought resistance. *Biol. Plant.* **2019**, *63*, 519–528. [CrossRef]
68. Waititu, J.K.; Zhang, X.E.; Chen, T.C.; Zhang, C.Y.; Zhao, Y.; Wang, H. Transcriptome analysis of tolerant and susceptible maize genotypes reveals novel insights about the molecular mechanisms underlying drought responses in leaves. *Int. J. Mol. Sci.* **2021**, *22*, 6980. [CrossRef] [PubMed]
69. Fracasso, A.; Trindade, L.M.; Amaducci, S. Drought stress tolerance strategies revealed by RNA-Seq in two sorghum genotypes with contrasting WUE. *BMC Plant Biol.* **2016**, *16*, 115. [CrossRef]
70. Varoquaux, N.; Cole, B.; Gao, C.; Pierroz, G.; Baker, C.R.; Patel, D.; Madera, M.; Jeffers, T.; Hollingsworth, J.; Sievert, J.; et al. Transcriptomic analysis of field-droughted sorghum from seedling to maturity reveals biotic and metabolic responses. *Proc. Natl. Acad. Sci. USA* **2019**, *116*, 27124–27132. [CrossRef]
71. Azzouz-Olden, F.; Hunt, A.G.; Dinkins, R. Transcriptome analysis of drought-tolerant sorghum genotype SC56 in response to water stress reveals an oxidative stress defense strategy. *Mol. Biol. Rep.* **2020**, *47*, 3291–3303. [CrossRef]
72. Kosova, K.; Vitamvas, P.; Prasil, I.T.; Klima, M.; Renaut, J. Plant proteoforms under environmental stress: Functional proteins arising from a single gene. *Front. Plant Sci.* **2021**, *12*, 793113. [CrossRef]
73. Nemati, M.; Piro, A.; Norouzi, M.; Vahed, M.M.; Nisticò, D.M.; Mazzuca, S. Comparative physiological and leaf proteomic analyses revealed the tolerant and sensitive traits to drought stress in two wheat parental lines and their F6 progenies. *Environ. Exp. Bot.* **2019**, *158*, 223–237. [CrossRef]
74. Chang, L.L.; Wang, L.M.; Peng, C.Z.; Tong, Z.; Wang, D.; Ding, G.H.; Xiao, J.H.; Guo, A.P.; Wang, X.C. The chloroplast proteome response to drought stress in cassava leaves. *Plant Physiol. Biochem.* **2019**, *142*, 351–362. [CrossRef] [PubMed]
75. Vineeth, T.V.; Kumar, P.; Krishna, G.K. Bioregulators protected photosynthetic machinery by inducing expression of photorespiratory genes under water stress in chickpea. *Photosynthetica* **2016**, *54*, 234–242. [CrossRef]
76. Barrs, H.; Weatherley, P. A re-examination of the relative turgidity technique for estimating water deficits in leaves. *Aust. J. Biol. Sci.* **1962**, *15*, 413–428. [CrossRef]
77. Rahman, S.; Shabeen, M.S.; Rahman, M.A.; Malik, T.A. Evaluation of excised leaf water loss and relative water content, as screening techniques for breeding drought resistant wheat. *Pak. J. Biol. Sci.* **2000**, *3*, 663–665.
78. Hiscox, J.D.; Israelstam, G.F. A method for the extraction of chlorophyll from leaf tissue without maceration. *Can. J. Bot.* **1979**, *57*, 1332–1334. [CrossRef]
79. Wellburn, A.R. The spectral determination of chlorophyll-a and chlorophhyll-b, as well as total carotenoids, using various solvents with spectrophotometers of different resolution. *J. Plant Physiol.* **1994**, *144*, 307–313. [CrossRef]
80. Gokul, A.; Carelse, M.F.; Niekerk, L.A.; Klein, A.; Ludidi, N.; Mendoza-Cozatl, D.; Keyster, M. Exogenous 3,3′-diindolylmethane improves vanadium stress tolerance in *Brassica napus* seedling shoots by modulating antioxidant enzyme activities. *Biomolecules* **2021**, *11*, 436. [CrossRef]
81. Velikova, V.; Yordanov, I.; Edreva, A. Oxidative stress and some antioxidant systems in acid rain treated bean plants—Protective role of exogenous polyamines. *Plant Sci.* **2000**, *151*, 59–66. [CrossRef]

82. Prinsen, H.; Schiebergen-Bronkhorst, B.G.M.; Roeleveld, M.W.; Jans, J.J.M.; de Sain-van der Velden, M.G.M.; Visser, G.; van Hasselt, P.M.; Verhoeven-Duif, N.M. Rapid quantification of underivatized amino acids in plasma by hydrophilic interaction liquid chromatography (HILIC) coupled with tandem mass-spectrometry. *J. Inherit. Metab. Dis.* **2016**, *39*, 651–660. [CrossRef] [PubMed]
83. Ngara, R. A Proteomic Analysis of Drought and Salt Stress Responsive Proteins of Different Sorghum Varieties. Ph.D. Thesis, University of the Western Cape, Cape Town, South Africa, 2009.
84. Moloi, M.J.; van der Merwe, R. Drought tolerance responses in vegetable-type soybean involve a network of biochemical mechanisms at flowering and pod-filling stages. *Plants* **2021**, *10*, 1502. [CrossRef]
85. UniProt, C. UniProt: The Universal Protein Knowledgebase in 2023. *Nucleic Acids Res.* **2023**, *51*, D523–D531.
86. Paysan-Lafosse, T.; Blum, M.; Chuguransky, S.; Grego, T.; Pinto, B.L.; Salazar, G.A.; Bileschi, M.L.; Bork, P.; Bridge, A.; Colwell, L.; et al. InterPro in 2022. *Nucleic Acids Res.* **2023**, *51*, D418–D427. [CrossRef] [PubMed]
87. Kanehisa, M.; Goto, S. KEGG: Kyoto Encyclopedia of Genes and Genomes. *Nucleic Acids Res.* **2000**, *28*, 27–30. [CrossRef]
88. Szklarczyk, D.; Kirsch, R.; Koutrouli, M.; Nastou, K.; Mehryary, F.; Hachilif, R.; Gable, A.L.; Fang, T.; Doncheva, N.T.; Pyysalo, S.; et al. The STRING database in 2023: Protein-protein association networks and functional enrichment analyses for any sequenced genome of interest. *Nucleic Acids Res.* **2023**, *51*, D638–D646. [CrossRef]
89. Ngwenya, S.P.; Moloi, S.J.; Shargie, N.G.; Brown, A.P.; Chivasa, S.; Ngara, R. Regulation of proline accumulation and protein secretion in sorghum under combined osmotic and heat stress. *Plants* **2024**, *13*, 1874. [CrossRef] [PubMed]
90. Ma, J.; Li, R.Q.; Wang, H.G.; Li, D.X.; Wang, X.Y.; Zhang, Y.C.; Zhen, W.C.; Duane, H.J.; Yan, G.J.; Li, Y.M. Transcriptomics analyses reveal wheat responses to drought stress during reproductive stages under field conditions. *Front. Plant Sci.* **2017**, *8*, 592. [CrossRef]
91. Kumar, J.; Gunapati, S.; Kianian, S.F.; Singh, S.P. Comparative analysis of transcriptome in two wheat genotypes with contrasting levels of drought tolerance. *Protoplasma* **2018**, *255*, 1487–1504. [CrossRef]
92. Ye, J.; Coulouris, G.; Zaretskaya, I.; Cutcutache, I.; Rozen, S.; Madden, T.L. Primer-BLAST: A tool to design target-specific primers for polymerase chain reaction. *BMC Bioinform.* **2012**, *13*, 134. [CrossRef]

plants

MDPI

Article

Biochemical and Epigenetic Regulation of Glutamate Metabolism in Maize (*Zea mays* L.) Leaves under Salt Stress

Alexander T. Eprintsev [1], Galina B. Anokhina [1], Polina S. Selivanova [1], Polina P. Moskvina [1] and Abir U. Igamberdiev [2,*]

[1] Department of Biochemistry and Cell Physiology, Voronezh State University, Voronezh 394018, Russia; bc366@bio.vsu.ru (A.T.E.); dowi2009@mail.ru (G.B.A.); lol221297@mail.ru (P.S.S.); polinamoskvina2001@gmail.com (P.P.M.)

[2] Department of Biology, Memorial University of Newfoundland, St. John's, NL A1C 5S7, Canada

* Correspondence: igamberdiev@mun.ca

Abstract: The effect of salt stress (150 mM NaCl) on the expression of genes, methylation of their promoters, and enzymatic activity of glutamate dehydrogenase (GDH), glutamate decarboxylase (GAD), and the 2-oxoglutarate (2-OG)–dehydrogenase (2-OGDH) complex was studied in maize (*Zea mays* L.). GDH activity increased continuously under salt stress, being 3-fold higher after 24 h. This was accompanied by the appearance of a second isoform with lower electrophoretic mobility. The expression of the *Gdh1* gene strongly increased after 6–12 h of incubation, which corresponded to the demethylation of its promoter, while *Gdh2* gene expression slightly increased after 2–6 h and then decreased. GAD activity gradually increased in the first 12 h, and then returned to the control level. This corresponded to the increase of *Gad* expression and its demethylation. Salt stress led to a 2-fold increase in the activity of 2-OGDH during the first 6 h of NaCl treatment, then the activity returned to the control level. Expression of the genes *Ogdh1* and *Ogdh3* peaked after 1–2 h of incubation. After 6–8 h with NaCl, the expression of these genes declined below the control levels, which correlated with the higher methylation of their promoters. We conclude that salt stress causes a redirection of the 2-OG flux to the γ-aminobutyric acid shunt via its amination to glutamate, by altering the expression of the *Gdh1* and *Gdh2* genes, which likely promotes the assembly of the native GDH molecule having a different subunit composition and greater affinity for 2-OG.

Keywords: glutamate dehydrogenase; mitochondria; NaCl; promoter methylation; 2-oxoglutarate dehydrogenase; salt stress; tricarboxylic acid cycle; *Zea mays* L.

check for updates

Citation: Eprintsev, A.T.; Anokhina, G.B.; Selivanova, P.S.; Moskvina, P.P.; Igamberdiev, A.U. Biochemical and Epigenetic Regulation of Glutamate Metabolism in Maize (*Zea mays* L.) Leaves under Salt Stress. *Plants* **2024**, *13*, 2651. https://doi.org/10.3390/plants13182651

Academic Editor: Pedro Diaz-Vivancos

Received: 16 July 2024
Revised: 17 September 2024
Accepted: 18 September 2024
Published: 21 September 2024

1. Introduction

Salinity affects vast areas of agricultural lands, and the most common factor causing salt stress is NaCl [1]. The salinity tolerance mechanisms in plants include limiting ion uptake and compartmentalizing ions, which leads to the prevention of negative osmotic effects on cells. The role of mitochondrial metabolism in the adaptation to stress conditions is essential for developing a salt tolerance via an increase in ATP generation, the scavenging of reactive oxygen species (ROS), and the regulation of ion transport across membranes [2]. The increase in the oxidation of mitochondrial substrates is an important characteristic of the salt stress response usually referred as a "salt respiration" [3]. It is caused by stress-induced changes in the modulation of redox balance in mitochondria resulting in the activation of respiratory enzymes and transporters at the genetic level, epigenetic level, and at the level of posttranscriptional modification of proteins [4–7].

The effect of salt stress includes two distinct phases, the early osmotic phase and the later ionic phase. The first phase refers to the rapid onset of osmotic stress within minutes due to the fast decrease in extracellular water potential, and the second phase refers to ionic and oxidative stress, as well as nutrient imbalance resulting from the accumulation of ions

(Na^+ and Cl^- in the case of NaCl stress) over time. The consequences of the second phase of stress can be observed over several days [8–11]. Salt stress affects the entire plant organism, including at the molecular genetic level. This occurs via specific signaling mechanisms that provide a cellular response to stress factors. Plants adapt to salt stress through the synthesis of osmolytes such as proline and soluble sugars and amines that possess the osmoprotective activity [12–14]. Proline biosynthesis occurs via the glutamate and ornithine pathways, and, in addition to reducing cytosolic osmotic potential, it is important for stabilizing protein structures, scavenging free radicals, and maintaining intracellular ionic homeostasis [14].

The tricarboxylic acid (TCA) cycle undergoes significant changes as a result of salt stress. Mitochondrial respiration and the active operation of the TCA cycle are important sources of ATP in stress conditions. Under the conditions of salinity, in the first six hours of stress, the intensification of the TCA cycle takes place [15–17]. The increase in the TCA cycle enzymes, including the 2-oxoglutarate–dehydrogenase (2-OGDH) complex [18], as well as the succinate dehydrogenase, which directly supplies electrons to the electron transport chains of mitochondria [17], is due to the need for an additional influx of energy to neutralize the negative effects of salts on cellular metabolic processes. The 2-OGDH complex (EC 1.2.1.105) is a complex multienzyme system that provides oxidation of 2-oxoglutarate (2-OG) to succinyl-CoA in mitochondria while simultaneously reducing NAD^+ to NADH. The 2-OGDH complex plays an important role in plant metabolism, catalyzing, firstly, the rate-limiting stage of mitochondrial respiration, and secondly, being a key participant in carbon–nitrogen interactions [19,20]. According to Che-Othman et al. [21], longer-term salt stress exhibits an inactivating effect on the activity of the 2-OGDH complex, which suggests that the operation of the TCA cycle is suppressed under these conditions. During the exposure to salinity, key metabolic enzymes necessary for the cyclic operation of the TCA cycle are inhibited by sodium chloride, which is overcome by the inclusion of an alternative pathway—the γ-aminobutyric acid (GABA) shunt—which provides an additional source of succinate for mitochondria [21,22]. This shunt bypasses the 2-OGDH complex via the decarboxylation of glutamate, which can be an alternative product of the 2-oxoglutarate (2-OG) conversion.

The connection between the TCA cycle and the compensatory stress response pathway, the GABA shunt, is ensured by the operation of GDH (EC 1.4.1.3), which catalyzes the reversible deamination of L-glutamate to 2-OG [23,24]. In addition to transamination, 2-OG can be converted to glutamate via the reverse reaction of glutamate dehydrogenase (GDH) [25,26]. In this reaction, the ammonium ion (NH_4^+) is utilized at a high concentration and at a high redox level [27]. These conditions are common in the case of salt stress [28]. GDH in plants is a simultaneous participant in both energy and biosynthetic metabolism, which is especially important from the point of view of cell adaptation to stress. A peculiarity of GDH in plants is that the catalytically active enzyme is formed via the assembly of a homo- or hetero-hexameric molecule consisting of α and/or β subunits. The predominance of α-subunits in the enzyme structure provides a higher affinity for glutamate, while the presence of β-subunits increases the affinity for 2-oxoglutarate [29–31].

The studies on cucumber leaves [13] showed that the activities of GDH (both in the amination and deamination directions), alanine aminotransferase, aspartate aminotransferase, and NADH-dependent isocitrate dehydrogenase changed after a prolonged (more than 24 h) exposure to salt stress. The activities of the above enzymes increased significantly throughout the experiment, which was associated with a high need for glutamate [13]. GDH induction at the genetic level in barley in the conditions of salt stress was demonstrated [32]. It is known that a high NaCl concentration stimulates the formation of reactive oxygen species (ROS), which in turn induce the synthesis of the α-subunit of GDH in tobacco and grapevines [33]. The important role of GDH in the adaptive response of plant cells to salt stress is confirmed by the report of Kumar et al. [34], which showed that in salt-tolerant varieties of rice (*Oryza sativa*), the activity of GDH in the direction of amination increases with the increasing salt stress, while in salt-sensitive varieties it decreases [34].

It has been shown that in pea (*Pisum sativum*) plants, which are resistant to ammonium excess, the aminating activity of GDH in roots is very high [35].

The glutamate formed during the amination of 2-oxoglutarate can be directed both to the synthesis of proline and to the formation of GABA, which in turn is a substrate for the GABA shunt, bypassing the 2-OGDH reaction. GDH isoforms that preferably operate in the direction of ammonia assimilation act as anti-stress enzymes in ammonia detoxification and glutamate production for proline synthesis [33]. In plants, GABA is formed primarily through the H^+-consuming α-decarboxylation of glutamate in an irreversible reaction catalyzed by the cytosol-localized glutamate decarboxylase (GAD, EC 4.1.1.15), which occurs under low pH conditions with the participation of pyridoxal phosphate. GABA synthesized in the cytosol enters mitochondria via the mitochondrial GABA transporter and is converted first to succinic semialdehyde (SSA) by GABA transaminase, and then to succinate, which is part of the TCA cycle [36]. This last step greatly influences the redox status of the cell because succinate bypasses the three TCA cycle reactions that produce NADH [37,38].

The adaptation of plants to salt stress includes changes in the content of metabolites, which are regulated at the biochemical and the genetic levels. However, another option for regulating the metabolic flows of a plant cell during an adaptive response to stress is the involvement of epigenetic mechanisms, which include DNA methylation. The methylation of various regions of the genome of the facultative halophyte plant *Mesembryanthemum crystallinum* to soil salinity was demonstrated [39]. A unique feature of this plant is its ability to switch from the C_3 pathway of photosynthesis to Crassulacean Acid Metabolism (CAM) under water deficiency and salinity. This switch is achieved through the activation of a large group of genes, among which the gene of the CAM-specific form of PEP carboxylase *Ppc1* is considered "diagnostic". The transcription level of this gene increases sharply during adaptation to salt stress; nevertheless, no changes in the methylation pattern of its promoter region were detected, and the methylation of repetitive rRNA genes did not change either. However, total DNA methylation at CpNpG sites increased 2-fold during salt stress. It turned out that this increase was associated mainly, if not exclusively, with a satellite DNA fraction. Apparently, adaptation to salt stress is achieved due to the global epigenetic reorganization of chromatin, which ensures the modulation of the expression of a large group of genes [39].

Despite the extensive knowledge about possible adaptive mechanisms of cellular metabolism to the effects of salinity caused by NaCl, the mode of regulation and functioning of the 2-OGDH complex and GDH under short-term salinity stress remains unclear. In this regard, the goal of this study was to identify the regulatory aspects of the operation of the 2-OGDH complex and GDH in maize leaves in the first 24 h of salt stress caused by NaCl. We studied the expression and epigenetic regulation of the genes *ZmOgdh1* and *ZmOgdh3* encoding the E1 subunit of 2-OGDH, of the genes *ZmGdh1* and *ZmGdh2* encoding α and β subunits of GDH, and of the gene *ZmGad* encoding GAD (further the names of genes are given without *Zm* for brevity). Our study demonstrates that salt stress affects the expression of these genes, and that this process is partially regulated via epigenetic mechanisms.

2. Results

2.1. Activity of the 2-OGDH Complex

The activity of the 2-OGDH complex increased in the first hour of incubation of the plants with the NaCl solution (Figure 1A). The maximum activity was more than twice that of the control plants at 2 h of incubation, and then it gradually decreased. After 12 h of incubation on the medium with NaCl, the activity became close to the control values.

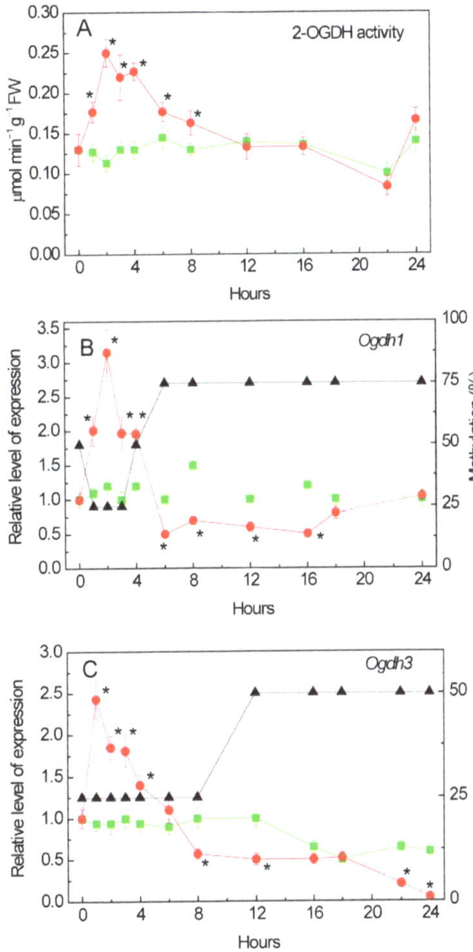

Figure 1. Effect of salt stress on the operation of the 2-oxoglutarate dehydrogenase (2-OGDH) complex. Changes in the activity of the 2-OGDH complex (**A**), in the relative levels of transcripts and the fraction of methylation of promoters (black triangles) of the genes *Ogdh1* (**B**) and *Ogdh3* (**C**) in maize leaves in the course of incubation of maize plants in 150 mM NaCl (red circles) as compared to the control plants (green squares). The data represent the means of three biological repeats ± SD. Statistically significant differences in activity and expression as compared to the control ($p \leq 0.05$) are shown by stars.

2.2. Changes in 2-Ogdh Gene Expression

We studied the transcriptional activity of the two genes encoding the subunit E1 of the 2-OGDH complex in the maize genome: *Ogdh1* (LOC100383579) and *Ogdh3* (LOC100383847) encoding the E1 subunit (2-OG–dehydrogenase) of the 2-OGDH complex [40,41] (the date of access to the NCBI database was 26 September 2023). Salt stress stimulated the expression in both genes. In the first hours of the experiment, an increase in the level of *Ogdh1* gene transcripts was observed compared to the control samples (Figure 1B). The maximum concentration of the *Ogdh1* gene mRNA was recorded after 2 h of incubation of the plants in the NaCl solution. Further incubation of the plants in a NaCl solution led to a decrease in the expression of *Ogdh1*. The maximum effect of NaCl incubation on *Ogdh3* transcript level was achieved in the first hour of stress exposure (Figure 1C). An increased level of expression of the *Ogdh3* gene mRNA in the first hours of the experiment (up to 6 h) was followed by its gradual decline. After 24 h, the relative level of *Ogdh3* transcripts in the NaCl incubated plants was 20 times lower than in the control plants.

2.3. Changes in the Degree of Methylation of Individual CpG Dinucleotides in Ogdh Promoters

Changes in the mRNA level of the maize *Ogdh1* gene when the seedlings were exposed to NaCl correlated with the changes in the methyl status of individual CpG dinucleotides in the promoter. An increase in gene transcription was accompanied by a decrease in the amount of methylated cytosines, while a decrease in the gene mRNA concentration correlated with an increase in the proportion of methylated CpG dinucleotides (Figure 1B). At the same time, in the control samples, the degree of methylation was 50% throughout the entire experiment (Figure S1).

An analysis of the NaCl induced changes in the *Ogdh3* gene methylation and its expression showed that in the first hours of salt stress there was an induction of the functioning of this gene. Moreover, there were no fluctuations in the degree of methylation of individual CpG dinucleotides—the proportion of methylated cytosines during a 4 h exposure to NaCl was 25% (relative to the number of CpG dinucleotides studied) (Figure 1C). The decrease in the transcriptional activity of the *Ogdh3* gene observed at 12 h of the experiment is apparently associated with an increase in the number of methylated cytosines in the promoter region to 50% (relative to the CpG dinucleotides under study). In the control group of plants, the degree of methylation remained at a constant level and amounted to 25% (Figure S1).

2.4. Analysis of 2-Ogdh Gene Promoters for the Presence of CpG Islands

An analysis of the *Ogdh* gene promoters for the presence of CpG islands showed that there is no single CpG islands in the promoter region of the *Ogdh1* gene (Figure 2A), while the promoter region of *Ogdh3* contains two CpG islands with a size of 116 bp and 591 bp (Figure 2B).

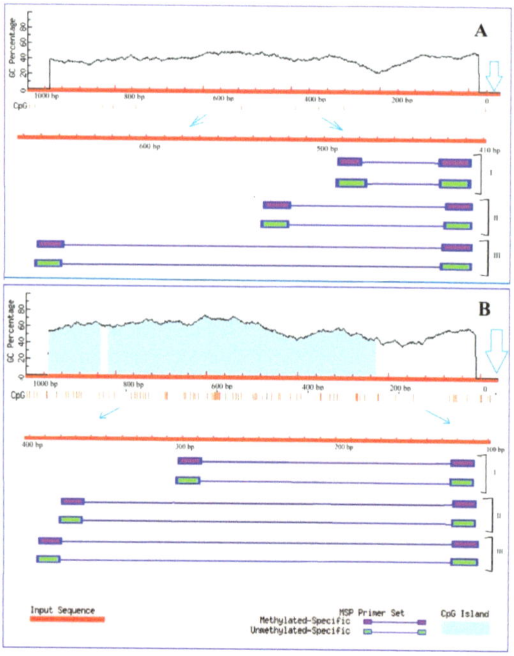

Figure 2. Results of analysis of the promoters of the genes *Ogdh1* (**A**) and *Ogdh3* (**B**) of *Zea mays* for the presence of CpG islands. Vertical lines indicate the positions of CpG dinucleotides. The outlined arrows indicate the position of the start codon. Thin blue arrows show the change of scale to outline the region used for designing three groups (I, II, III) of primers to the different CpG dinucleotides.

Due to the fact that CpG islands are present in the promoter region of the gene, it can be assumed that the operation of this gene can be regulated through methylation. However, the absence of CpG islands does not exclude the possibility of controlling gene expression by changing the methyl status, since in plant organisms, unlike animals, cytosine methylation is possible not only at CpG sites, but also at CpNpG and CpNpN sites, where N is A, T, or C [42]. The analysis of the promoter region of the *Ogdh1* gene for the presence of CpNpG and CpNpN sites resulted in the estimation that the studied promoter region of 1000 bp in size contains 24.9% CpNpN sites and 11.4% CpNpG sites, which indicates the possibility of regulating gene transcription by changing the methyl status of its promoter.

2.5. Effect of Salt Stress on Glutamate Dehydrogenase Activity

Salinity induced the overall activity of GDH in the amination reaction (Figure 3A). An almost 1.5-fold increase in GDH activity was recorded after 1 h of NaCl incubation, then it further continuously increased, and at 24 h of incubation, the total GDH activity was three times higher than in the control.

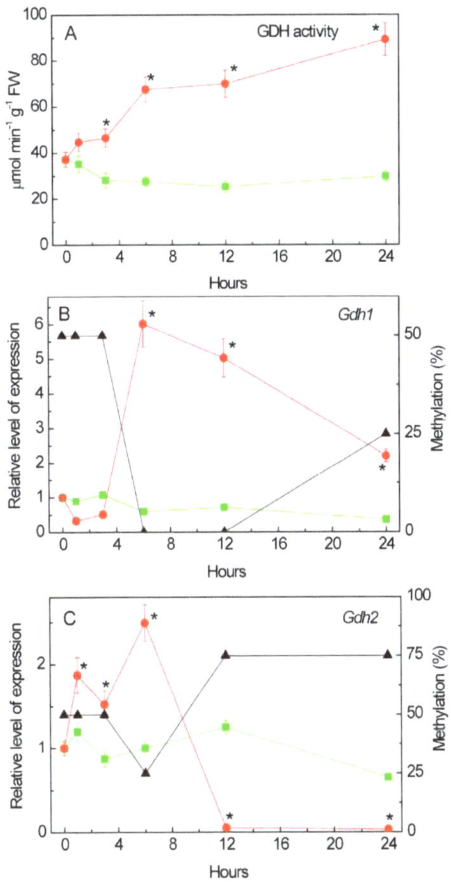

Figure 3. Effect of salt stress on glutamate dehydrogenase activity and expression in maize leaves. Changes in the activity of glutamate dehydrogenase (GDH) (**A**), in the relative levels of transcripts and the fraction of methylation of promoters (black triangles) of the genes *Gdh1* (**B**) and *Gdh2* (**C**) in maize leaves in the course of incubation of maize plants in 150 mM NaCl (red circles) as compared to the control plants (green squares). The data represent the means of three biological repeats ± SD. Statistically significant differences in activity and expression as compared to the control ($p \leq 0.05$) are shown by stars.

2.6. Isoenzyme Composition of GDH under Salt Stress Conditions

The increase in the overall enzymatic activity of GDH was accompanied by the appearance of an additional isoform with lower electrophoretic mobility on electropherograms (Figure 4). This second isoform was detectable after 1 h of incubation in the NaCl solution, and after 12 h of the experiment, it disappeared from electropherograms. Before the start of the experiment, as well as in the plants from the control group, only one form of GDH with an electrophoretic mobility value of 0.17 was observed; however, starting from the first hour of incubation in a saline solution, the appearance of a second isoform with a mobility value of 0.13 was noted. It is known that the α-subunit has greater electrophoretic mobility than the β-subunit [43], which suggests that the appearance of the second isoenzyme is due to the occurrence in the structure of the polypeptide of the less mobile β-subunit.

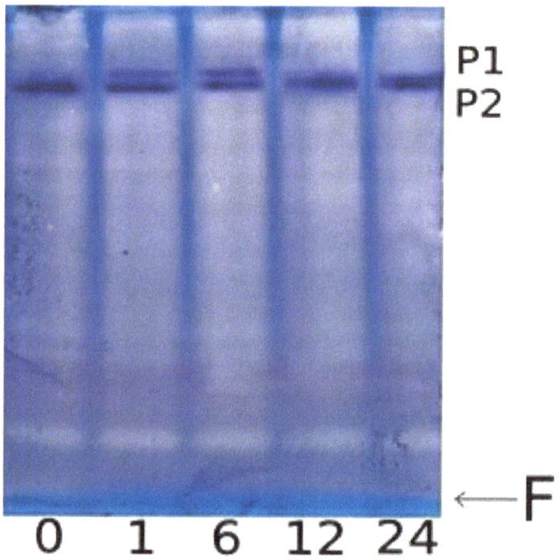

Figure 4. Effect of salt stress on the isoenzyme composition of GDH in maize leaves. PAGE electropherogram of GDH from maize leaves under salt stress: 0, 1, 6, 12, 24—incubation time in the NaCl solution (hours); P1, P2—protein bands representing native GDH protein molecules (isoenzymes) stained by the tetrazolium method; and F—dye front.

2.7. Changes in the Expression of Gdh Genes under Salt Stress

The induction of GDH activity under salt stress can be associated with changes in the functioning of the *Gdh1* (LOC542220) and *Gdh2* (LOC100193614) genes [44,45] (Figure 3B,C). The relative level of *Gdh1* gene transcripts, despite the initial reduction in the first 5 h of the experiment, increased at 6 h of salt stress. The low level of expression of this gene in the first hours of salinity was compensated by the operation of the *Gdh2* gene, for which transcriptional activity increased after 1 h of incubation in the solution of NaCl (Figure 3C). This corresponds to the appearance of an isoform with lower electrophoretic mobility (Figure 4). Starting from 6 h, the expression of the *Gdh1* gene increased.

An analysis of the promoters of the *Gdh* family genes for the presence of CpG islands showed that the promoter of the *Gdh1* gene does not contain a single CpG island (Figure 5A), while two islands of 404 and 383 bp in size were found in the promoter region of the *Gdh2* gene (Figure 5B). A study of the promoter region of the *Gdh1* gene for the presence of CpNpG and CpNpN sites revealed that the structure of the 1000 bp promoter region contains 24.9% CpNpN sites and 13.8% CpNpG sites. A study of changes in the degree of methylation of individual CpG dinucleotides in the *Gdh1* gene promoter revealed that the increase in the mRNA concentration at 6 h of stress exposure was associated with a complete demethylation of the dinucleotides under study, which were 50% methylated before the experiment and in the first 3 h of the experiment (Figure 3B). An increase in the concentration of the *Gdh2* gene mRNA at 6 h of salt stress was accompanied by a decrease in the proportion of methylated cytosines (from 50 to 25%) in its promoter, and further inactivation of the gene corresponded to an increase in the degree of methylation to 75% (Figure 3C). In the control group of plants, the degree of methylation of the promoters of both genes was 50% throughout the entire duration of the experiment (Figure S2).

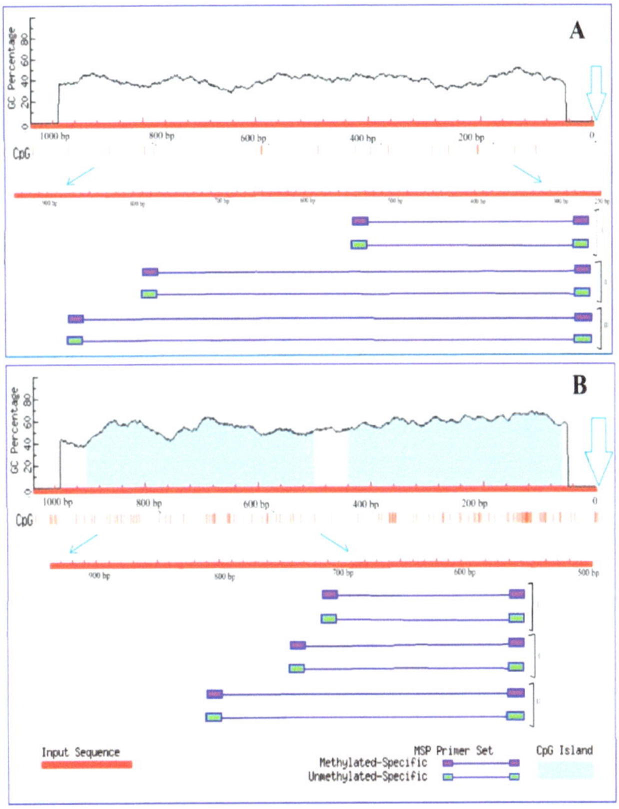

Figure 5. Results of analysis of the promoters of the genes *Gdh1* (**A**) and *Gdh2* (**B**) of *Zea mays* for the presence of CpG islands. Vertical lines indicate the positions of CpG dinucleotides. The outlined arrows indicate the position of the start codon. Thin blue arrows show the change of scale to outline the region used for designing three groups (I, II, III) of primers to the different CpG dinucleotides.

2.8. Effect of Salt Stress on Glutamate Decarboxylase Activity

An analysis of the total enzymatic activity of glutamate decarboxylase (GAD) in corn leaves under salinity conditions caused by 150 mM of sodium chloride showed that from the first hour of stress exposure there already was a gradual increase in the studied parameter relative to the values recorded in the control group of plants (Figure 6A). The maximum increase in GAD activity was observed at 12 h of the experiment. In the following hours, a gradual decrease in the values of the catalytic activity of GAD was recorded.

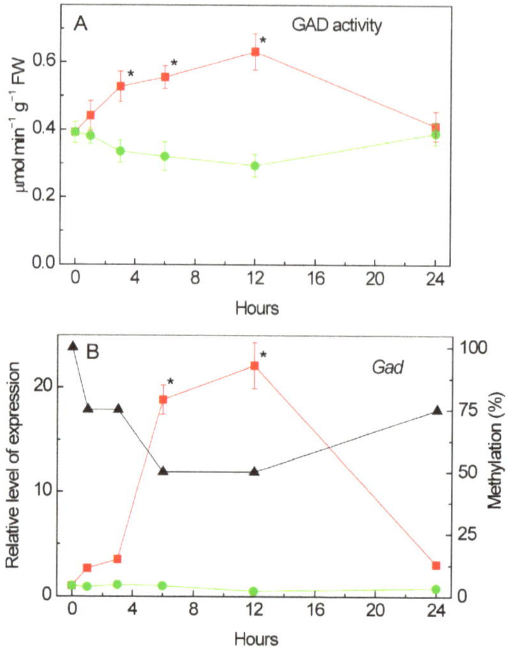

Figure 6. Effect of salt stress on glutamate decarboxylase (GAD) activity and expression in maize leaves. Changes in the total enzymatic activity of GAD (**A**), and in the relative levels of transcripts and the fraction of methylation of the promoters (black triangles) of the gene *Gad* (**B**) in maize leaves in the course of incubation of maize plants in 150 mM NaCl (red circles) as compared to the control plants (green squares). The data represent the means of three biological repeats ± SD. Statistically significant differences in activity and expression as compared to the control ($p \leq 0.05$) are shown by stars.

2.9. Changes in the Expression of *Gad* under Salt Stress

An analysis of the transcriptional activity of the *Gad* gene (LOC100284394) [46], which encodes glutamate decarboxylase in the maize genome, showed an increase in the relative level of transcripts from the first hour of stress exposure (Figure 6B). The maximum increase in the mRNA level was observed at 12 h of the experiment and exceeded the control values by more than 20 times. By 24 h of the experiment, a significant decrease in the transcriptional activity of the *Gad* gene was noted; however, the relative level of transcripts in the group exposed to salt stress was more than three times higher than in the control group of plants. An analysis of the nucleotide sequence of the promoter region of the *Gad* gene showed the presence in its structure of two CpG islands measuring 130 and 543 bp (Figure 7).

A comparative analysis of the changes in the relative level of gene transcripts and the degree of methylation of its promoter showed that an increase in the transcriptional activity of the *Gad* gene is accompanied by a decrease in the degree of promoter methylation to 50%. Before the start of the experiment, all dinucleotides under study were methylated; however, by the 6th hour of the experiment, the degree of methylation of the individual CpG dinucleotides promoter was 50%. The decrease in the mRNA level at 24 h of the experiment was associated with an increase in the methyl status of the promoter region of the *Gad* gene. Thus, changes in the transcriptional activity of the *Gad* gene are likely due to changes in the methyl status of its promoter, which indicates the epigenetic nature of the regulation of this gene in maize leaves under salinity conditions caused by 150 mM of sodium chloride. In the control group of plants, 50% of all studied CpG dinucleotides were methylated throughout the entire experiment (Figure S3).

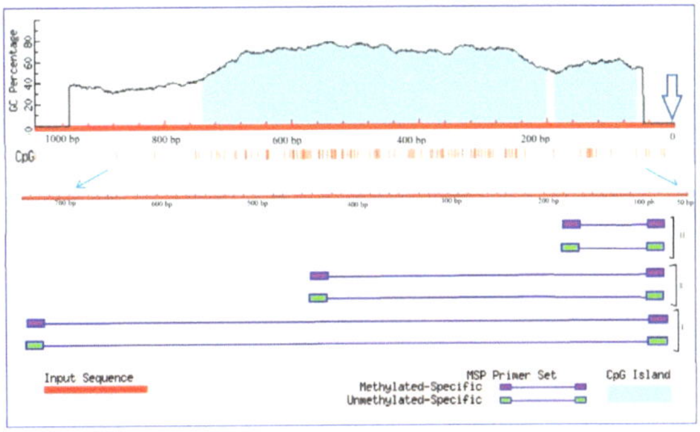

Figure 7. Analysis of *Gad* gene promoters of *Zea mays* for the presence of CpG islands. Vertical lines indicate the positions of CpG dinucleotides. The outlined arrow indicates the position of the start codon. Thin blue arrows show the change of scale to outline the region used for designing three groups (I, II, III) of primers to the different CpG dinucleotides.

3. Discussion

The obtained data reveal significant changes in the activity and expression of the enzymes involved in 2-OG conversion under the conditions of salt stress. The 2-OGDH multienzyme complex catalyzes the rate-limiting stage of the TCA cycle [18]. The incubation of maize seedlings in the NaCl solution in the first six hours leads to the intensification of the TCA cycle due to an increase in the activity of the 2-OGDH complex in leaves. The operation of this complex is regulated at the biochemical [20], genetic [47,48], and epigenetic (in this study) levels. An additional influx of 2-OG to the TCA cycle due to an increase in the deamination activity of GDH leads to further intensification of the formation of succinyl-CoA. This is evidenced by the appearance of an additional GDH isoenzyme, which is caused by an increase in the expression profile of the *Gdh2* gene. These assumptions are confirmed by data obtained from Tercé-Laforgue et al. [43], who showed that a polypeptide containing in its structure the β-subunit encoded by the *Gdh2* gene has lower electrophoretic mobility during PAGE electrophoresis than the GDH enzyme consisting of α-subunits. GDH, consisting predominantly of β-subunits, has high deamination activity, thus the direction of the reaction is shifted towards the formation of 2-OG, while the polypeptide consisting of α-subunits is characterized by a high affinity for 2-OG, which contributes to the redirection of the reaction towards the formation of glutamate in the amination reaction [43].

The initial rapid induction of the TCA cycle under conditions of salt stress, caused by an increase in the catalytic activity of its enzymes in the first hours of incubation, is known as "salt respiration" [6,17]. The observed increase in the activity of the 2-OGDH complex in the first hours of salinity indicates an increase in the rate of functioning of the entire TCA cycle. Since the 2-OGDH complex provides the oxidation of 2-OG through the interaction of its three components, while the slowest reaction in this complex multi-step process is catalyzed by 2-OG dehydrogenase (E1 subunit; E.C. 1.2.4.2), a study was carried out on the dynamics of the transcriptional activity of genes encoding the subunit E1 (*Ogdh1* and *Ogdh3*).

A study of the mRNA level of genes encoding 2-OGDH showed that salt stress stimulates their expression: in the first hours of the experiment, an increase in the level of *Ogdh1* gene transcripts was observed compared to the control samples (Figure 1A,B). Further incubation of the seedlings in a NaCl solution led to a decrease in the expression of this gene, which corresponded to the decline in the enzymatic activity of the 2-OGDH complex compared to the control level. The level of *Ogdh3* gene transcripts increased even earlier, and it was also followed by a gradual decline in its transcriptional activity (Figure 1C). The decline in the relative level of *Ogdh3* transcripts in maize leaves after 24 h of salt stress was substantial, being ~20 times lower than in the leaves of the control group; thus, the activity of 2-OGDH in the conditions of salt stress was supported by the *Ogdh1* gene.

There is compelling evidence that epigenetic modifications may be an important mechanism of adaptation to different habitats. Moreover, there is a large amount of data indicating changes in the degree of methylation of both the entire genome and its individual regions are induced by stress [49]. The transcriptional activity of genes can be regulated by changing the methyl status of both individual CpG dinucleotides and CpG dinucleotides within the CpG island [50]. In addition, it is known that DNA methylation regulates the activity of enhancers through changes in the methyl status of the transcription factor binding sites [49]. An analysis of the OGDH gene promoters for the presence of CpG islands showed that there is not a single CpG island in the promoter region of the *Ogdh1* gene (Figure 2A), while the promoter region of *Ogdh3* contains two CpG islands (Figure 2B). The presence of CpG islands in the promoter region suggests that the operation of the corresponding gene is regulated through methylation. However, the absence of CpG islands does not exclude the possibility of controlling gene expression by changing the methyl status, since in plant organisms, unlike animals, cytosine methylation is possible not only at CpG sites, but also at CpNpG and CpNpN sites [51]. Changes in the methyl status of the CpG dinucleotides of the *Ogdh1* and *Ogdh3* gene promoters, correlating with the transformation of the relative level of their transcripts, indicate the important role of DNA methylation in the regulation of the operation of the 2-OGDH complex during the adaptive response of cellular metabolism to salinity conditions.

The disappearance of the effect of salt respiration at the level of the 2-OGDH complex during the long-term (more than 6 h) incubation of seedlings in a 150 mM NaCl solution (Figure 1) indicates a possibility of a switch in the 2-OG flow in the mitochondria using the reverse reaction of GDH by directing it to the GABA shunt [17]. This redistribution of the key metabolite of the TCA cycle is due to the fact that the 2-OGDH activity is reduced after 6 h of salt stress, which corresponds to an increase in the proportion of methylated cytosines in the promoters of the *Ogdh1* and *Ogdh3* genes (Figure 1B,C). If 2-OGDH does not utilize all accumulated 2-OG, this may lead to the activation of a GDH isoform with a higher capacity in the amination reaction.

The induction of GDH activity under salt stress is associated with the changes in the functioning of the *Gdh1* and *Gdh2* genes (Figure 3). In the *Zea mays* L. genome, GDH is encoded by two genes, each of which contains information about different types of GDH subunits, α and β [52]. The predominance of one or another type of subunit in the structure of the polypeptide leads to a shift in the equilibrium of the reaction towards the formation of 2-OG, or, conversely, towards the synthesis of glutamate. The activation of glutamate conversion to 2-OG corresponds with the appearance of an isoform with lower electrophoretic mobility (Figure 4), which possibly contains in its structure the α-subunit of GDH, providing the displacement of the reaction towards the formation of 2-OG, which subsequently enters the TCA cycle, to maintain energy metabolism. Then, starting from 6 h, the *Gdh1* gene is activated (Figure 3), encoding the β-subunit of GDH, which promotes the reaction towards the formation of glutamate, which can be used for the synthesis of proline [29,34,35] or decarboxylated to form GABA [22,37]. The observed changes in the expression of the *Gdh1* gene under salt stress conditions are associated with a transformation of the methyl status, indicating an epigenetic mechanism for the regulation of its transcription. An increase in the expression of the *Gdh1* gene, caused by a decrease in the degree of cytosine methylation of this gene, as well as an increase in the proportion of methylated cytosines in the promoter of the *Gdh2* gene, provides an alternative pathway for the conversion of 2-OG, which results in the activation of the GABA shunt.

A change in the expression of the GDH genes "switches" the flow of 2-OG from the TCA cycle to amino acid synthesis. It should be noted that GDH activity both in the first 6 h and at the second stage of stress exposure (after 6 h) was regulated due to the differential expression of the *Gdh1* and *Gdh2* genes. At the same time, the processes of methylation/demethylation of CpG dinucleotides within their promoters play a key role in the regulation of the GDH activity (Figures 3 and 5). Glutamate entering the GABA shunt is decarboxylated to GABA, which is converted via succinic semialdehyde to succinate, and the latter subsequently enters the TCA cycle in the consequent reaction. This bypasses the limiting step of the TCA cycle at the level of the 2-OGDH complex and provides efficient adaptation to the conditions of salt stress.

Thus, the incubation of maize seedlings in a NaCl solution results in a long-term induction of GDH, which includes a short-term induction of the second isoform. Both genes (*Gdh1* and *Gdh2*) participate in the regulation of the GDH activity under salt stress. The appearance of an additional isoform in the first hours of the experiment is associated with the formation of a form of GDH consisting of α-subunits, the predominance of which in the structure of the polypeptide ensures the occurrence of an enzymatic reaction towards the formation of 2-OG, and contributes to the maintenance of the TCA cycle during the short period of "salt respiration". Long-term exposure of maize seedlings to NaCl (6 h or more) induces the transcription of the *Gdh1* gene mRNA, resulting in the synthesis of β-subunits of the enzyme. This leads the 2-OG from the TCA cycle to its amination, thereby turning it into glutamate, which can then be used for the synthesis of proline and GABA, which is confirmed by the data on changes in the expression and activity of GAD (Figure 6). The regulation of the glutamate dehydrogenase genes *Gdh1* and *Gdh2* under salt stress was carried out epigenetically due to changes in the methyl status of the CpG dinucleotides in their promoters.

The initial activation of expression of the subunit E1 of the 2-OGDH complex provides the conversion of an additional pool of 2-OG formed from glutamate. After 6 h of salt stress, the limitations in 2-OG conversion would result in its amination to glutamate, which is directed to proline and GABA (Figure 8). A possibility of metabolite channeling in the TCA cycle [53] can facilitate these conversions and result in an efficient rearrangement of the mitochondrial metabolism for its adjustment to the conditions of salt stress.

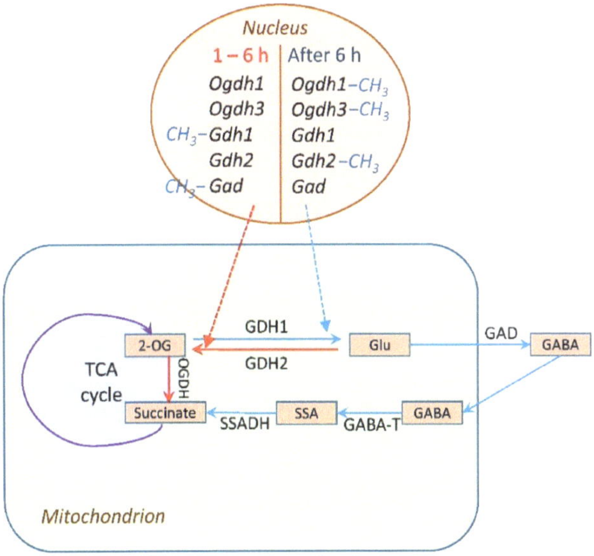

Figure 8. Regulation of glutamate metabolism in plant cells under salt stress. Abbreviations: TCA cycle, tricarboxylic acid cycle; 2-OG, 2-oxoglutarate; OGDH, 2-oxoglutarate dehydrogenase complex; GDH1 and GDH2, polypeptides encoded by the *Gdh1* and *Gdh2* genes, respectively; Glu, glutamate; GABA, γ-aminobutyric acid; SSA, succinic acid semialdehyde; Suc, succinate. Salt stress affects plant cell metabolism in two stages: 1. During the first 6 h of salt stress, 2-OGDH activity increases, while GDH, due to the induction of *Gdh2* gene expression, acts as a supplier of 2-OG, the source of which is glutamate (red arrow). 2. After 6 h, 2-OGDH is inhibited, GDH redirects the flow of 2-OG to glutamate due to the induction of the *Gdh1* gene, 2-OG is converted into glutamate, which is used for the synthesis of GABA (blue arrow). $-CH_3$ indicates the methylation of gene promoter.

4. Material and Methods

4.1. Object of Investigation

The object of research was two-weeks-old maize seedlings (*Zea mays* L., cv Voronezhskaya-76, obtained from the Voronezh branch of the All-Russian Research Institute of Maize), grown hydroponically under 10 h of daylight with a light intensity of 90 μmol quanta m^{-2} s^{-1} (climatic chamber "LabTech", Namyangju, Republic of Korea) and an ambient temperature of 25 °C.

4.2. Induction of Salt Stress

The induction of salt stress was achieved by placing the plants, with the roots removed, in a 0.15 M NaCl solution for 24 h. Plants placed in water for the duration of the experiment were used as a control.

4.3. Determination of 2-OGDH, GDH and GAD Activities

4.3.1. Extraction of 2-OGDH and GDH

The activities of 2-OGDH and GDH were determined in the mitochondrial fraction. For this purpose, the plant material was homogenized in the grinding medium containing 0.15 M potassium phosphate buffer (pH 7.4), 0.4 M sucrose, 2.5 mM EDTA, 1 mM KCl, and 4 mM $MgCl_2$ in a ratio of 1 g plant material to 10 mL grinding medium. The homogenate was filtered through four layers of cheesecloth and centrifuged for 3 min at 3000× *g*. The supernatant was centrifuged for 10 min at 20,000× *g*. The pellet was resuspended in 50 mM potassium phosphate buffer (pH 7.4). Thiamine pyrophosphate (final concentration 0.2 mM) was added to the mitochondrial suspension when resuspending the sediment to measure 2-OGDH and GDH activity. The degree of mitochondrial breakage was more than 90%, which was controlled by microscopy on an Olympus CX41RF ("Olympus", Tokyo, Japan). The cross-contamination of mitochondria by the cytoplasmic fraction did not exceed 5%, by the chloroplast fraction was 10%, and by the peroxisomal fraction was 3%; this was determined by measuring the enzymes specific to each fraction as described earlier [54].

4.3.2. Extraction of GAD

The extraction of GAD was carried out by homogenizing the plant material in a 20 mM acetate buffer (pH 4.8) with the addition of 10 mM of pyridoxal phosphate, after which it was centrifuged at 12,000× *g* for 30 min. The supernatant was used for the GAD activity assay. All procedures were carried out at 4 °C.

4.3.3. Determination of 2-OGDH Activity

2-Oxoglutarate–dehydrogenase complex (2-OGDH; EC 1.2.1.105) activity was determined in the mitochondrial fraction at 25 °C on an Evolution 260 Bio spectrophotometer (Thermo Fisher Scientific, Waltham, MA, USA) by the rate of NADH formation at 340 nm in a 0.1 M Tris-HCl buffer (pH 7.5) containing 1 mM potassium 2-oxoglutarate, 2 mM NAD^+, 0.5 mM $MgCl_2$, 0.12 mM CoA (lithium salt), 0.2 mM TPP, 2.5 mM cysteine, 1 mM AMP, and 0.05% Triton X-100 [18].

4.3.4. Determination of GDH Activity

For the L-Glutamate, NAD(P)-oxidoreductase (GDH; EC 1.4.1.3) activity in the amination reaction was determined at 340 nm in 0.1 M Tris-HCl buffer (pH 8.0) containing 13 mM 2-oxoglutarate, 0.25 mM NADH, 1 mM $CaCl_2$, and 50 mM $(NH_4)_2SO_4$ on an Evolution 260 Bio spectrophotometer (Thermo Fisher Scientific, Waltham, MA, USA) [55].

4.3.5. Determination of GAD Activity

The determination of glutamate decarboxylase (GAD; EC 4.1.1.15) activity was carried out using an Evolution 260 Bio spectrophotometer (Thermo Fisher Scientific, Waltham, MA, USA) by measuring changes in optical density at 620 nm for a solution containing a 20 mM acetate buffer (pH 4.8), 70 μM bromocresol green, 10 mM pyridoxal-5-phosphate, and 2 mM sodium glutamic acid [56]. The method is based on recording changes in the pH of the medium in the course of the reaction, which is reflected in the appearance of the protonated form of bromocresol green. The protonated form of bromocresol green has an absorption maximum at the wavelength of 620 nm.

One unit of enzymatic activity of GDH, GAD, and 2-OGDH corresponds to the conversion of 1 μmol of the substrate per minute at 25 °C and the optimal pH value.

4.3.6. Electrophoretic Separation of GDH Isozymes

The isoenzyme patterns of GDH were determined by PAGE electrophoresis at 2–4 °C, according to Maurer [57]. The concentration of acrylamide in the upper gel was 4%, and in the lower gel was 7%. Equal amounts of protein (determined by the Lowry method [58]) were added to each pocket of the gel. The specific development of GDH was carried out using the tetrazolium method by incubating the gel in the dark in a 0.1 M Tris-HCl buffer (pH 8.0) containing 1 mM phenazine methosulfate (PMS), 1.8 mM nitroblue tetrazolium (NBT), 2 mM NAD^+, and 10 mM sodium glutamate [59].

4.4. Expression of Gdh, Gad, and Ogdh Genes

To assess changes in the transcriptional activity of the *Gdh*, *Gad*, and *Ogdh* genes in maize leaves, we analyzed changes in the relative levels of transcripts of the *Gdh1* (*LOC542220*) and *Gdh2* (*LOC100193614*) genes encoding the α and β subunits of GDH, the *Gad* (*LOC 100284394*) gene encoding GAD, and the *Ogdh1* (*LOC100383579*) and *Ogdh3* (*LOC100383847*) genes encoding the subunit E1 of 2-OGDH in real-time PCR using specific primers (Tables S1–S3). The efficiency of the primers was estimated in the range of 0.2–0.8 μM by performing a series of dilutions followed by amplification by real-time PCR. The range of primer efficiency for the genes *Gdh1*, *Gdh2*, *Gad*, *Ogdh1*, and *Ogdh3* was 0.3–0.5 μM.

The progress in sequencing of the maize genome [60,61] made it possible to identify the genes encoding the studied enzymes. As a template for RT-PCR, we used cDNA obtained through a reverse transcription reaction with the MMLV-RT Kit (JSC Evrogen, Moscow, Russia) in accordance with the manufacturer's protocol. RNA for the reverse transcription reaction was obtained through a phenol–chloroform extraction using LiCl to remove DNA [62,63]. The plant material (100 mg tissue per 1 mL medium) was ground in a porcelain mortar with an extraction medium containing 4 M guanidine thiocyanate, 30 mM sodium citrate, 30 mM β-mercaptoethanol, and a pH between 7.0 and 7.5; then 2 M of sodium acetate (pH 4.0) was added (1/10 of the volume) and mixed, followed by adding 1 volume of phenol saturated with water. After vortexing, a mixture of chloroform and isoamyl alcohol (49:1) was added, mixed (20 s), and incubated on ice for 15–20 min. After centrifugation for 20 min at 10,000× *g* at 4 °C, the upper aqueous phase was transferred to a new Eppendorf tube and 1 mL of isopropanol was added, after which the mixture was incubated at −20 °C for 1 h. The RNA was precipitated by centrifugation for 20 min at 10,000× *g* at 4 °C. The supernatant was removed, and the pellet was washed twice with 500 μL of 80% ethanol, dried and dissolved in RNase-free water (100 μL).

To purify the RNA from DNA contamination, 1 volume of 12 M lithium chloride was added, after which the sample was incubated at −20 °C for 30 min and centrifuged at 10,000× *g* for 15 min. The supernatant containing DNA was removed, and the RNA sediment was washed twice with 80% ethanol, then dried and dissolved in 40 μL of RNase-free water.

A real-time PCR was carried out on a LightCycler96 device (Roche, Solna, Sweden), using SYBR Green I as an intercalating dye. Amplification was carried out according to the following parameters: initial denaturation at 95 °C, for 5 min; then 35 cycles, including the stages of denaturation at 95 °C for 20 s, and at 72 °C for 20 s. Finally, a 10 min final elongation was performed at 72 °C. Quantitative matrix control was performed using gene-specific primers for housekeeping genes (Ef-1α elongation factors) (Table S4). Total RNA without the RT-PCR step was used as a negative control. Calculation of the relative levels of transcripts of the studied genes was carried out using the $2^{-\Delta\Delta Ct}$ method [64].

4.5. Analysis of the Promoters of the Studied Genes for the Presence of CpG Islands

An analysis of the promoters of the studied genes for the presence of CpG islands and selection of primers for methyl-specific PCR (Tables S5–S7) were carried out using the MethPrimer online service. The choice of the region for which the primers were selected was determined by the presence of individual CpG dinucleotides, as well as of plant-specific methylation sites (CNN, CNG). The analysis of potential targets was performed manually considering the microenvironment of the individual CpG dinucleotides, the presence of promoter regulatory elements near the studied CpG dinucleotides, and the methylation/demethylation of CpG dinucleotides (as well as CNN and CNG nucleotides) near the start codon.

4.6. Analysis of the Degree of Methylation of CpG Dinucleotides by Methyl-Specific PCR

To study the methyl status of individual CpG dinucleotides in the promoters of the genes under study, methyl-specific PCR was performed using ScreenMix (JSC Evrogen, Moscow, Russia), using DNA modified with sodium bisulfite as a template with some modifications [65]. In the first stage, pre-denatured DNA (incubation at 55 °C in the presence of 0.3 M NaOH) was modified by adding 500 µL of a mixture of 2.3 M $NaHSO_3$ solution (Sigma-Aldrich, St. Louis, MO, USA) and 0.5 M hydroquinone (Sigma-Aldrich, St. Louis, MO, USA), followed by incubation in the dark at 55 °C for 4 h. At the second stage, the modified DNA was purified from sodium bisulfite using the DNeasy Plant Pro Kit (Qiagen, Hilden, Germany).

Desulfonation of the modified DNA was carried out by incubation with 0.3 M of NaOH (final concentration) for 20 min at 37 °C. To precipitate DNA from the solution, 2 µL of glycogen solution (20 mg/mL), 35 µL of 10 M ammonium acetate, and 3 volumes of 80% ethanol were added followed by incubation for 20 min at −20 °C. Then, it was centrifuged for 30 min at 13,000× *g*, washed twice with cooled 80% ethanol, and dried and dissolved in DNase-free water.

Amplification was carried out on a Mini AMP Thermal cycler (Thermo Fisher, Waltham, MA, USA) according to the following scheme: initial denaturation (10 min, 95 °C), 35 cycles of amplification, and final elongation (10 min, 72 °C). The cycles of amplification included denaturation (30 s, 95 °C), amplification (30 s, 52 °C), and elongation (30 s, 72 °C).

A calculation of the percent of methylation of the promoters of the studied genes was performed based on the results of the electrophoresis of the methyl-specific PCR amplicons. A visualization of the PCR products was carried out in a 2% agarose gel (Helicon, Moscow, Russia) with the addition of ethidium bromide (Sigma-Aldrich, St. Louis, MO, USA) [66]. For visualization, a Serva Blue cube 300 transilluminator (SERVA Electrophoresis GmbH, Heidelberg, Germany) with a wavelength of 312 nm was used. The degree of methylation of the promoter region is the total value of the results of the methyl-specific PCR of the studied CpG dinucleotides of a particular gene. To assess the degree of promoter methylation (the fraction of methylated CpG dinucleotides), the following criteria were used: 0% methylation—all three studied CpG dinucleotides were unmethylated; 25% methylation—1 or 2 CpG dinucleotides were partially methylated or 1 CpG dinucleotide was completely methylated; 50% methylation—1 or 2 CpG dinucleotides were completely methylated or all three examined CpG dinucleotides were partially methylated; 75% methylation—1 or 2 CpG dinucleotides were partially methylated, and the rest were fully methylated; and 100% methylation—all three CG dinucleotides were methylated [67]. This method provides an estimation of the degree of methylation in our experiments with a high precision ($p < 0.01$) within the frame of the four defined degrees of methylation.

4.7. Statistical Analysis

The experiments were performed in three biological replicates; analytical determinations for each sample were carried out in triplicate. The STATISTICA 12.0 program was used. The quantitative indicators were assessed for compliance with normal distribution using the Shapiro–Wilk test. The data distribution was assessed as normal distribution.

The differences were analyzed for statistical significance using the Student's *t*-test. The results in the graphs were expressed as mean ± SD. Additionally, a one-way analysis of variance ANOVA was used [68]. The differences discussed in this study are statistically significant ($p \leq 0.05$).

5. Conclusions

In the first six hours (stage 1 of exposure) of salt stress, the TCA cycle is intensified due to an increase in the catalysis of the limiting reaction catalyzed by the 2-OGDH complex. The rate of operation of this metabolon is regulated at the biochemical, genetic, and epigenetic levels. Salt stress results in a short-term increase in the expression of the genes *Ogdh1* and *Ogdh3* encoding the E1 subunit of the 2-OGDH complex, due to a decrease in the proportion of methylated cytosines in the promoters of these genes. This provides the conversion of 2-OG formed from glutamate in the reaction of GDH. After 6 h of salt stress, the limitations in 2-OG conversion would result in its amination to glutamate, which is directed to GABA via GAD and proline. This rearrangement of the mitochondrial metabolism is important for the adjustment of plant cells to the conditions of salt stress. The process of coordinated changes in the activities of the 2-OGDH complex, GAD, and GDH is partially regulated via epigenetic mechanisms.

Supplementary Materials: The following supporting information can be downloaded at: https://www.mdpi.com/article/10.3390/plants13182651/s1, Figure S1: Changes in the relative levels of transcripts of the genes *Ogdh1* and *Ogdh3* and of the fraction of methylation of their promoters in the control (non-stressed) maize plants.; Figure S2: Changes in the relative levels of transcripts of the genes *Gdh1* and *Gdh2* and of the fraction of methylation of their promoters in the control (non-stressed) maize plants; Figure S3: Changes in the relative levels of transcripts of the gene *Gad* and of the fraction of methylation of its promoters in the control (non-stressed) maize plants. Table S1: Primers to the genes encoding glutamate dehydrogenaseTable S2: Primers to the genes of glutamate decarboxylase ; Table S3: Primers to the genes of 2-oxoglutarate dehydrogenase; Table S4: Primers to the genes encoding Ef-1α; Table S5: Oligonucleotides for methyl-specific PCR for the primers of *Gdh1* and *Gdh2*; Table S6: Oligonucleotides for methyl-specific PCR for the primers of *Gad*; Table S7: Oligonucleotides for methyl-specific PCR for the primers of *Ogdh1* and *Ogdh3*.

Author Contributions: Conceptualization, A.T.E. and A.U.I.; investigation, A.T.E., G.B.A., P.S.S. and P.P.M.; methodology, A.T.E., G.B.A., P.S.S. and P.P.M.; writing—original draft preparation, A.T.E., G.B.A. and P.S.S.; writing—review and editing, A.T.E. and A.U.I.; project administration, A.T.E. and A.U.I.; funding acquisition, A.T.E. All authors have read and agreed to the published version of the manuscript.

Funding: This research was funded (to A.T.E.) by the Ministry of Science and Higher Education of the Russian Federation as part of the state task for universities in the field of scientific activity for 2023–2025, project No FZGU-2023-0009.

Data Availability Statement: The datasets generated for this study are available upon request from the corresponding author.

Conflicts of Interest: The authors declare no conflicts of interest.

Abbreviations

GABA	γ-aminobutyric acid
GAD	glutamate decarboxylase
GDH	glutamate dehydrogenase
NBT	nitroblue tetrazolium
2-OG	2-oxoglutarate
2-OGDH	2-oxoglutarate dehydrogenase
PMS	phenazine methosulfate
TCA	cycle, tricarboxylic acid cycle

References

1. Jacoby, R.P.; Taylor, N.L.; Millar, A.H. The role of mitochondrial respiration in salinity tolerance. *Trends Plant Sci.* **2011**, *16*, 614–623. [CrossRef] [PubMed]
2. Schwarzländer, M.; Fricker, M.D.; Sweetlove, L.J. Monitoring the in vivo redox state of plant mitochondria: Effect of respiratory inhibitors, abiotic stress and assessment of recovery from oxidative challenge. *Biochim. Biophys. Acta* **2009**, *1787*, 468–475. [CrossRef]
3. Robertson, R.N.; Wilkins, M. Quantitative Relation between Salt Accumulation and Salt Respiration in Plant Cells. *Nature* **1948**, *161*, 101. [CrossRef] [PubMed]
4. Bandehagh, A.; Taylor, N.L. Can alternative metabolic pathways and shunts overcome salinity-induced inhibition of central carbon metabolism in crops? *Front. Plant Sci.* **2020**, *11*, 1072. [CrossRef] [PubMed]
5. Møller, I.M.; Igamberdiev, A.U.; Bykova, N.V.; Finkemeier, I.; Rasmusson, A.G.; Schwarzländer, M. Matrix Redox Physiology Governs the Regulation of Plant Mitochondrial Metabolism through Posttranslational Protein Modifications. *Plant Cell* **2020**, *32*, 573–594. [CrossRef] [PubMed]
6. Van Zelm, E.; Zhang, Y.; Testerink, C. Salt Tolerance Mechanisms of Plants. *Annu. Rev. Plant Biol.* **2020**, *71*, 403–433. [CrossRef]
7. Igamberdiev, A.U.; Bykova, N.V. Mitochondria in photosynthetic cells: Coordinating redox control and energy balance. *Plant Physiol.* **2023**, *191*, 2104–2119. [CrossRef]
8. Hasanuzzaman, M.; Nahar, K.; Fujita, M. Plant response to salt stress and role of exogenous protectants to mitigate salt-induced damages. In *Ecophysiology and Responses of Plants under Salt Stress*; Springer: New York, NY, USA, 2013; pp. 25–87.
9. Isayenkov, S.V.; Maathuis, F.J.M. Plant salinity stress: Many unanswered questions remain. *Front. Plant Sci.* **2019**, *10*, 80. [CrossRef]
10. Saddhe, A.A.; Malvankar, M.R.; Karle, S.B.; Kumar, K. Reactive nitrogen species: Paradigms of cellular signaling and regulation of salt stress in plants. *Env. Exp. Bot.* **2019**, *161*, 86–97. [CrossRef]
11. Qiu, X.M.; Sun, Y.Y.; Ye, X.Y.; Li, Z.G. Signaling Role of Glutamate in Plants. *Front. Plant Sci.* **2020**, *10*, 1743. [CrossRef]
12. Negrão, S.; Schmöckel, S.M.; Tester, M. Evaluating physiological responses of plants to salinity stress. *Ann. Bot.* **2017**, *119*, 1–11. [CrossRef] [PubMed]
13. Naliwajski, M.R.; Skłodowska, M. The relationship between carbon and nitrogen metabolism in cucumber leaves acclimated to salt stress. *Peer J.* **2018**, *6*, e6043. [CrossRef] [PubMed]
14. El Moukhtari, A.; Cabassa-Hourton, C.; Farissi, M.; Savouré, A. How Does Proline Treatment Promote Salt Stress Tolerance During Crop Plant Development? *Front. Plant Sci.* **2020**, *11*, 1127. [CrossRef] [PubMed]
15. Eprintsev, A.T.; Fedorin, D.N.; Cherkasskikh, M.V.; Igamberdiev, A.U. Effect of salt stress on the expression and promoter methylation of the genes encoding the mitochondrial and cytosolic forms of aconitase and fumarase in maize. *Int. J. Mol. Sci.* **2021**, *22*, 6012. [CrossRef] [PubMed]
16. Epintsev, A.T.; Gataullina, M.O. Regulation of the Activity of Decarboxylating Malate Dehydrogenases in Corn Leaves during an Adaptive Response to Salt Stress. *Biol. Bull.* **2023**, *50*, 39–46. [CrossRef]
17. Fedorin, D.N.; Eprintsev, A.T.; Florez Caro, O.J.; Igamberdiev, A.U. Effect of Salt Stress on the Activity, Expression, and Promoter Methylation of Succinate Dehydrogenase and Succinic Semialdehyde Dehydrogenase in Maize (*Zea mays* L.) Leaves. *Plants* **2022**, *12*, 68. [CrossRef]
18. Araújo, W.L.; Nunes-Nesi, A.; Trenkamp, S.; Bunik, V.I.; Fernie, A.R. Inhibition of 2-oxoglutarate dehydrogenase in potato tuber suggests the enzyme is limiting for respiration and confirms its importance in nitrogen assimilation. *Plant Physiol.* **2008**, *148*, 1782–1796. [CrossRef]
19. Nunes-Nesi, A.; Araújo, W.L.; Obata, T.; Fernie, A.R. Regulation of the mitochondrial tricarboxylic acid cycle. *Curr. Opin. Plant Biol.* **2013**, *16*, 335–343. [CrossRef]
20. Millar, A.H.; Hill, S.A.; Leaver, C.J. Plant mitochondrial 2-oxoglutarate dehydrogenase complex: Purification and characterization in potato. *Biochem. J.* **1999**, *343*, 327–334. [CrossRef]
21. Che-Othman, M.H.; Jacoby, R.P.; Millar, A.H.; Taylor, N.L. Wheat mitochondrial respiration shifts from the tricarboxylic acid cycle to the GABA shunt under salt stress. *New Phytol.* **2020**, *225*, 1166–1180. [CrossRef]
22. Shelp, B.J.; Bozzo, G.G.; Trobacher, C.P.; Chiu, G.; Bajwa, V.S. Strategies and tools for studying the metabolism and function of γ-aminobutyrate in plants. I. Pathway structure. *Botany* **2012**, *90*, 651–668. [CrossRef]
23. Grzechowiak, M.; Sliwiak, J.; Jaskolski, M.; Ruszkowski, M. Structural and functional studies of *Arabidopsis thaliana* glutamate dehydrogenase isoform 2 demonstrate enzyme dynamics and identify its calcium binding site. *Plant Physiol. Biochem.* **2023**, *201*, 107895. [CrossRef] [PubMed]
24. Grzechowiak, M.; Sliwiak, J.; Jaskolski, M.; Ruszkowski, M. Structural Studies of Glutamate Dehydrogenase (Isoform 1) From *Arabidopsis thaliana*, an Important Enzyme at the Branch-Point Between Carbon and Nitrogen Metabolism. *Front. Plant Sci.* **2020**, *11*, 754. [CrossRef] [PubMed]
25. Tercé-Laforgue, T.; Dubois, F.; Ferrario-Méry, S.; de Crecenzo, M.A.; Sangwan, R.; Hirel, B. Glutamate dehydrogenase of tobacco is mainly induced in the cytosol of phloem companion cells when ammonia is provided either externally or released during photorespiration. *Plant Physiol.* **2004**, *136*, 4308–4317. [CrossRef]
26. Skopelitis, D.S.; Paranychianakis, N.V.; Kouvarakis, A.; Spyros, A.; Stephanou, E.G.; Roubelakis-Angelakis, K.A. The isoenzyme 7 of tobacco NAD(H)-dependent glutamate dehydrogenase exhibits high deaminating and low aminating activities in vivo. *Plant Physiol.* **2007**, *145*, 1726–1734. [CrossRef]

27. Melo-Oliveira, R.; Oliveira, I.C.; Coruzzi, G.M. Arabidopsis mutant analysis and gene regulation define a nonredundant role for glutamate dehydrogenase in nitrogen assimilation. *Proc. Natl. Acad. Sci. USA* **1996**, *93*, 4718–4723. [CrossRef]
28. Tercé-Laforgue, T.; Clément, G.; Marchi, L.; Restivo, F.M.; Lea, P.J.; Hirel, B. Resolving the Role of Plant NAD-Glutamate Dehydrogenase: III. Overexpressing Individually or Simultaneously the Two Enzyme Subunits Under Salt Stress Induces Changes in the Leaf Metabolic Profile and Increases Plant Biomass Production. *Plant Cell Physiol.* **2015**, *56*, 1918–1929. [CrossRef]
29. Purnell, M.P.; Botella, J.R. Tobacco isoenzyme 1 of NAD (H)-dependent glutamate dehydrogenase catabolizes glutamate in vivo. *Plant Physiol.* **2007**, *143*, 530–539. [CrossRef]
30. Purnell, M.P.; Skopelitis, D.S.; Roubelakis-Angelakis, K.A.; Botella, J.R. Modulation of higher-plant NAD(H)-dependent glutamate dehydrogenase activity in transgenic tobacco via alteration of beta subunit levels. *Planta* **2005**, *222*, 167–180. [CrossRef]
31. Tercé-Laforgue, T.; Bedu, M.; Dargel-Grafin, C.; Dubois, F.; Gibon, Y.; Restivo, F.M.; Hirel, B. Resolving the role of plant glutamate dehydrogenase: II. Physiological characterization of plants overexpressing the two enzyme subunits individually or simultaneously. *Plant Cell Physiol.* **2013**, *54*, 1635–1647. [CrossRef]
32. Barqawi, A.A.; Abulfaraj, A.A. Salt Stress-Related Mechanisms in Leaves of the Wild Barley Hordeum spontaneum Generated from RNA-Seq Datasets. *Life* **2023**, *13*, 1454. [CrossRef] [PubMed]
33. Skopelitis, D.S.; Paranychianakis, N.V.; Paschalidis, K.A.; Pliakonis, E.D.; Delis, I.D.; Yakoumakis, D.I.; Kouvarakis, A.; Papadakis, A.K.; Stephanou, E.G.; Roubelakis-Angelakis, K.A. Abiotic stress generates ROS that signal expression of anionic glutamate dehydrogenases to form glutamate for proline synthesis in tobacco and grapevine. *Plant Cell* **2006**, *18*, 2767–2781. [CrossRef] [PubMed]
34. Kumar, R.G.; Shah, K.; Dubey, R.S. Salinity induced behavioural changes in malate dehydrogenase and glutamate dehydrogenase activities in rice seedlings of differing salt tolerance. *Plant Sci.* **2000**, *156*, 23–34. [CrossRef] [PubMed]
35. Lasa, B.; Frechilla, S.; Aparicio-Tejo, P.M.; Lamsfus, C. Role of glutamate dehydrogenase and phosphoenolpyruvate carboxylase activity in ammonium nutrition tolerance in roots. *Plant Physiol. Biochem.* **2002**, *40*, 969–976. [CrossRef]
36. Ji, J.; Shi, Z.; Xie, T.; Zhang, X.; Chen, W.; Du, C.; Sun, J.; Yue, J.; Zhao, X.; Jiang, Z.; et al. Responses of GABA shunt coupled with carbon and nitrogen metabolism in poplar under NaCl and CdCl2 stresses. *Ecotoxicol. Environ. Saf.* **2020**, *193*, 110322. [CrossRef] [PubMed]
37. Shelp, B.J.; Bown, A.W.; McLean, M.D. Metabolism and functions of gamma-aminobutyric acid. *Trends Plant Sci.* **1999**, *4*, 446–452. [CrossRef]
38. Eprintsev, A.T.; Selivanova, N.V.; Igamberdiev, A.U. Enzymatic Conversions of Glutamate and γ-Aminobutyric Acid as Indicators of Plant Stress Response. *Methods Mol. Biol.* **2020**, *2057*, 71–78. [CrossRef]
39. Dyachenko, O.V.; Zakharchenko, N.S.; Shevchuk, T.V.; Bohnert, H.J.; Cushman, J.C.; Buryanov, Y.I. Effect of hypermethylation of CCWGG sequences in DNA of *Mesembryanthemum crystallinum* plants on their adaptation to salt stress. *Biochemistry* **2006**, *71*, 570–575. [CrossRef]
40. Gardiner, J.; Schroeder, S.; Polacco, M.L.; Sanchez-Villeda, H.; Fang, Z.; Morgante, M.; Landewe, T.; Fengler, K.; Useche, F.; Hanafey, M.; et al. Anchoring 9371 maize expressed sequence tagged unigenes to the bacterial artificial chromosome contig map by two-dimensional overgo hybridization. *Plant Physiol.* **2004**, *134*, 1317–1326. [CrossRef]
41. Soderlund, C.; Descour, A.; Kudrna, D.; Bomhoff, M.; Boyd, L.; Currie, J.; Angelova, A.; Collura, K.; Wissotski, M.; Ashley, E.; et al. Sequencing, mapping, and analysis of 27,455 maize full-length cDNAs. *PLoS Genet.* **2009**, *5*, e1000740. [CrossRef]
42. Vanyushin, B.F.; Kirnos, M.D. DNA methylation in plants. *Gene* **1988**, *74*, 117–121. [CrossRef] [PubMed]
43. Tercé-Laforgue, T.; Lothier, J.; Limami, A.M.; Rouster, J.; Lea, P.J.; Hirel, B. The Key Role of Glutamate Dehydrogenase 2 (GDH2) in the Control of Kernel Production in Maize (*Zea mays* L.). *Plants* **2023**, *12*, 2612. [CrossRef] [PubMed]
44. Sakakibara, H.; Fujii, K.; Sugiyama, T. Isolation and characterization of a cDNA that encodes maize glutamate dehydrogenase. *Plant Cell Physiol.* **1995**, *36*, 789–797. [CrossRef] [PubMed]
45. Alexandrov, N.N.; Brover, V.V.; Freidin, S.; Troukhan, M.E.; Tatarinova, T.V.; Zhang, H.; Swaller, T.J.; Lu, Y.P.; Bouck, J.; Flavell, R.B.; et al. Insights into corn genes derived from large-scale cDNA sequencing. *Plant Mol. Biol.* **2009**, *69*, 179–194. [CrossRef]
46. Schnable, P.S.; Ware, D.; Fulton, R.S.; Stein, J.C.; Wei, F.; Pasternak, S.; Liang, C.; Zhang, J.; Fulton, L.; Graves, T.A.; et al. The B73 maize genome: Complexity, diversity, and dynamics. *Science* **2009**, *326*, 1112–1115. [CrossRef]
47. Condori-Apfata, J.A.; Batista-Silva, W.; Medeiros, D.B.; Vargas, J.R.; Valente, L.M.L.; Heyneke, E.; Pérez-Diaz, J.L.; Fernie, A.R.; Araújo, W.L.; Nunes-Nesi, A. The Arabidopsis E1 subunit of the 2-oxoglutarate dehydrogenase complex modulates plant growth and seed production. *Plant Mol. Biol.* **2019**, *101*, 183–202. [CrossRef]
48. Condori-Apfata, J.A.; Batista-Silva, W.; Medeiros, D.B.; Vargas, J.R.; Valente, L.M.L.; Pérez-Díaz, J.L.; Fernie, A.R.; Araújo, W.L.; Nunes-Nesi, A. Downregulation of the E2 Subunit of 2-Oxoglutarate Dehydrogenase Modulates Plant Growth by Impacting Carbon-Nitrogen Metabolism in *Arabidopsis thaliana*. *Plant Cell Physiol.* **2021**, *62*, 798–814. [CrossRef]
49. Ashapkin, V.V.; Kutueva, L.I.; Aleksandrushkina, N.I.; Vanyushin, B.F. Epigenetic Mechanisms of Plant Adaptation to Biotic and Abiotic Stresses. *Int. J. Mol. Sci.* **2020**, *21*, 7457. [CrossRef]
50. Ashapkin, V.V.; Kutueva, L.I.; Vanyushin, B.F. Aging as an Epigenetic Phenomenon. *Curr. Genom.* **2017**, *18*, 385–407. [CrossRef]
51. Kreibich, E.; Kleinendorst, R.; Barzaghi, G.; Kaspar, S.; Krebs, A.R. Single-molecule footprinting identifies context-dependent regulation of enhancers by DNA methylation. *Mol. Cell* **2023**, *83*, 787–802.e9. [CrossRef]
52. Lehmann, T.; Ratajczak, L. The pivotal role of glutamate dehydrogenase (GDH) in the mobilization of N and C from storage material to asparagine in germinating seeds of yellow lupine. *J. Plant. Physiol.* **2008**, *165*, 149–158. [CrossRef] [PubMed]

53. Zhang, Y.; Beard, K.F.M.; Swart, C.; Bergmann, S.; Krahnert, I.; Nikoloski, Z.; Graf, A.; Ratcliffe, R.G.; Sweetlove, L.J.; Fernie, A.R.; et al. Protein-protein interactions and metabolite channelling in the plant tricarboxylic acid cycle. *Nat. Commun.* **2017**, *8*, 15212. [CrossRef] [PubMed]
54. Eprintsev, A.T.; Fedorin, D.N.; Nikitina, M.V.; Igamberdiev, A.U. Expression and properties of the mitochondrial and cytosolic forms of aconitase in maize scutellum. *J. Plant Physiol.* **2015**, *181*, 14–19. [CrossRef]
55. Sarasketa, A.; González-Moro, M.B.; González-Murua, C.; Marino, D. Nitrogen Source and External Medium pH Interaction Differentially Affects Root and Shoot Metabolism in Arabidopsis. *Front. Plant Sci.* **2016**, *7*, 29. [CrossRef]
56. Yu, K.; Hu, S.; Huang, J.; Mei, L.H. A high-throughput colorimetric assay to measure the activity of glutamate decarboxylase. *Enzym. Microb. Technol.* **2011**, *49*, 272–276. [CrossRef]
57. Maurer, H.R. *Disc Electrophoresis and Related Techniques of Polyacrylamide Gel Electrophoresis*; De Gruyter: Berlin, Germany, 1971.
58. Lowry, O.H.; Rosebrough, N.J.; Farr, A.L.; Randall, R.J. Protein measurement with the Folin phenol reagent. *J. Biol. Chem.* **1951**, *193*, 265–275. [CrossRef] [PubMed]
59. Thurman, D.A.; Palin, C.; Laycock, M.V. Isoenzymatic nature of L-glutamic dehydrogenase of higher plants. *Nature* **1965**, *207*, 193–194. [CrossRef]
60. Martienssen, R.A.; Rabinowicz, P.D.; O'Shaughnessy, A.; McCombie, W.R. Sequencing the maize genome. *Curr. Opin. Plant Biol.* **2004**, *7*, 102–107. [CrossRef]
61. Messing, J.; Dooner, H.K. Organization and variability of the maize genome. *Curr. Opin. Plant Biol.* **2006**, *9*, 157–163. [CrossRef]
62. Chomczynski, P.; Sacchi, N. Single-step method of RNA isolation by acid guanidinium thiocyanate-phenol-chloroform extraction. *Anal. Biochem.* **1987**, *162*, 156–159. [CrossRef]
63. Matz, M.V. Amplification of representative cDNA samples from microscopic amounts of invertebrate tissue to search for new genes. *Methods Mol. Biol.* **2002**, *183*, 3–18. [CrossRef] [PubMed]
64. Livak, K.J.; Schmittgen, T.D. Analysis of relative gene expression data using real-time quantitative PCR and the $2^{-\Delta\Delta Ct}$ Method. *Methods* **2001**, *25*, 402–408. [CrossRef] [PubMed]
65. Hsieh, C.L. Evidence that protein binding specifies sites of DNA demethylation. *Mol. Cell. Biol.* **1999**, *19*, 46–56. [CrossRef] [PubMed]
66. Green, M.R.; Sambrook, J. Analysis of DNA by Agarose Gel Electrophoresis. *Cold Spring Harb. Protoc.* **2019**, *1*, pdb-top100388. [CrossRef] [PubMed]
67. Eprintsev, A.T.; Fedorin, D.N.; Karabutova, L.A.; Igamberdiev, A.U. Expression of genes encoding subunits A and B of succinate dehydrogenase in germinating maize seeds is regulated by methylation of their promoters. *J. Plant Physiol.* **2016**, *205*, 33–40. [CrossRef]
68. Zar, J.H. *Biostatistical Analysis*, 5th ed.; Pearson: London, UK, 2010.

plants

MDPI

Article

Deciphering the Mechanism of Melatonin-Induced Enhancement of Photosystem II Function in Moderate Drought-Stressed Oregano Plants

Julietta Moustaka [1,†], Ilektra Sperdouli [2], Sumrunaz İşgören [3], Begüm Şaş [4] and Michael Moustakas [1,*]

1 Department of Botany, Aristotle University of Thessaloniki, 54124 Thessaloniki, Greece; ioumoustaka@gmail.com
2 Institute of Plant Breeding and Genetic Resources, Hellenic Agricultural Organisation-Demeter (ELGO-Demeter), 57001 Thessaloniki, Greece; esperdouli@elgo.gr
3 Department of Molecular Biology and Genetics, Istanbul Kültür University, Ataköy 7-8-9-10, 34158 Bakırköy, Turkey; isgorensn@gmail.com
4 School of Life Sciences, Faculty of Biotechnology, ITMO University, Kronverkskiy Prospekt 49, 197101 Saint-Petersburg, Russia; begum.sas99@gmail.com
* Correspondence: moustak@bio.auth.gr
† Current address: Department of Food Science, Aarhus University, 8200 Aarhus, Denmark.

Abstract: Melatonin (MT) is considered as an antistress molecule that plays a constructive role in the acclimation of plants to both biotic and abiotic stress conditions. In the present study, we assessed the impact of 10 and 100 μM MT foliar spray, on chlorophyll content, and photosystem II (PSII) function, under moderate drought stress, on oregano (*Origanum vulgare* L.) plants. Our aim was to elucidate the molecular mechanism of MT action on the photosynthetic electron transport process. Foliar spray with 100 μM MT was more effective in mitigating the negative impact of moderate drought stress on PSII function, compared to 10 μM MT. MT foliar spray significantly improved the reduced efficiency of the oxygen-evolving complex (OEC), and PSII photoinhibition (Fv/Fm), which were caused by drought stress. Under moderate drought stress, foliar spray with 100 μM MT, compared with the water sprayed (WA) leaves, increased the non-photochemical quenching (NPQ) by 31%, at the growth irradiance (GI, 205 μmol photons m^{-2} s^{-1}), and by 13% at a high irradiance (HI, 1000 μmol photons m^{-2} s^{-1}). However, the lower NPQ increase at HI was demonstrated to be more effective in decreasing the singlet-excited oxygen ($^1O_2$) production at HI (−38%), in drought-stressed oregano plants sprayed with 100 μM MT, than the corresponding decrease in $^1O_2$ production at the GI (−20%), both compared with the respective WA-sprayed leaves under moderate drought. The reduced $^1O_2$ production resulted in a significant increase in the quantum yield of PSII photochemistry (Φ_{PSII}), and the electron transport rate (ETR), in moderate drought-stressed plants sprayed with 100 μM MT, compared with WA-sprayed plants, but only at the HI (+27%). Our results suggest that the enhancement of PSII functionality, with 100 μM MT under moderate drought stress, was initiated by the NPQ mechanism, which decreased the $^1O_2$ production and increased the fraction of open PSII reaction centers (qp), resulting in an increased ETR.

Keywords: chlorophyll content; non-photochemical quenching; electron transport; quantum yield of PSII photochemistry; singlet-excited oxygen; redox state; excess excitation energy; photoinhibition

Citation: Moustaka, J.; Sperdouli, I.; İşgören, S.; Şaş, B.; Moustakas, M. Deciphering the Mechanism of Melatonin-Induced Enhancement of Photosystem II Function in Moderate Drought-Stressed Oregano Plants. *Plants* **2024**, *13*, 2590. https://doi.org/10.3390/plants13182590

Academic Editors: Carmen Arena and Bernhard Huchzermeyer

Received: 1 August 2024
Revised: 12 September 2024
Accepted: 14 September 2024
Published: 16 September 2024

1. Introduction

Despite constant improvements, increase in crop yields is currently not sufficient to avoid food shortages by 2050 [1,2]. Worldwide food production is estimated to be reduced by 11–25% by the end of the century [3]. Drought stress, when compared to other stresses, is the major problem associated with climate change [4–8], lowering drastically global crop production [9,10], and resulting, e.g., in 21% yield decreases in wheat and in even more, in 40% in maize [11].

Drought causes numerous morphophysiological disorders in plants [12], resulting in reduction in leaf area and plant biomass, in leaf chlorosis and wilting, and in abortion of leaves and flowers, with a subsequent decline in productivity [13–15]. The closure of stomata under drought, which reduces transpiration in order to avoid dehydration, decreases CO_2 disposal and, as a result, the use of electrons for carbon fixation [16–18]. Subsequently, the electrons in the light reactions divert to oxygen, thus creating the superoxide anion radical ($O_2^{\bullet -}$) [18–20]. Moreover, the transfer of the excess absorbed light energy from the triplet-excited state of chlorophylls ($^3Chl^*$) to molecular oxygen (O_2) creates the singlet oxygen ($^1O_2$), which can further produce $O_2^{\bullet -}$ and hydrogen peroxide (H_2O_2) [21–23].

The reactive oxygen species (ROS), which are produced in the light reactions ($^1O_2$, H_2O_2, and $O_2^{\bullet -}$), are formed persistently, but are kept in homeostasis by enzymatic and non-enzymatic antioxidants under non-stress conditions [24–27]. However, under drought stress, the homeostasis between the creation and elimination of ROS breakdowns [7], and the generation of ROS rises remarkably [18], initiating oxidative stress that triggers membrane damages, protein degradation and enzyme inactivation, causing damage to cellular components [28,29]. Thus, in drought stress conditions, the excess absorbed light energy that cannot be used in the Calvin–Benson–Bassham cycle results in the disruption of ROS homeostasis [20,21] and, even with the presence of photoprotective mechanisms, the generation of ROS that rises results in photooxidative damage [30–33]. ROS-provoked oxidative damage under drought stress is one of the key factors that limit plant growth [34–36]. Osmolyte accumulation under drought stress conditions confers plant tolerance if it is supplemented by a controlled drought-induced ROS generation [37–39]. A non-photochemical quenching (NPQ) mechanism protects the chloroplasts from the damaging effects of ROS by dissipating the excess light energy as heat [40–42].

Drought stress impairs cell division, elongation and differentiation, and the osmotic adjustment, triggering loss of turgor [43,44]. It impairs photosynthetic efficiency and plant growth, disturbing light energy partitioning, and finally decreasing plant productivity [43,44]. The primary process in increasing plant productivity is photosynthesis; thus, enhancing photosynthetic efficiency can improve crop yield [45,46]. The use of bio-stimulants against drought stress is the newly adopted strategy to improve photosynthetic efficiency in enhancing crop production, and to overcome threatening agricultural challenges [47,48].

Over the last decade, the agricultural industry has used bio-stimulants as a key approach to stimulate plant growth and enhance crop production under both non-stress and stress conditions [49–53]. The increasing demand for ecological agricultural solutions has positioned bio-stimulants as valuable tools for reinventing farming and sustaining resilience to environmental stressors [54,55]. Applying bio-stimulants mitigates the harmful effects of abiotic stresses, enhancing growth performance and plant tolerance [56]. Bio-stimulants reduce the negative effects of drought stress by lessening damage from oxidative stress [15] and offer a promising and ecological approach to addressing environmental concerns and promoting sustainable agriculture [52]. However, the precise mechanisms by which plant bio-stimulants enhance nutrient uptake and utilization, improve yield, increase photosynthetic function, and confer resistance to both biotic and abiotic stresses have not yet been fully elucidated [50,51,53]. To implement successfully bio-stimulants in the field, an understanding of their action mechanism in coping with diverse abiotic stresses is needed [52].

Melatonin (MT) is a versatile molecule that functions as a natural antioxidant, boosting plants' tolerance to environmental stresses, and considered as a promising bio-stimulant for agricultural use [57]. MT is considered as a new plant hormone that plays key roles in a broad range of physiological processes, like photosynthesis, leaf senescence, osmoregulation, stomatal closure, circadian cycle, secondary metabolism, germination, shoot and root growth, flowering and fruit setting, and as an anti-stressor [58–62]. As an anti-stress molecule, it plays a constructive role in numerous biotic and abiotic stresses, like drought, salinity, heavy metals, high and low temperatures, plant–pathogen interactions, fungal diseases, and UV radiation [63–66].

MT can penetrate cell membranes, raising endogenous MT levels [67,68], and acting as an antioxidant under stress conditions [69]. By improving antioxidant capacity and increasing the xanthophyll pool size, MT enhances photosynthetic function and photosynthetic enzyme activities under salt stress [65]. Exogenous MT, under chilling stress, enhanced NPQ, i.e., the photoprotective heat dissipation of excess excitation energy, mitigating photoinhibition [70]. In the medicinal plant *Mentha spicate*, under non-stress conditions, MT stimulated PSII functionality under low light [62]. MT enhanced tolerance to high light in *Arabidopsis thaliana* through improving the anti-oxidative system and photosynthesis [71]. In tomato seedlings MT improved drought tolerance by enhancing soluble sugar accumulation [72]. Soluble sugars, under drought and salt stress, act as osmo-protectants, reducing ROS production and maintaining the cellular redox state [18,39]. MT alleviated abscisic acid deficiency and improved the antioxidant system in rice under salt stress [73]. In maize seeds under salt stress, MT enhanced the antioxidant ability of the seeds, reducing salt-induced damage, and thus promoting seed germination [74]. Under salt stress, exogenous MT led to an increase in endogenous levels, activating the antioxidant system and playing an essential role as ROS scavenger [75]. In cotton, MT, by regulating the expression of photosynthetic genes and the antioxidant system, enhanced cold tolerance [76].

It has been shown that MT preserves chlorophyll molecules and the photosynthetic function [77], regulates the expression of photosynthetic genes [78], and interacts with Ca^{2+}, nitric oxide (NO), and ROS, to regulate the redox network [78–81]. ROS signaling and MT have been shown to be interconnected [65]. Melatonin is related also to the plant hormones, e.g., ethylene (ETH), indole acetic acid (IAA), cyto-kinins (CTK), gibberellins (GAs), abscisic acid (ABA), brassino-steroids (BR), jasmonic acid (JA), strigolactone (SL), and salicylic acid (SA) [82,83].

Origanum vulgare L. is a perennial polymorphic species native to the Mediterranean basin, but occurring almost all over Europe and West and Central Asia [84]. It is used both as a medicinal and culinary herb, especially in the Spanish, Mexican, French, Turkish, Italian, and Greek cuisines [62]. In this research work, we tested whether MT foliar spray can enhance PSII function of *Origanum vulgare* L. plants under moderate drought stress and tried to disentangle the underlying mechanisms of MT action on the electron transport process under moderate drought stress conditions.

2. Results

2.1. Volumetric Soil Water Content and Leaf Water Content

Withholding the irrigation for 6 days resulted in moderate drought-stressed plants, having a soil water content of 50 ± 2% of that of the well-watered plants. At the same time, the leaf water content in drought-stressed water sprayed plants was 87.6 ± 0.4^{b}%; in drought-stressed plants sprayed with 10 μM MT, this was 88.1 ± 0.4^{b}%, and in drought-stressed plants sprayed with 100 μM MT, t his was 89.6 ± 0.3^{a}% (different lower-case letters show significant differences at $p < 0.05$).

2.2. Impact of Melatonin on Chlorophyll Content of Drought-Stressed Oregano Leaves

A decrease in volumetric soil water content of about 50%, resulted in a significant decrease (47%) in the chlorophyll content of the drought-stressed water sprayed plants (WA) (Figure 1a). In drought-stressed oregano plants sprayed with 10 μM MT (10 μM MT), the chlorophyll content decreased to 80% of that of the well-watered plants (CK)while, in drought-stressed oregano plants sprayed with 100 μM MT (100 μM MT), the chlorophyll content remained at the level of the well-watered plants (CK) (Figure 1a).

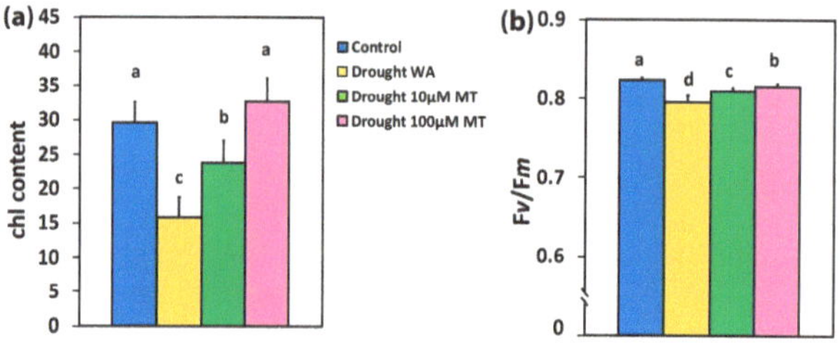

Figure 1. The chlorophyll content, expressed in relative units (**a**), and the maximum efficiency of PSII photochemistry (*Fv/Fm*), in dark adapted leaves (**b**) of control oregano plants, of drought-stressed oregano plants sprayed with water (WA), of drought-stressed oregano plants sprayed with 10 µM MT, and of drought-stressed oregano plants sprayed with 100 µM MT. Error bars are standard deviations (SDs). Significant difference at $p < 0.05$ is shown by different lower-case letters.

2.3. Impact of Melatonin on the Maximum Efficiency of Photosystem II Photochemistry and the Efficiency of the Oxygen-Evolving Complex

In WA plants, the maximum efficiency of PSII photochemistry (*Fv/Fm*) decreased compared to CK plants by 4%, while in 10 µM MT-treated plants this decreased by 2%, and in 100 µM MT plants by 1% (Figure 1b). The efficiency of the oxygen-evolving complex (OEC) in WA plants decreased compared to CK by 17%, while in 10 µM MT this decreased by 10% and in 100 µM MT by 6% (Figure 2a).

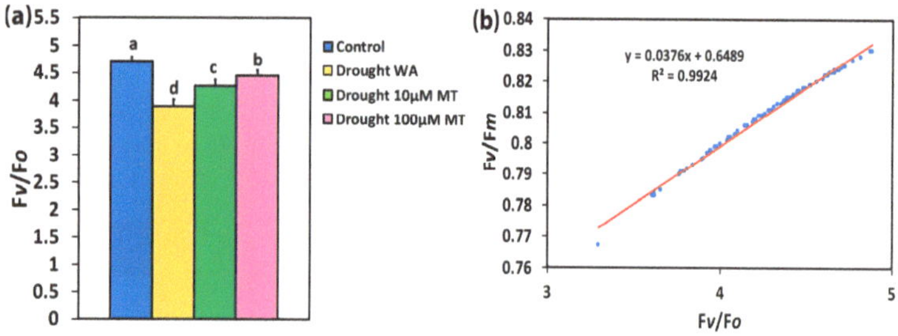

Figure 2. The efficiency of the oxygen-evolving complex (*Fv/Fo*) in dark adapted leaves (**a**), of control oregano plants, of drought-stressed oregano plants sprayed with water (WA), of drought-stressed oregano plants sprayed with 10 µM MT, and of drought-stressed oregano plants sprayed with 100 µM MT. Error bars are standard deviations (SDs). Significant difference at $p < 0.05$ is shown by different lower-case letters. In (**b**), the relationship between the *Fv/Fm*, and the *Fv/Fo* is depicted based on the data from Figures 1b and 2a. A positive significant correlation exists ($R^2 = 0.9924$, $p < 0.001$).

The response patterns of *Fv/Fm* and of *Fv/Fo* were similar and, as documented by regression analysis, they showed a positive significant correlation ($R^2 = 0.9924$, $p < 0.001$) (Figure 2b).

2.4. Impact of Melatonin on Light Energy Use Efficiency of Photosystem II

The absorbed light energy by PSII antenna is either distributed to photochemistry (Φ_{PSII}), or dissipated as heat (Φ_{NPQ}), or is misplaced in a nonregulatory way (Φ_{NO}) [85]. All have a sum equal to 1 [85].

At the growth irradiance of 205 µmol photons m^{-2} s^{-1} (GI), the quantum yield for the photochemistry (Φ_{PSII}) of WA oregano plants decreased compared to CK plants by 27%, while the MT-treated plants at both concentrations showed a decrease of 24% (Figure 3a). Similarly, at the high irradiance of 1000 µmol photons m^{-2} s^{-} (HI), Φ_{PSII} in WA plants decreased compared to CK plants by 32% while, in 10 µM MT and 100 µM MT, there was a lower decrease compared to CK plants, by 18% and 13%, respectively (Figure 3b).

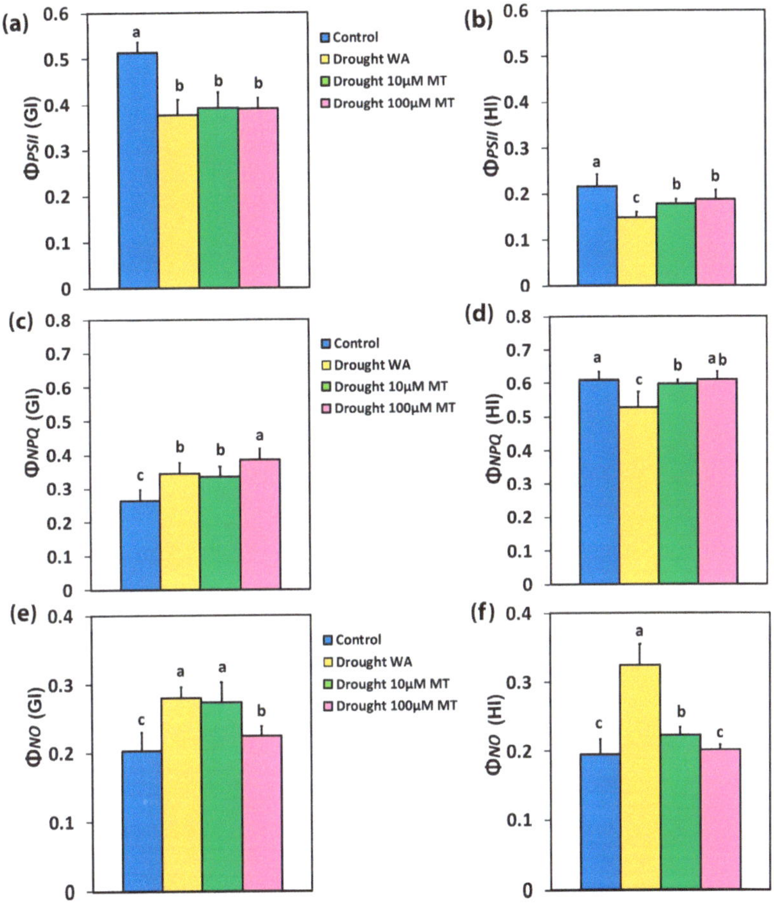

Figure 3. Distribution of the absorbed light energy at PSII. The effective quantum yield of the PSII photochemistry (Φ_{PSII}) at the growth irradiance (GI) (**a**), and at a high irradiance (HI) (**b**), the quantum yield of regulated non-photochemical energy loss in PSII (Φ_{NPQ}) at the GI (**c**), and at a HI (**d**), and the quantum yield of non-regulated energy loss in PSII (Φ_{NO}) at the GI (**e**), and at a HI (**f**); of control oregano plants, of drought-stressed oregano plants sprayed with water (WA), of drought-stressed oregano plants sprayed with 10 µM MT, and of drought-stressed oregano plants sprayed with 100 µM MT. Error bars are standard deviations (SDs). Significant difference at $p < 0.05$ is shown by different lower-case letters.

The regulated non-photochemical energy loss in PSII (Φ_{NPQ}) of WA oregano plants increased (31%) at the GI (Figure 3c), but at HI Φ_{NPQ} decreased (13%) (Figure 3d), compared to CK plants. In 10 µM MT-treated oregano plants, Φ_{NPQ} increased (by 27%) at the GI (Figure 3c), but at HI Φ_{NPQ} decreased (by 2%) (Figure 3d), compared to CK plants. At the GI in 100 µM MT-treated oregano plants, Φ_{NPQ} increased (by 46%) (Figure 3c), compared to CK plants, but at HI Φ_{NPQ} remained at the same level as CK plants (Figure 3d).

In WA oregano plants, and in 10 µM MT plants, the yield of non-regulated energy loss in PSII (Φ_{NO}) increased at the GI by 38%, and 35%, respectively (Figure 3e), while at the HI Φ_{NO} increased by 66% and 15%, respectively (Figure 3f), compared to CK plants. In 100 µM MT plants, Φ_{NO} increased (by 11%) at the GI (Figure 3e), compared to CK plants, while at the HI Φ_{NO} remained at the same level as CK plants (Figure 3e).

2.5. Impact of Melatonin on the Photoprotective Heat Dissipation, the Fraction of Open PSII Reaction Centers and the Electron Transport Rate

The non-photochemical quenching (NPQ), at the GI, in WA plants, and in 10 µM MT plants, remained at the same level as CK plants (Figure 4a). In 100 µM MT-treated plants, NPQ increased (by 46%) at the GI (Figure 4a), compared to CK plants, but at HI NPQ de-creased (by 14%) (Figure 4b). At the HI, NPQ in WA oregano plants, and in 10 µM MT plants, decreased in both by 24%, compared to CK plants (Figure 4b).

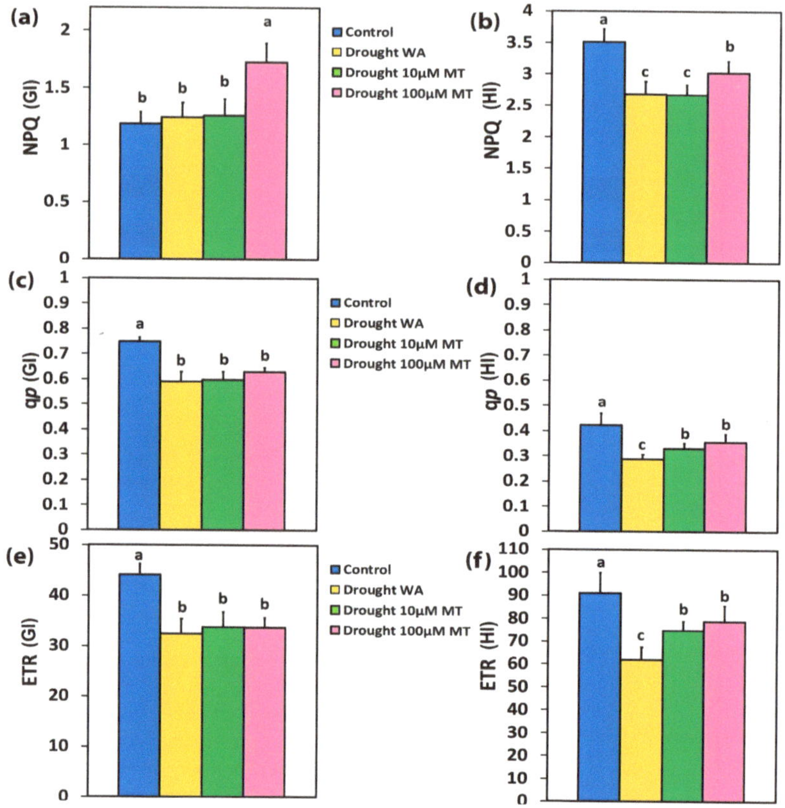

Figure 4. Non-photochemical quenching (NPQ), at the growth irradiance (GI) (**a**), and at high irradiance (HI) (**b**), the fraction of open PSII rection centers (RCs) (*qp*), at the GI (**c**), and at a HI (**d**), and the electron transport rate (ETR) at the GI (**e**), and at a HI (**f**), of control oregano plants, of drought-stressed oregano plants sprayed with water (WA), of drought-stressed oregano plants sprayed with 10 µM MT, and of drought-stressed oregano plants sprayed with 100 µM MT. Error bars are standard deviations (SDs). Significant difference at $p < 0.05$ is shown by different lower-case letters.

The percentage of open PSII rection centers (RCs) (*qp*), which indicates also the redox state of quinone A (Q_A), in WA plants at 10 µM MT and 100 µM MT, decreased by 21%, 20%, and 16%, respectively at the GI, compared to CK plants (Figure 4c). At the HI, in WA, for 10 µM MT, and 100 µM MT-treated plants, the redox state of Q_A decreased by 32%, 22%, and 16%, respectively, compared to CK plants (Figure 4d).

The electron transport rate (ETR), at the GI, in WA plants, in 10 µM MT and 100 µM MT plants, decreased by 27%, 24%, and 24%, respectively, compared to CK plants (Figure 4e). At the HI, ETR in WA plants decreased by 32%, compared to CK plants while, in 10 µM MT, and 100 μM MT-treated plants, ETR decreased by 18%, and 13%, respectively, compared to CK plants (Figure 4f).

2.6. Impact of Melatonin on the Efficiency of the Open PSII Reaction Centers and the Excess Excitation Energy at PSII

The efficiency of the open PSII RCs (Fv'/Fm'), at the GI, in WA oregano plants and in 100 μM MT plants, decreased compared to CK plants, but did not change in 10 μM MT plants (Figure 5a). At the HI, Fv'/Fm' did not change with any treatment compared to CK plants (Figure 5b).

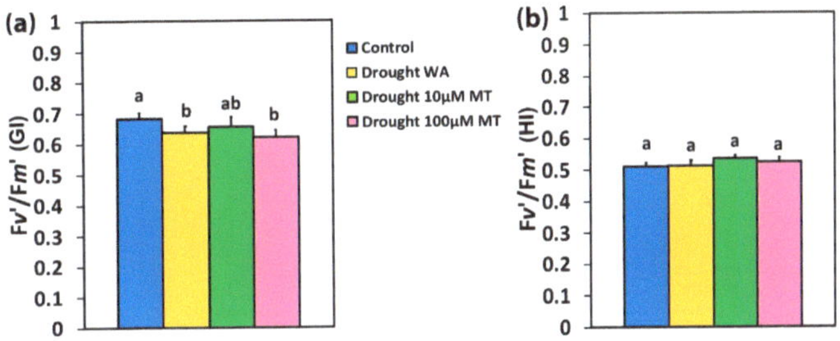

Figure 5. The efficiency of the open PSII RCs (Fv'/Fm'), at the growth irradiance (GI) (**a**), and at a high irradiance (HI) (**b**), of control oregano plants, of drought-stressed oregano plants sprayed with water (WA), of drought-stressed oregano plants sprayed with 10 μM MT, and of drought-stressed oregano plants sprayed with 100 μM MT. Error bars are standard deviations (SDs). Significant difference at $p < 0.05$ is shown by different lower-case letters.

In WA oregano plants, in 10 μM MT and 100 μM MT plants, the excess excitation energy at PSII (EXC), increased by 52%, 54%, and 35%, respectively, at the GI (Figure 6a) compared to CK plants while, at the HI, the EXC increased by 25%, 22%, and 15%, respectively, compared to CK plants (Figure 6b).

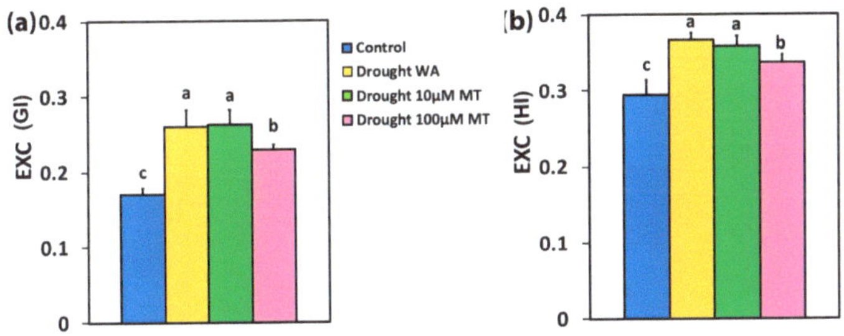

Figure 6. The excess excitation energy at PSII (EXC), at the growth irradiance (GI) (**a**), and at a high irradiance (HI) (**b**), of control oregano plants, of drought-stressed oregano plants sprayed with water (WA), of drought-stressed oregano plants sprayed with 10 μM MT, and of drought-stressed oregano plants sprayed with 100 μM MT. Error bars are standard deviations (SDs). Significant difference at $p < 0.05$ is shown by different lower-case letters.

2.7. Impact of Melatonin on PSII Excitation Pressure

At the GI, the excitation pressure at PSII ($1 - qL$), in WA plants, in 10 μM MT and 100 μM MT plants, increased by 27%, 29%, and 19%, respectively, compared to CK plants (Figure S1a) while, at the HI, $1 - qL$, this increased by 14%, 11%, and 8%, respectively, compared to CK plants (Figure S1b).

2.8. *Impact of Drought and Melatonin on the Spatiotemporal Heterogeneity of PSII Photochemistry*

Representative color pictures of the whole oregano leaf area that were acquired by chlorophyll fluorescence imaging analysis, for the parameters Fv/Fm (captured in dark adapted leaves), and of Φ_{PSII}, Φ_{NPQ}, Φ_{NO}, and qp (captured at the growth irradiance, GI), from control oregano plants (CK), drought-stressed oregano plants sprayed with water (WA), drought-stressed oregano plants sprayed with 10 μM MT, and drought-stressed oregano plants sprayed with 100 μM MT are presented in Figure 7. The leaf area of WA plants decreased drastically, while the treatment with 10 μM MT and 100 μM MT mitigated this leaf area decrease (Figure 7). A higher spatial heterogeneity, mainly for the parameters Φ_{PSII}, Φ_{NPQ}, and Φ_{NO}, was noticed in CK oregano plants, compared to all treatments of moderate drought-stressed plants (Figure 7). Moderate drought stress seemed to mask this spatial heterogeneity under all treatments (Figure 7). The negative impact of moderate drought stress, in all parameters, was significantly mitigated mainly by the 100 μM MT treatment (Figure 7).

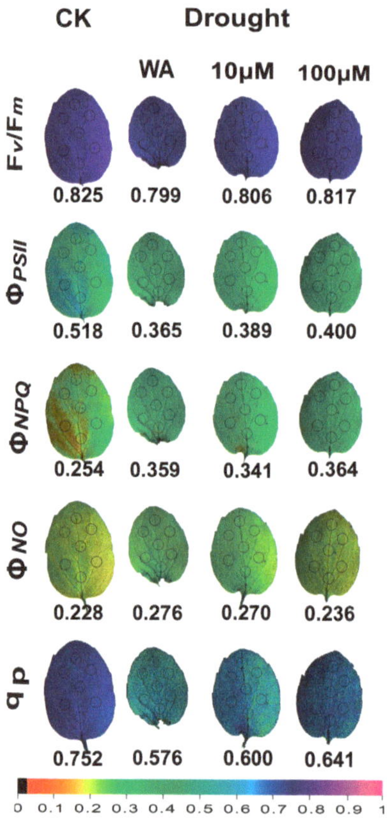

Figure 7. Leaf color-coded pictures of the parameters: Fv/Fm captured in dark adapted plants, Φ_{PSII}, Φ_{NPQ}, Φ_{NO}, and qp, captured at the growth irradiance (GI), of the control oregano plants (CK), of the drought-stressed oregano plants sprayed with water (WA), of the drought-stressed oregano plants sprayed with 10 μM MT, and of the drought-stressed oregano plants sprayed with 100 μM MT. The seven areas of interest (AOI) that were selected in each leaf are shown by circles. The average whole leaf value for each parameter is shown. At the bottom, the color code indicates the corresponding color values.

At high irradiance, (HI), the representative color pictures, which were acquired by chlorophyll fluorescence imaging analysis, of the whole oregano leaf area, for the parameters Φ_{PSII}, Φ_{NPQ}, Φ_{NO}, and qp, from CK plants, WA, 10 µM MT and 100 µM MT-treated, are shown in Figure 8. The higher spatial heterogeneity in CK oregano plants, compared to all other treatments (Figure 8), mainly for the parameters Φ_{PSII}, Φ_{NPQ}, and qp, was still noticed, but was significantly less than that observed at the GI (Figure 7). It seems that the increased dissipation of the absorbed light energy as heat (Φ_{NPQ}), at the HI compared to the GI, reduced the spatial PSII heterogeneity. The negative impact of moderate drought stress, in all the parameters, was also significantly mitigated at the HI, mainly by the application of 100 µM MT (Figure 8).

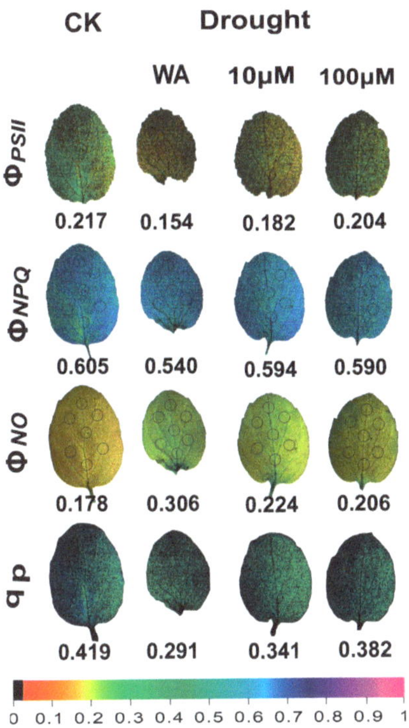

Figure 8. Leaf color-coded pictures of the parameters Φ_{PSII}, Φ_{NPQ}, Φ_{NO}, and qp, captured at high irradiance (HI), of the control oregano plants (CK), of the drought-stressed oregano plants sprayed with water (WA), of the drought-stressed oregano plants sprayed with 10 µM MT, and of the drought-stressed oregano plants sprayed with 100 µM MT. The seven areas of interest (AOI) selected in each leaf are shown by circles. The average whole leaf value for each parameter is shown. At the bottom, the color code indicates the corresponding color values.

3. Discussion

Drought stress triggers a decline in CO_2 availability owing to stomatal closure, leading to the over-reduction of the electron transport process and to increased production of ROS [86–90]. Under moderate drought stress, the percentage of open PSII reaction centers (qp) decreased (Figure 4c,d), resulting in a reduced quantum yield of photochemistry (Φ_{PSII}) (Figure 3a,b), and to increased non-photochemical energy loss in PSII (Φ_{NPQ}) (Figure 3c) [91–95]. The inability of Φ_{NPQ} to increase under HI in WA oregano plants (Figure 3d) resulted in an increase in Φ_{NO} at HI by 66% (Figure 3f), indicating that the photoprotective regulatory mechanisms were insufficient, and that the plants had severe problems in coping with the absorbed light energy [94]. The non-photochemical quenching (NPQ) mechanism assists in dissipating the excess light energy as heat, preventing overexcitation of the chlorophyll molecules, and protecting the photosynthetic apparatus against photodamage under drought stress conditions [96–98]. Drought stress impairs photosynthetic efficiency by disturbing the light energy partitioning [7,44].

In our experiments, the decreased Φ_{PSII} in WA or 10 μM MT plants, compared to CK plants (Figure 3a,b), could not be compensated for by the increased Φ_{NPQ} (Figure 3c,d) and, as a result, the quantum yields of non-regulated energy loss in PSII (Φ_{NO}) increased (Figure 3e,f). However, in 100 μM MT plants, the decreased Φ_{PSII} was balanced by the increase in Φ_{NPQ}, partially at the GI (Figure 3c), and fully balanced at the HI (Figure 3d), compared to CK plants. As a result, Φ_{NO} at HI remained at the level of that of the CK plants (Figure 3f). Non-regulated energy dissipated in PSII (Φ_{NO}) is the result of $^1O_2$ formation [99–102]. Thus, the reduced $^1O_2$ production, which remained at the level of CK plants with 100 μM MT, resulted in a significant increase in Φ_{PSII} (Figure 3b) and ETR (Figure 4f) by 27%, only at the HI, both parameters being compared with the respective WA ones. This increase in Φ_{PSII} at the HI, in 100 μM MT, compared with the respective WA (Figure 3b), was due to the 25% increase in the open PSII RCs (Figure 4d), since the excitation capture efficiency of the open PSII RCs (Fv'/Fm') at the HI did not differ (Figure 5b).

The leaf water content of drought-stressed plants sprayed with 100 μM MT was significantly higher from drought-stressed water sprayed plants and drought-stressed plants sprayed with 10 μM MT, suggesting the induction of osmolyte accumulation [72]. In accordance with our results, foliar application of 100 μM MT reversed drought-induced impairment of the leaf water content in tomato seedlings, by increasing the sugar accumulation, particularly sucrose content [72]. Soluble sugars, together with proline, act as osmo-protectants in subsequent periods of drought stress [39,72]. Exogenous MT application under non-stressed conditions in *Chara australis* [103] and in *Mentha spicata* [68] amplified the number of open PSII RCs, thus improving Φ_{PSII}. In tall fescue under high-temperature stress, MT also increased Φ_{PSII} by increasing the fraction of open RCs [104]. Under non-stressed conditions, application of MT in *Camellia sinensis* stimulated photosynthesis and alleviated PSII photoinhibition, displaying a direct antioxidant effect [105]. Drought stress is documented to harm the oxygen-evolving complex (OEC) of PSII [106], decreasing the abundance of OEC proteins [89]. Malfunction of the OEC is related to donor side photoinhibition [107–110].

In our experiments, application of MT relieved photoinhibition (judged from Fv/Fm), which was caused by moderate drought stress (Figure 1b). The reduced Fv/Fm from drought stress (Figure 1b) was due to donor-side photoinhibition due to malfunction of the OEC [109–111]. The malfunction of the OEC, caused by moderate drought stress (Figure 2a), was judged from the ratio Fv/Fo [112–118]. The higher efficiency of the OEC in MT-sprayed oregano plants, compared to WA-sprayed ones (Figure 2a), matched the respective improved maximum efficiency of PSII photochemistry (Fv/Fm) (Figure 1b). Reduced efficiency of the OEC leads to a decline in Fv/Fm [119,120], as documented by the positive significant correlation in our regression analysis (Figure 2b). A donor-side photoinhibition is often linked with ROS production [111,121,122]. The observed increased $^1O_2$ generation in moderate drought-stressed plants sprayed either with WA or with 10 μM MT (Figure 3e,f) was possible due to the donor-side photoinhibition (Figure 1b), owing to the malfunction of the OEC (Figure 2a). Application of 100 μM MT significantly alleviated the decreased efficiency of the OEC and PSII photoinhibition caused by moderate drought stress. In contrast, under non-stress conditions, MT reduced the efficiency of the OEC, causing donor-side photoinhibition [62].

The reduction of the excess excitation energy, which was detected with 100 μM MT, at both GI (Figure 6a), and HI (Figure 6b), compared to WA and 10 μM MT, indicates enhanced efficiency of PSII. Enhancing photosynthesis is a major challenge for plant scientists, particularly given the growing global demand for food [123–125]. The key to improving photosynthetic efficiency lies in optimizing the distribution of absorbed light energy [94,126].

A decreased excitation pressure at PSII (1-qL), at both GI (Figure S1a) and HI (Figure S1b), was detected with 100 μM MT, compared to WA plants. A decreased excitation pressure has been corelated to increased expression level of photosystem II subunit S (PsbS) [127,128], which is associated with increased levels of NPQ [129]. A higher NPQ level is a characteristic of drought-tolerant cultivars [95]. Foliar spray with 100 μM MT, compared with WA leaves, resulted in increased NPQ by 31% at the GI (Figure 4a), and by 13% at the HI (Figure 4b). The lower NPQ increase at HI, with 100 μM MT, compared with WA leaves, was proved to be more effective in decreasing $^1O_2$ production at the HI by 38% (Figure 3f), than the corresponding decrease in $^1O_2$ production at the GI by 20% (Figure 3e), both compared with the respective WA leaves.

Whole leaf evaluation of PSII functionality by chlorophyll fluorescence imaging methodology revealed a higher spatial heterogeneity in control oregano plants at the GI, compared to all treatments of drought-stressed plants (Figure 7). However, at HI (Figure 8), this spatial heterogeneity was significantly lower than that observed at the GI (Figure 7). It seems that the increased dissipation of the absorbed light energy as heat (Φ_{NPQ}), at the HI compared to the GI, reduced the spatial heterogeneity. However, in *Arabidopsis thaliana,* the observed order of the higher spatial PSII heterogeneity was first at mild drought stress, next at moderate drought stress, and with least heterogeneity for control plants [130]. The spatial PSII heterogeneity observed in control oregano plants suggests that water potential and stomatal function differ in cells from different regions of the leaf, contributing to spatial differences in photochemical activity and reflecting different zones of the leaf anatomy and mesophyll development [131,132]. Moderate drought stress in oregano plants seemed to mask the differences, resulting in less spatial heterogeneity (Figure 7).

Foliar application of MT in mint plants, under non-stress conditions, improved PSII functionality by triggering the NPQ mechanism that stimulated ROS production, which enhanced the photosynthetic function [62]. In the present experiment, oregano plants that were sprayed with MT before the moderate drought stress show an enhancement of PSII functionality, initiated by the NPQ mechanism, which decreased $^1O_2$ production and increased the ETR. It seems that, under different environmental growth conditions, MT application triggers differentially the NPQ mechanism, which activates differential ROS regulatory networks of light energy partitioning signaling to improve PSII function.

4. Materials and Methods

4.1. Plant Material and Cultivation

Six-week-old oregano (*Origanum vulgare* L.) plants obtained from the Garden Center Vaseiliadis were transported to a growth chamber and transplanted into a 1.5 L pot. Throughout the experimental period the oregano plants were grown at a day/night temperature of 21 ± 1/19 ± 1 °C, a relative humidity during day/night of 60 ± 5/70 ± 5%, and a 14-h photoperiod provided by white fluorescent light lamps with a photosynthetic photon flux density (PPFD) of 200 ± 10 μmol photons m^{-2} s^{-1} [133]. Eight to ten plants were measured from each treatment (n = 8–10).

4.2. Melatonin Treatments

In the experiments, we used MT (N-acetyl-5-methoxytryptamine) obtained from Sigma-Aldrich (St. Louis, MO, USA). Twenty mg of MT was dissolved in 1 mL ethanol, before being further diluted with ultra-pure water [62,134] to a final concentration of 10 μM or 100 μM MT [62]. Oregano plants used as control were sprayed with distilled water, with an equal amount of ethanol to that in MT-sprayed plants, as before [62].

4.3. Drought Stress Treatment

All oregano plants, after being each transplanted to a 1.5 L pot, were irrigated at full soil water capacity, proceeding to measurements that served as controls. After measurement, oregano plants were foliar-sprayed with 15 mL per plant (until full wetting), with either distilled water (dH_2O), or 10 μM MT, or 100 μM MT. The surface of the soil was isolated by aluminum foil to prevent MT from dropping into the soil. All plants were sprayed during the dark cycle, since MT may be photo-responsive [135]. The aluminum foil was removed after spraying, and irrigation was withheld in all three groups for 6 days, until soil water content was maintained at 50 ± 2% of the well-watered plants. Four to five plants from each group were measured and two independent experiments were performed (n = 8–10).

4.4. *Soil Water Content*

The soil moisture sensor (5TE; Decagon Devices, Pullman, WA, USA), jointly with the ProCheck device (Decagon Devices), was used for measuring the volumetric soil water content [130]. The results are presented as percentage of the full soil water capacity of the well-watered oregano plants.

4.5. *Leaf Water Content*

The water content of oregano leaves was determined by the electronic moisture balance (MOC120H, Shimadzu, Tokyo, Japan) with the formula: (FW–DW)/DW × 100%, where FW is fresh weight and DW is dry weight [136].

4.6. *Chlorophyll Content*

The chlorophyll content in oregano leaves was measured with a portable chlorophyll content meter (Model Cl-01, Hansatech Instruments Ltd., Norfolk, UK), as described in detail [137]. Results were expressed in relative units [137].

4.7. *Chlorophyll Fluorescence Imaging Analysis*

The modulated Imaging-PAM Fluorometer M-Series (Heinz Walz GmbH, Effeltrich, Germany) was used for chlorophyll fluorescence imaging analysis, performed as described in detail previously [99]. Seven areas of interest (AOI) were selected in each leaf to cover the whole leaf area. The actinic light (AL) used for estimating PSII function was 205 μmol photons m^{-2} s^{-1} (corresponding to growth irradiance, GI) and 1000 μmol photons m^{-2} s^{-1} (corresponding to high irradiance, HI). The measured chlorophyll fluorescence parameters (described in Supplementary Table S1) were estimated using the Win software version 2.32 (Heinz Walz GmbH, Effeltrich, Germany). Color-coded images of selected chlorophyll fluorescence parameters for control (well-watered, water-sprayed, CK), moderate drought-stressed, water sprayed (WA), moderate drought-stressed sprayed with 10 μM MT (10 μM MT) and moderate drought-stressed sprayed with 100 μM MT (100 μM MT) are also presented.

4.8. *Statistical Analysis*

Statistical analysis was performed using the IBM SPSS Statistics for Windows version 28. A one-way ANOVA test was performed to evaluate the effect of the treatment on each photosynthetic parameter, followed by Tukey post-hoc test for pairwise comparisons. Significance was set at a $p < 0.05$ level. A linear regression analysis was also performed. Eight to ten plants were used for statistical analysis for each treatment (n = 8–10).

5. Conclusions

Foliar spraying of oregano plants with 100 μM MT was documented to be more effective than 10 μM MT, by retaining higher leaf water content and preserving the chlorophyll content under moderate drought stress, thus mitigating the negative impact on PSII function. MT significantly improved the malfunction of the OEC and the resulting PSII photoinhibition caused by moderate drought stress. It is suggested that, under moderate drought stress, MT exerts its action on oregano plants, by triggering the NPQ mechanism to decrease $^1O_2$ production, mainly at HI. The reduced $^1O_2$ production resulted in ameliorating PSII photochemistry and, by increasing the percentage of open PSII reaction centers, ETR was increase. It is concluded that MT may reduce the excess excitation energy by reducing $^1O_2$ formation, and may also enhance the photosynthetic function of moderate drought-stressed oregano plants. Consequently, it can be regarded as a promising agent for improving the ability of crop plants to face drought episodes in combination with the high light conditions of the Mediterranean area, which influence crop production detrimentally. However, since there is a differential impact of MT on the light energy use efficiency at PSII, depending on the light intensity and the plant species, more experiments must be performed with different crop species to establish the extensive use of MT in agriculture, in order to accomplish sustainable crop production to meet the challenge of drought stress due to climate change.

Supplementary Materials: The following supporting information can be downloaded at: https://www.mdpi.com/article/10.3390/plants13182590/s1, Figure S1. The excitation pressure at PSII ($1 - qL$). Table S1: Definitions of the chlorophyll fluorescence parameters used in the experiments.

Author Contributions: Conceptualization, M.M.; methodology, J.M., I.S. and M.M.; validation, J.M. and M.M.; formal analysis, J.M., I.S., S.İ., B.Ş. and M.M.; investigation, I.S., S.İ. and B.Ş.; resources, M.M.; data curation, J.M., I.S., S.İ., B.Ş. and M.M.; writing—original draft preparation, J.M. and M.M.; writing—review and editing, J.M., I.S., S.İ., B.Ş. and M.M.; visualization, J.M., I.S. and M.M.; supervision, M.M.; project administration, M.M. All authors have read and agreed to the published version of the manuscript.

Funding: This research received no external funding.

Data Availability Statement: The data presented in this study are available in this article.

Conflicts of Interest: The authors declare no conflicts of interest.

References

1. Ray, D.K.; Ramankutty, N.; Mueller, N.D.; West, P.C.; Foley, J.A. Recent patterns of crop yield growth and stagnation. *Nat. Commun.* **2012**, *3*, 1293. [CrossRef] [PubMed]
2. Sahay, S.; Grzybowski, M.; Schnable, J.C.; Głowacka, K. Genetic control of photoprotection and photosystem II operating efficiency in plants. *New Phytol.* **2023**, *239*, 1068–1082. [CrossRef] [PubMed]
3. Wing, I.S.; De Cian, E.; Mistry, M.N. Global vulnerability of crop yields to climate change. *J. Environ. Econ. Manag.* **2021**, *109*, 102462. [CrossRef]
4. Placide, R.; Hirut, G.B.; Stephan, N.; Fekadu, B. Assessment of drought stress tolerance in root and tuber crops. *Afr. J. Plant Sci.* **2014**, *8*, 214–224. [CrossRef]
5. Sperdouli, I.; Mellidou, I.; Moustakas, M. Harnessing chlorophyll fluorescence for phenotyping analysis of wild and cultivated tomato for high photochemical efficiency under water deficit for climate change resilience. *Climate* **2021**, *9*, 154. [CrossRef]
6. Seleiman, M.F.; Al-Suhaibani, N.; Ali, N.; Akmal, M.; Alotaibi, M.; Refay, Y.; Dindaroglu, T.; Abdul-Wajid, H.H.; Battaglia, M.L. Drought stress impacts on plants and different approaches to alleviate its adverse effects. *Plants* **2021**, *10*, 259. [CrossRef]
7. Moustakas, M.; Sperdouli, I.; Moustaka, J. Early drought stress warning: Color pictures of photosystem II photochemistry. *Climate* **2022**, *10*, 179. [CrossRef]
8. Moustaka, J.; Moustakas, M. Early-stage detection of biotic and abiotic stress on plants by chlorophyll fluorescence imaging analysis. *Biosensors* **2023**, *13*, 796. [CrossRef]
9. Hussain, M.; Farooq, S.; Hasan, W.; Ul-Allah, S.; Tanveer, M.; Farooq, M.; Nawaz, A. Drought stress in sunflower: Physiological effects and its management through breeding and agronomic alternatives. *Agr. Water Manag.* **2018**, *201*, 152–166. [CrossRef]
10. Muktadir, M.A.; Adhikari, K.N.; Ahmad, N.; Merchant, A. Chemical composition and reproductive functionality of contrasting faba bean genotypes in response to water deficit. *Physiol. Plant.* **2021**, *172*, 540–551. [CrossRef]
11. Daryanto, S.; Wang, L.; Jacinthe, P.A. Global synthesis of drought effects on maize and wheat production. *PLoS ONE* **2016**, *11*, e0156362. [CrossRef] [PubMed]
12. Dietz, K.J.; Zörb, C.; Geilfus, C.M. Drought and crop yield. *Plant Biol.* **2021**, *23*, 881–893. [CrossRef] [PubMed]
13. Fathi, A.; Tari, D.B. Effect of drought stress and its mechanism in plants. *Int. J. Life Sci.* **2016**, *10*, 1–6. [CrossRef]
14. Sah, R.P.; Chakraborty, M.; Prasad, K.; Pandit, M.; Tudu, V.K.; Chakravarty, M.K.; Moharana, D. Impact of water deficit stress in maize: Phenology and yield components. *Sci. Rep.* **2020**, *10*, 2944. [CrossRef]
15. Melo, G.B.; da Silva, A.G.; da Costa, A.C.; Alves da Silva, A.; Rosa, M.; Bessa, L.A.; Rodrigues, C.R.; Castoldi, G.; Vitorino, L.C. Foliar application of biostimulant mitigates water stress effects on soybean. *Agronomy* **2024**, *14*, 414. [CrossRef]
16. Bray, E.A. Plant responses to water deficit. *Trends Plant Sci.* **1997**, *2*, 48–54. [CrossRef]
17. Flexas, J.; Barón, M.; Bota, J.; Ducruet, J.-M.; Gallé, A.; Galmés, J.; Jiménez, M.; Pou, A.; Ribas-Carbó, M.; Sajnani, C. Photosynthesis limitations during water stress acclimation and recovery in the drought-adapted *Vitis* hybrid Richter-110 (*V. berlandieri* x *V. rupestris*). *J. Exp. Bot.* **2009**, *60*, 2361–2377. [CrossRef]

18. Miller, G.; Suzuki, N.; Ciftci-Yilmaz, S.; Mittler, R. Reactive oxygen species homeostasis and signalling during drought and salinity stresses. *Plant Cell Environ.* **2010**, *33*, 453–467. [CrossRef]
19. Niyogi, K.K. Safety valves for photosynthesis. *Curr. Opin. Plant Biol.* **2000**, *3*, 455–460. [CrossRef]
20. Asada, K. Production and scavenging of reactive oxygen species in chloroplasts and their functions. *Plant Physiol.* **2006**, *141*, 391–396. [CrossRef]
21. Moustakas, M. Plant Photochemistry, Reactive Oxygen Species, and Photoprotection. *Photochem* **2022**, *2*, 5–8. [CrossRef]
22. Müller, P.; Li, X.P.; Niyogi, K.K. Non-photochemical quenching. A response to excess light energy. *Plant Physiol.* **2001**, *125*, 1558–1566. [CrossRef] [PubMed]
23. Wilson, K.E.; Ivanov, A.G.; Öquist, G.; Grodzinski, B.; Sarhan, F.; Huner, N.P.A. Energy balance, organellar redox status, and acclimation to environmental stress. *Can. J. Bot.* **2006**, *84*, 1355–1370. [CrossRef]
24. Apel, K.; Hirt, H. Reactive oxygen species: Metabolism, oxidative stress, and signal transduction. *Annu. Rev. Plant Biol.* **2004**, *55*, 373–399. [CrossRef] [PubMed]
25. Moustaka, J.; Tanou, G.; Adamakis, I.D.; Eleftheriou, E.P.; Moustakas, M. Leaf age dependent photoprotective and antioxidative mechanisms to paraquat-induced oxidative stress in *Arabidopsis thaliana*. *Int. J. Mol. Sci.* **2015**, *16*, 13989–14006. [CrossRef]
26. Ruban, A.V. Nonphotochemical chlorophyll fluorescence quenching: Mechanism and effectiveness in protecting plants from photodamage. *Plant Physiol.* **2016**, *170*, 1903–1916. [CrossRef]
27. Sachdev, S.; Ansari, S.A.; Ansari, M.I.; Fujita, M.; Hasanuzzaman, M. Abiotic stress and reactive oxygen species: Generation, signaling, and defense mechanisms. *Antioxidants* **2021**, *10*, 277. [CrossRef]
28. Munné-Bosch, S.; Peñuelas, J. Photo- and antioxidative protection, and a role for salicylic acid during drought and recovery in field-grown *Phillyrea angustifolia* plants. *Planta* **2003**, *217*, 758–766. [CrossRef]
29. Hajiboland, R.; Cheraghvareh, L.; Poschenrieder, C. Improvement of drought tolerance in tobacco (*Nicotiana rustica* L.) plants by silicon. *J. Plant Nutr.* **2017**, *40*, 1661–1676. [CrossRef]
30. Kasajima, I.; Ebana, K.; Yamamoto, T.; Takahara, K.; Yano, M.; Kawai-Yamada, M.; Uchimiya, H. Molecular distinction in genetic regulation of nonphotochemical quenching in rice. *Proc. Natl. Acad. Sci. USA* **2011**, *108*, 13835–13840. [CrossRef]
31. Noctor, G.; Veljovic-Jovanovic, S.O.N.J.A.; Driscoll, S.; Novitskaya, L.; Foyer, C.H. Drought and oxidative load in the leaves of C3 plants: A predominant role for photorespiration? *Ann. Bot.* **2002**, *89*, 841–850. [CrossRef] [PubMed]
32. Hasanuzzaman, M.; Bhuyan, M.H.M.B.; Zulfiqar, F.; Raza, A.; Mohsin, S.M.; Mahmud, J.A.; Fujita, M.; Fotopoulos, V. Reactive oxygen species and antioxidant defense in plants under abiotic stress: Revisiting the crucial role of a universal defense regulator. *Antioxidants* **2020**, *9*, 681. [CrossRef] [PubMed]
33. Sperdouli, I.; Moustaka, J.; Ouzounidou, G.; Moustakas, M. Leaf age-dependent photosystem II photochemistry and oxidative stress responses to drought stress in *Arabidopsis thaliana* are modulated by flavonoid accumulation. *Molecules* **2021**, *26*, 4157. [CrossRef] [PubMed]
34. Betteridge, D.J. What is oxidative stress? *Metabolism* **2000**, *49*, 3–8. [CrossRef]
35. Jain, M.; Kataria, S.; Hirve, M.; Prajapati, R. Water deficit stress effects and responses in maize. In *Plant Abiotic Stress Tolerance*; Hasanuzzaman, M., Hakeem, K., Nahar, K., Alharby, H., Eds.; Springer: Cham, Switzerland, 2019; pp. 129–151.
36. Moustakas, M.; Sperdouli, I.; Moustaka, J.; Şaş, B.; İşgören, S.; Morales, F. Mechanistic insights on salicylic acid mediated enhancement of photosystem II function in oregano seedlings subjected to moderate drought stress. *Plants* **2023**, *12*, 518. [CrossRef]
37. Sperdouli, I.; Panteris, E.; Moustaka, J.; Aydın, T.; Bayçu, G.; Moustakas, M. Mechanistic insights on salicylic acid-induced enhancement of photosystem II function in basil plants under non-stress or mild drought stress. *Int. J. Mol. Sci.* **2024**, *25*, 5728. [CrossRef]
38. Anjum, S.A.; Xie, X.Y.; Wang, L.C.; Saleem, M.F.; Man, C.; Lei, W. Morphological, physiological and biochemical responses of plants to drought stress. *Afr. J. Agric. Res.* **2011**, *6*, 2026–2032.
39. Sperdouli, I.; Moustakas, M. Interaction of proline, sugars, and anthocyanins during photosynthetic acclimation of *Arabidopsis thaliana* to drought stress. *J. Plant Physiol.* **2012**, *169*, 577–585. [CrossRef]
40. Foyer, C.H.; Shigeoka, S. Understanding oxidative stress and antioxidant functions to enhance photosynthesis. *Plant Physiol.* **2011**, *155*, 93–100. [CrossRef]
41. Niyogi, K.K.; Wolosiuk, R.A.; Malkin, R. Photosynthesis. In *Biochemistry & Molecular Biology of Plants*, 2nd ed.; Buchanan, B.B., Gruissem, W., Jones, R.L., Eds.; John Wiley & Sons, Ltd.: Hoboken, NJ, USA, 2015; pp. 508–566.
42. Moustakas, M.; Malea, P.; Zafeirakoglou, A.; Sperdouli, I. Photochemical changes and oxidative damage in the aquatic macrophyte *Cymodocea nodosa* exposed to paraquat-induced oxidative stress. *Pest. Biochem. Physiol.* **2016**, *126*, 28–34. [CrossRef]
43. Zhu, J.K. Abiotic stress signaling and responses in plants. *Cell* **2016**, *167*, 313–324. [CrossRef] [PubMed]
44. Hou, X.; Zhang, W.; Du, T.; Kang, S.; Davies, W.J. Responses of water accumulation and solute metabolism in tomato fruit to water scarcity and implications for main fruit quality variables. *J. Exp. Bot.* **2020**, *71*, 1249–1264. [CrossRef] [PubMed]
45. López-Calcagno, P.E.; Brown, K.L.; Simkin, A.J.; Fisk, S.J.; Vialet-Chabrand, S.; Lawson, T.; Raines, C.A. Stimulating photosynthetic processes increases productivity and water-use efficiency in the field. *Nat. Plants* **2020**, *6*, 1054–1063. [CrossRef]
46. Zhu, X.G.; Hasanuzzaman, M.; Jajoo, A.; Lawson, T.; Lin, R.; Liu, C.M.; Liu, L.N.; Liu, Z.; Lu, C.; Moustakas, M.; et al. Improving photosynthesis through multidisciplinary efforts: The next frontier of photosynthesis research. *Front. Plant Sci.* **2022**, *13*, 967203. [CrossRef] [PubMed]

47. Khalid, M.F.; Huda, S.; Yong, M.; Li, L.; Li, L.; Chen, Z.H.; Ahmed, T. Alleviation of drought and salt stress in vegetables: Crop responses and mitigation strategies. *Plant Growth Regul.* **2023**, *99*, 177–194. [CrossRef]
48. Zhang, F.; Rosental, L.; Ji, B.; Brotman, Y.; Dai, M. Metabolite-mediated adaptation of crops to drought and the acquisition of tolerance. *Plant J.* **2024**, *118*, 626–644. [CrossRef]
49. Van Oosten, M.J.; Pepe, O.; De Pascale, S.; Silletti, S.; MAggio, A. The role of biostimulants and bioeffectors as alleviators of abiotic stress in crop plants. *Chem. Biol. Technol. Agric.* **2014**, *4*, 5. [CrossRef]
50. Yakhin, O.I.; Lubyanov, A.A.; Yakhin, I.A.; Brown, P.H. Biostimulants in plant science: A global perspective. *Front. Plant Sci.* **2017**, *7*, 2049. [CrossRef]
51. Nephali, L.; Piater, L.A.; Dubery, I.A.; Patterson, V.; Huyser, J.; Burgess, K.; Tugizimana, F. Biostimulants for plant growth and mitigation of abiotic stresses: A metabolomics perspective. *Metabolites* **2020**, *10*, 505. [CrossRef]
52. Ma, Y.; Freitas, H.; Dias, M.C. Strategies and prospects for biostimulants to alleviate abiotic stress in plants. *Front. Plant Sci.* **2022**, *13*, 1024243. [CrossRef]
53. Monteiro, E.; Gonçalves, B.; Cortez, I.; Castro, I. The role of biostimulants as alleviators of biotic and abiotic stresses in grapevine: A review. *Plants* **2022**, *11*, 396. [CrossRef] [PubMed]
54. Li, J.; Lardon, R.; Mangelinckx, S.; Geelen, D. Practical guide toward discovery of biomolecules with biostimulant activity. *J. Exp. Bot.* **2024**, *75*, 3797–3817. [CrossRef] [PubMed]
55. Moustakas, M.; Panteris, E.; Moustaka, J.; Aydın, T.; Bayçu, G.; Sperdouli, I. Modulation of photosystem II function in celery via foliar-applied salicylic acid during gradual water deficit stress. *Int. J. Mol. Sci.* **2024**, *25*, 6721. [CrossRef] [PubMed]
56. Gedeon, S.; Ioannou, A.; Balestrini, R.; Fotopoulos, V.; Antoniou, C. Application of biostimulants in tomato plants (*Solanum lycopersicum*) to enhance plant growth and salt stress tolerance. *Plants* **2022**, *11*, 3082. [CrossRef] [PubMed]
57. Karumannil, S.; Khan, T.A.; Kappachery, S.; Gururani, M.A. Impact of exogenous melatonin application on photosynthetic machinery under abiotic stress conditions. *Plants* **2023**, *12*, 2948. [CrossRef]
58. Arnao, M.B.; Hernández-Ruiz, J. Melatonin: A new plant hormone and/or a plant master regulator? *Trends Plant Sci.* **2019**, *24*, 38–48. [CrossRef]
59. Li, D.; Wei, J.; Peng, Z.; Ma, W.; Yang, Q.; Song, Z.; Sun, W.; Yang, W.; Yuan, L.; Xu, X.; et al. Daily rhythms of phytomelatonin signaling modulate diurnal stomatal closure via regulating reactive oxygen species dynamics in *Arabidopsis*. *J. Pineal Res.* **2020**, *68*, e12640. [CrossRef]
60. Wang, K.; Xing, Q.; Ahammed, G.J.; Zhou, J. Functions and prospects of melatonin in plant growth, yield, and quality. *J. Exp. Bot.* **2022**, *73*, 5928–5946. [CrossRef]
61. Khan, D.; Cai, N.; Zhu, W.; Li, L.; Guan, M.; Pu, X.; Chen, Q. The role of phytomelatonin receptor 1-mediated signaling in plant growth and stress response. *Front. Plant Sci.* **2023**, *14*, 1142753. [CrossRef]
62. Moustakas, M.; Sperdouli, I.; Adamakis, I.-D.S.; Şaş, B.; İşgören, S.; Moustaka, J.; Morales, F. Mechanistic approach on melatonin-induced hormesis of photosystem II function in the medicinal plant *Mentha spicata*. *Plants* **2023**, *12*, 4025. [CrossRef]
63. Arnao, M.B.; Hernãndez-Ruiz, J. Functions of melatonin in plants: A review. *J. Pineal Res.* **2015**, *59*, 133–150. [CrossRef] [PubMed]
64. Wang, Y.; Reiter, R.J.; Chan, Z. Phytomelatonin: A universal abiotic stress regulator. *J. Exp. Bot.* **2018**, *69*, 963–974. [CrossRef] [PubMed]
65. Yan, F.; Zhang, J.; Li, W.; Ding, Y.; Zhong, Q.; Xu, X.; Wei, H.; Li, G. Exogenous melatonin alleviates salt stress by improving leaf photosynthesis in rice seedlings. *Plant Physiol. Biochem.* **2021**, *163*, 367–375. [CrossRef] [PubMed]
66. Sati, H.; Chinchkar, A.V.; Kataria, P.; Pareek, S. The role of phytomelatonin in plant homeostasis, signaling, and crosstalk in abiotic stress mitigation. *Physiol. Plant.* **2024**, *176*, e14413. [CrossRef] [PubMed]
67. Zhang, H.J.; Zhang, N.; Yang, R.C.; Wang, L.; Sun, Q.Q.; Li, D.B.; Cao, Y.Y.; Weeda, S.; Zhao, B.; Ren, S.; et al. Melatonin promotes seed germination under high salinity by regulating antioxidant systems, ABA and GA_4 interaction in cucumber (*Cucumis sativus* L.). *J. Pineal Res.* **2014**, *57*, 269–279. [CrossRef]
68. Sanie Khatam, A.; Rastegar, S.; Aboutalebi Jahromi, A.; Hassanzadeh Khankahdani, H.; Akbar Bagherian, S.A. Biochemical and physiological mechanism induced by melatonin in Mexican lime (*Citrus aurantifolia* Swingle) plants: Cold and freezing stress. *Acta Physiol. Plant.* **2023**, *45*, 98. [CrossRef]
69. Song, R.; Ritonga, F.N.; Yu, H.; Ding, C.; Zhao, X. Effects of exogenous antioxidant melatonin on physiological and biochemical characteristics of *Populus cathayana* x *canadansis* 'Xin Lin 1' under salt and alkaline stress. *Forests* **2022**, *13*, 1283. [CrossRef]
70. Ding, F.; Wang, M.; Liu, B.; Zhang, S. Exogenous melatonin mitigates photoinhibition by accelerating non-photochemical quenching in tomato seedlings exposed to moderate light during chilling. *Front. Plant Sci.* **2017**, *8*, 244. [CrossRef]
71. Yang, S.J.; Huang, B.; Zhao, Y.Q.; Hu, D.; Chen, T.; Ding, C.B.; Chen, Y.E.; Yuan, S.; Yuan, M. Melatonin enhanced the tolerance of *Arabidopsis thaliana* to high light through improving anti-oxidative system and photosynthesis. *Front. Plant Sci.* **2021**, *12*, 752584. [CrossRef]
72. Jahan, M.S.; Yang, J.Y.; Althaqafi, M.M.; Alharbi, B.M.; Wu, H.Y.; Zhou, X.B. Melatonin mitigates drought stress by increasing sucrose synthesis and suppressing abscisic acid biosynthesis in tomato seedlings. *Physiol. Plant.* **2024**, *176*, e14457. [CrossRef]
73. Yan, F.; Chen, X.; Wang, Z.; Xia, Y.; Zheng, D.; Xue, S.; Zhao, H.; Huang, Z.; Niu, Y.; Zhang, G. Melatonin alleviates abscisic acid deficiency inhibition on photosynthesis and antioxidant systems in rice under salt stress. *Phyton-Int. J. Exp. Bot.* **2024**, *93*, 1421–1440. [CrossRef]

74. Wang, J.; Yan, D.; Liu, R.; Wang, T.; Lian, Y.; Lu, Z.; Hong, Y.; Wang, Y.; Li, R. The physiological and molecular mechanisms of exogenous melatonin promote the seed germination of maize (*Zea mays* L.) under salt stress. *Plants* **2024**, *13*, 2142. [CrossRef] [PubMed]
75. Zhan, H.; Nie, X.; Zhang, T.; Li, S.; Wang, X.; Du, X.; Tong, W.; Song, W. Melatonin: A small molecule but important for salt stress tolerance in plants. *Int. J. Mol. Sci.* **2019**, *20*, 709. [CrossRef]
76. Zhu, J.; Lou, H.; Yan, C.; Zhang, W.; Li, Z. Exogenous melatonin enhances cold tolerance by regulating the expression of photosynthetic performance, antioxidant system, and related genes in cotton. *Plants* **2024**, *13*, 2010. [CrossRef]
77. Bose, S.K.; Howlader, P. Melatonin plays multifunctional role in horticultural crops against environmental stresses: A review. *Environ. Exp. Bot.* **2020**, *176*, 104063. [CrossRef]
78. Arnao, M.; Hernández-Ruiz, J. Melatonin and reactive oxygen and nitrogen species: A model for the plant redox network. *Melatonin Res.* **2019**, *2*, 152–168. [CrossRef]
79. Ahmad, I.; Song, X.; Hussein Ibrahim, M.E.; Jamal, Y.; Younas, M.U.; Zhu, G.; Zhou, G.; Adam Ali, A.Y. The role of melatonin in plant growth and metabolism, and its interplay with nitric oxide and auxin in plants under different types of abiotic stress. *Front. Plant Sci.* **2023**, *14*, 1108507. [CrossRef]
80. Khan, M.S.S.; Ahmed, S.; Ikram, A.U.; Hannan, F.; Yasin, M.U.; Wang, J.; Zhao, B.; Islam, F.; Chen, J. Phytomelatonin: A key regulator of redox and phytohormones signaling against biotic/abiotic stresses. *Redox Biol.* **2023**, *64*, 102805. [CrossRef]
81. Hussain, A.; Faheem, B.; Jang, H.-S.; Lee, D.-S.; Mun, B.-G.; Rolly, N.K.; Yun, B.-W. Melatonin nitric oxide crosstalk in plants and the prospects of NOMela as a nitric oxide donor. *Int. J. Mol. Sci.* **2024**, *25*, 8535. [CrossRef]
82. Tan, D.X.; Hardeland, R.; Manchester, L.C.; Korkmaz, A.; Ma, S.; Rosales-Corral, S.; Reiter, R.J. Functional roles of melatonin in plants, and perspectives in nutritional and agricultural science. *J. Exp. Bot.* **2012**, *63*, 577–597. [CrossRef]
83. Arnao, M.B.; Hernãndez-Ruiz, J. Melatonin and its relationship to plant hormones. *Ann. Bot.* **2018**, *121*, 195–207. [CrossRef] [PubMed]
84. Kofidis, G.; Bosabalidis, A.M.; Moustakas, M. Contemporary seasonal and altitudinal variations of leaf structural features in oregano (*Origanum vulgare* L.). *Ann. Bot.* **2003**, *92*, 635–645. [CrossRef] [PubMed]
85. Kramer, D.M.; Johnson, G.; Kiirats, O.; Edwards, G.E. New fluorescence parameters for the determination of QA redox state and excitation energy fluxes. *Photosynth. Res.* **2004**, *79*, 209–218. [CrossRef] [PubMed]
86. Zhou, Y.; Lam, H.M.; Zhang, J. Inhibition of photosynthesis and energy dissipation induced by water and high light stresses in rice. *J. Exp. Bot.* **2007**, *58*, 1207–1217. [CrossRef]
87. Murata, N.; Allakhverdiev, S.I.; Nishiyama, Y. The mechanism of photoinhibition in vivo: Re-evaluation of the roles of catalase, -tocopherol, non-photochemical quenching, and electron transport. *Biochim. Biophys. Acta* **2012**, *1817*, 1127–1133. [CrossRef]
88. Tripathy, B.C.; Oelmüller, R. Reactive oxygen species generation and signaling in plants. *Plant Sig. Behav.* **2012**, *7*, 1621–1633. [CrossRef]
89. Dalal, V.K.; Tripathy, B.C. Water-stress induced downsizing of light-harvesting antenna complex protects developing rice seedlings from photo-oxidative damage. *Sci. Rep.* **2018**, *8*, 5955. [CrossRef] [PubMed]
90. Qiao, M.; Hong, C.; Jiao, Y.; Hou, S.; Gao, H. Impacts of drought on photosynthesis in major food crops and the related mechanisms of plant responses to drought. *Plants* **2024**, *13*, 1808. [CrossRef] [PubMed]
91. Alberte, R.S.; Fiscus, E.L.; Naylor, A.W. The effects of water stress on the development of the photosynthetic apparatus in greening leaves. *Plant Physiol.* **1975**, *55*, 317–321. [CrossRef]
92. He, J.X.; Wang, J.; Liang, H.G. Effects of water-stress on photochemical function and protein-metabolism of photosystem-II in wheat leaves. *Physiol. Plant.* **1995**, *93*, 771–777. [CrossRef]
93. Giardi, M.T.; Cona, A.; Geiken, B.; Kučera, T.; Masojidek, J.; Mattoo, A.K. Long-term drought stress induces structural and functional reorganization of photosystem II. *Planta* **1996**, *199*, 118–125. [CrossRef]
94. Sperdouli, I.; Moustakas, M. A better energy allocation of absorbed light in photosystem II and less photooxidative damage contribute to acclimation of *Arabidopsis thaliana* young leaves to water deficit. *J. Plant Physiol.* **2014**, *171*, 587–593. [CrossRef] [PubMed]
95. Bano, H.; Athar, H.R.; Zafar, Z.U.; Ogbaga, C.C.; Ashraf, M. Peroxidase activity and operation of photo-protective component of NPQ play key roles in drought tolerance of mung bean [*Vigna radiata* (L.) Wilcziek]. *Physiol. Plant.* **2021**, *172*, 603–614. [CrossRef] [PubMed]
96. Ruban, A.V.; Wilson, S. The mechanism of non-photochemical quenching in plants: Localization and driving forces. *Plant Cell Physiol.* **2021**, *62*, 1063–1072. [CrossRef] [PubMed]
97. Johnson, M.P.; Ruban, A.V. Photoprotective energy dissipation in higher plants involves alteration of the excited state energy of the emitting chlorophyll(s) in the light harvesting antenna II (LHCII). *J. Biol. Chem.* **2009**, *284*, 23592–23601. [CrossRef]
98. Ivanov, A.G.; Sane, P.V.; Hurry, V.; Oquist, G.; Huner, N.P. Photosystem II reaction centre quenching: Mechanisms and physiological role. *Photosynth. Res.* **2008**, *98*, 565–574. [CrossRef]
99. Moustaka, J.; Panteris, E.; Adamakis, I.D.S.; Tanou, G.; Giannakoula, A.; Eleftheriou, E.P.; Moustakas, M. High anthocyanin accumulation in poinsettia leaves is accompanied by thylakoid membrane unstacking, acting as a photoprotective mechanism, to prevent ROS formation. *Environ. Exp. Bot.* **2018**, *154*, 44–55. [CrossRef]
100. Vass, I. Role of charge recombination processes in photodamage and photoprotection of the photosystem II complex. *Physiol. Plant.* **2011**, *142*, 6–16. [CrossRef]

101. Prasad, A.; Sedlářová, M.; Pospíšil, P. Singlet oxygen imaging using fluorescent probe Singlet Oxygen Sensor Green in photosynthetic organisms. *Sci. Rep.* **2018**, *8*, 13685. [CrossRef]
102. Gawroński, P.; Witoń, D.; Vashutina, K.; Bederska, M.; Betliński, B.; Rusaczonek, A.; Karpiński, S. Mitogen-activated protein kinase 4 is a salicylic acid-independent regulator of growth but not of photosynthesis in Arabidopsis. *Mol. Plant* **2014**, *7*, 1151–1166. [CrossRef]
103. Lazár, D.; Murch, S.J.; Beilby, M.J.; Al Khazaaly, S. Exogenous melatonin affects photosynthesis in characeae *Chara australis*. *Plant Signal. Behav.* **2013**, *8*, e23279. [CrossRef] [PubMed]
104. Wang, G.; Xing, M.; Hu, T.; Ji, M.; Li, X.; Amombo, E.; Shao, A.; Xu, X.; Fu, J. Photosystem II photochemical adjustment of tall fescue against heat stress after melatonin priming. *J. Plant Physiol.* **2022**, *275*, 153758. [CrossRef] [PubMed]
105. Yang, N.; Han, M.H.; Teng, R.M.; Yang, Y.Z.; Wang, Y.H.; Xiong, A.S.; Zhuang, J. Exogenous melatonin enhances photosynthetic capacity and related gene expression in a dose-dependent manner in the tea plant (*Camellia sinensis* (L.) Kuntze). *Int. J. Mol. Sci.* **2022**, *23*, 6694. [CrossRef]
106. Toivonen, P.; Vidaver, W. Variable chlorophyll a fluorescence and CO_2 uptake in water stressed white spruce seedlings. *Plant Physiol.* **1988**, *86*, 744–748. [CrossRef] [PubMed]
107. Callahan, F.E.; Becker, D.W.; Cheniae, G.M. Studies on the photo-inactivation of the water-oxidizing enzyme. II. Characterization of weak light photoinhibition of PSII and its light-induced recovery. *Plant Physiol.* **1986**, *82*, 261–269. [CrossRef]
108. Chen, G.X.; Kazimir, J.; Cheniae, G.M. Photoinhibition of hydroxylamine-extracted photosystem II membranes: Studies of the mechanism. *Biochemistry* **1992**, *31*, 11072–11083. [CrossRef]
109. Anderson, J.M.; Park, Y.I.; Chow, W.S. Unifying model for the photoinactivation of photosystem II in vivo: A hypothesis. *Photosynth. Res.* **1998**, *56*, 1–13. [CrossRef]
110. Sarvikas, P.; Hakala, M.; Pätsikkä, E.; Tyystjärvi, T.; Tyystjärvi, E. Action spectrum of photoinhibition in leaves of wild type and npq1-2 and npq4-1 mutants of *Arabidopsis thaliana*. *Plant Cell Physiol.* **2006**, *47*, 391–400. [CrossRef]
111. Moustakas, M.; Dobrikova, A.; Sperdouli, I.; Hanć, A.; Adamakis, I.S.; Moustaka, J.; Apostolova, E. A hormetic spatiotemporal photosystem II response mechanism of salvia to excess zinc exposure. *Int. J. Mol. Sci.* **2022**, *23*, 11232. [CrossRef]
112. Govindachary, S.; Bukhov, N.G.; Joly, D.; Carpentier, R. Photosystem II inhibition by moderate light under low temperature in intact leaves of chilling-sensitive and -tolerant plants. *Physiol. Plant.* **2004**, *121*, 322–333. [CrossRef]
113. Pellegrini, E.; Carucci, M.G.; Campanella, A.; Lorenzini, G.; Nali, C. Ozone stress in *Melissa officinalis* plants assessed by photosynthetic function. *Environ. Exp. Bot.* **2011**, *73*, 94–101. [CrossRef]
114. Siddiqui, H.; Ahmed, K.B.M.; Hayat, S. Comparative effect of 28-homobrassinolide and 24-epibrassinolide on the performance of different components influencing the photosynthetic machinery in *Brassica juncea* L. *Plant Physiol. Biochem.* **2018**, *129*, 198–212. [CrossRef]
115. Mosadegh, H.; Trivellini, A.; Lucchesini, M.; Ferrante, A.; Maggini, R.; Vernieri, P.; Mensuali Sodi, A. UV-B physiological changes under conditions of distress and eustress in sweet basil. *Plants* **2019**, *8*, 396. [CrossRef] [PubMed]
116. Gohari, G.; Farhadi, H.; Panahirad, S.; Zareei, E.; Labib, P.; Jafari, H.; Mahdavinia, G.; Hassanpouraghdam, M.B.; Ioannou, A.; Kulak, M.; et al. Mitigation of salinity impact in spearmint plants through the application of engineered chitosan-melatonin nanoparticles. *Int. J. Biol. Macromol.* **2023**, *224*, 893–907. [CrossRef]
117. Kalisz, A.; Kornaś, A.; Skoczowski, A.; Oliwa, J.; Jurkow, R.; Gil, J.; Sękara, A.; Sałata, A.; Caruso, G. Leaf chlorophyll fluorescence and reflectance of oakleaf lettuce exposed to metal and metal(oid) oxide nanoparticles. *BMC Plant Biol.* **2023**, *23*, 329. [CrossRef] [PubMed]
118. Zia, A.; Farrag, E.S.; Mahmoud, S.Y. Dieback of royal poinciana (*Delonix regia*) trees induced by *Alternaria tenuissima* and its impact on photochemical efficiency of photosystem II. *Physiol. Mol. Plant Pathol.* **2024**, *133*, 102357. [CrossRef]
119. Tóth, S.Z.; Nagy, V.; Puthur, J.T.; Kovács, L.; Garab, G. The physiological role of ascorbate as photosystem II electron donor: Protection against photoinactivation in heat-stressed leaves. *Plant Physiol.* **2011**, *156*, 382–392. [CrossRef]
120. Széles, E.; Kuntam, S.; Vidal-Meireles, A.; Nagy, V.; Nagy, K.; Ábrahám, Á.; Kovács, L.; Tóth, S.Z. Single-cell microfluidics in combination with chlorophyll a fluorescence measurements to assess the lifetime of the Chlamydomonas PSBO protein. *Photosynthetica* **2023**, *61*, 417–424. [CrossRef]
121. Hamdani, S.; Khan, N.; Perveen, S.; Qu, M.; Jiang, J.; Govindjee; Zhu, X.G. Changes in the photosynthesis properties and photoprotection capacity in rice (*Oryza sativa*) grown under red, blue, or white light. *Photosynth. Res.* **2019**, *139*, 107–121. [CrossRef]
122. Tryfon, P.; Sperdouli, I.; Adamakis, I.-D.S.; Mourdikoudis, S.; Dendrinou-Samara, C.; Moustakas, M. Modification of tomato photosystem II photochemistry with engineered zinc oxide nanorods. *Plants* **2023**, *12*, 3502. [CrossRef]
123. Ort, D.R.; Merchant, S.S.; Alric, J.; Barkan, A.; Blankenship, R.E.; Bock, R.; Croce, R.; Hanson, M.R.; Hibberd, J.M.; Long, S.P.; et al. Redesigning photosynthesis to sustainably meet global food and bioenergy demand. *Proc. Natl. Acad. Sci. USA* **2015**, *112*, 8529–8536. [CrossRef] [PubMed]
124. Paul, M.J. Improving photosynthetic metabolism for crop yields: What is going to work? *Front. Plant Sci.* **2021**, *12*, 743862. [CrossRef] [PubMed]
125. Long, S.P.; Ainsworth, E.A.; Leakey, A.D.B.; Nosberger, J.; Ort, D.R. Food for thought: Lower-than-expected crop yield stimulation with rising CO_2 concentrations. *Science* **2006**, *312*, 1918–1921. [CrossRef] [PubMed]

126. Yin, X.; Struik, P.C. Constraints to the potential efficiency of converting solar radiation into phytoenergy in annual crops: From leaf biochemistry to canopy physiology and crop ecology. *J. Exp. Bot.* **2015**, *66*, 6535–6549. [CrossRef]
127. Busch, F.A. Opinion: The red-light response of stomatal movement is sensed by the redox state of the photosynthetic electron transport chain. *Photosynth. Res.* **2014**, *119*, 131–140. [CrossRef]
128. Głowacka, K.; Kromdijk, J.; Kucera, K.; Xie, J.; Cavanagh, A.P.; Leonelli, L.; Leakey, A.D.B.; Ort, D.R.; Niyogi, K.K.; Long, S.P. Photosystem II Subunit S overexpression increases the efficiency of water use in a field-grown crop. *Nat. Commun.* **2018**, *9*, 868. [CrossRef] [PubMed]
129. Niyogi, K.K.; Li, X.P.; Rosenberg, V.; Jung, H.S. Is PsbS the site of nonphotochemical quenching in photosynthesis? *J. Exp. Bot.* **2005**, *56*, 375–382. [CrossRef]
130. Sperdouli, I.; Ouzounidou, G.; Moustakas, M. Hormesis responses of photosystem II in *Arabidopsis thaliana* under water deficit stress. *Int. J. Mol. Sci.* **2023**, *24*, 9573. [CrossRef]
131. Terashima, I. Anatomy of non-uniform leaf photosynthesis. *Photosynth. Res.* **1992**, *31*, 195–212. [CrossRef]
132. Meng, Q.; Siebke, K.; Lippert, P.; Baur, B.; Mukherjee, U.; Weis, E. Sink–source transition in tobacco leaves visualized using chlorophyll fluorescence imaging. *New Phytol.* **2001**, *151*, 585–595. [CrossRef]
133. Moustakas, M.; Sperdouli, I.; Adamakis, I.-D.S.; Moustaka, J.; İşgören, S.; Şaş, B. Harnessing the role of foliar applied salicylic acid in decreasing chlorophyll content to reassess photosystem II photoprotection in crop plants. *Int. J. Mol. Sci.* **2022**, *23*, 7038. [CrossRef] [PubMed]
134. Zahedi, S.M.; Hosseini, M.S.; Fahadi Hoveizeh, N.; Gholami, R.; Abdelrahman, M.; Tran, L.P. Exogenous melatonin mitigates salinity-induced damage in olive seedlings by modulating ion homeostasis, antioxidant defense, and phytohormone balance. *Physiol. Plant.* **2021**, *173*, 1682–1694. [CrossRef] [PubMed]
135. Zhang, N.; Zhao, B.; Zhang, H.J.; Weeda, S.; Yang, C.; Yang, Z.C.; Ren, S.; Guo, Y.D. Melatonin promotes water-stress tolerance, lateral root formation, and seed germination in cucumber (*Cucumis sativus* L.). *J. Pineal Res.* **2013**, *54*, 15–23. [CrossRef] [PubMed]
136. Sperdouli, I.; Moustakas, M. Spatio-temporal heterogeneity in *Arabidopsis thalian*a leaves under drought stress. *Plant Biol.* **2012**, *14*, 118–128. [CrossRef]
137. Borek, M.; Bączek-Kwinta, R.; Rapacz, M. Photosynthetic activity of variegated leaves of *Coleus* × *hybridus* hort. cultivars characterised by chlorophyll fluorescence techniques. *Photosynthetica* **2016**, *54*, 331–339. [CrossRef]

plants

MDPI

Article

Comparative Physiological and Gene Expression Analyses Reveal Mechanisms Involved in Maintaining Photosynthesis Capacity, Alleviating Ion Toxicity and Oxidative Stress of Kentucky Bluegrass under NaCl Treatment

Rong Wang †, Shi-Jie Yan †, Chao Liu, Huan Guo * and Yan-Nong Cui *

College of Grassland Agriculture, Northwest A&F University, Yangling 712100, China; wr00142023@163.com (R.W.); yjay@nwafu.edu.cn (S.-J.Y.); liuchao3314463317@163.com (C.L.)

* Correspondence: huan.guo@nwafu.edu.cn (H.G.); cuiyn@nwafu.edu.cn (Y.-N.C.)

† These authors contributed equally to this work.

Abstract: Kentucky bluegrass (*Poa pratensis* L.), a widely used cool-season turfgrass, shows a high sensitivity to soil salinity. Clarifying the adaptative mechanisms of Kentucky bluegrass that serve to improve its salt tolerance in saline environments is urgent for the application of this turfgrass in salt-affected regions. In this study, physiological responses of the Kentucky bluegrass cultivars "Explorer" and "Blue Best" to NaCl treatment, as well as gene expressions related to photosynthesis, ion transport, and ROS degradation, were analyzed. The results showed that the growth of "Explorer" was obviously better compared to "Blue Best" under 400 mM NaCl treatment. "Explorer" exhibited a much stronger photosynthetic capacity than "Blue Best" under NaCl treatment, and the expression of key genes involved in chlorophyll biosynthesis, photosystem II, and the Calvin cycle in "Explorer" was greatly induced by salt treatment. Compared with "Blue Best", "Explorer" could effectively maintain Na^+/K^+ homeostasis in its leaves under NaCl treatment, which can be attributed to upregulated expression of genes, such as *HKT1;5*, *HAK5*, and *SKOR*. The relative membrane permeability and contents of O_2^- and H_2O_2 in "Explorer" were significantly lower than those in "Blue Best" under NaCl treatment, and, correspondingly, the activities of SOD and POD in the former were significantly higher than in the latter. Moreover, the expression of genes involved in the biosynthesis of enzymes in the ROS-scavenging system of "Explorer" was immediately upregulated after NaCl treatment. Additionally, free proline and betaine are important organic osmolytes for maintaining hydration status in Kentucky bluegrass under NaCl treatment, as the contents of these metabolites in "Explorer" were significantly higher than in "Blue Best". This work lays a theoretical basis for the improvement of salt tolerance in Kentucky bluegrass.

Keywords: salt stress; photosynthesis; Na^+ transport; ROS; turfgrass

Citation: Wang, R.; Yan, S.-J.; Liu, C.; Guo, H.; Cui, Y.-N. Comparative Physiological and Gene Expression Analyses Reveal Mechanisms Involved in Maintaining Photosynthesis Capacity, Alleviating Ion Toxicity and Oxidative Stress of Kentucky Bluegrass under NaCl Treatment. *Plants* **2024**, *13*, 2107. https://doi.org/10.3390/plants13152107

Academic Editor: Emilia Apostolova

Received: 10 July 2024
Revised: 24 July 2024
Accepted: 25 July 2024
Published: 30 July 2024

1. Introduction

In recent decades, soil salinity has been a major threat to turf establishment and management, especially in arid regions, as the cultivation of turfgrasses in these regions highly depends on reclaimed salt-containing water [1–3]. Kentucky bluegrass (*Poa pratensis* L.) is widely utilized in lawns and water and soil conservation, as well as slope protection, owing to its eminent traits, such as winter hardiness, a fine texture, a developed rhizome, and a great capacity for ramet generation [2]. This species is also employed in athletic fields like golf courses [3]. Nevertheless, Kentucky bluegrass displays a high sensitivity to soil salinity [4]. Therefore, elucidating the adaptative mechanisms of Kentucky bluegrass with respect to salinity and improving its salt tolerance through genetic approaches are high priorities for the development of the turf industry.

Salinity directly exerts osmotic stress and ion toxicity on plants or initiates secondary stresses (e.g., oxidative stress), ultimately influencing many aspects of biological processes, such as root activity, tissue hydration, photosynthesis, nutrient balance, membrane integrity, and cellular metabolism [1,5]. Plants deal with osmotic stresses mainly by accumulating large quantities of organic or inorganic osmolytes in cells to enhance water influx under salt conditions [1]. For the majority of glycophytes, the biosynthesis of organic metabolites, including amino acids, soluble sugars, and alkaloids, is the most effective way to decrease tissue osmotic potential and maintain hydration status under salt stresses [6,7]. Na^+ in saline soils is metabolically toxic to plants and disturbs K^+ homeostasis in plant tissues [8]. It has been found that restricting Na^+ transport into photosynthetic organs is essential for Poaceae plants [9,10]. The formation rate of reactive oxygen species (ROS) also accelerates substantially under salt stresses, which leads to membrane dysfunction [11]. The activities of key antioxidases in the ROS-scavenging system can increase the degradation of ROS under salt stresses [12,13]. However, the accumulation patterns of organic osmolytes, as well as the strategies for alleviating ion toxicity and oxidative stresses in Kentucky bluegrass under saline conditions, still require further investigation.

Na^+/K^+ transporters are employed by plants to alleviate ion toxicity; for instance, the Na^+/H^+ antiporter NHX1 mediates the sequestration of Na^+ into vacuoles, high-affinity K^+ transporters (HKTs) mediate Na^+ and/or K^+ uptake into cells, and shaker-type K^+ channels, such as AKT1, are responsible for root K^+ uptake [14–17]. Many researchers have reported that the expression of genes associated with Na^+ and K^+ transport undergoes significant changes after salt treatments, and these changes play significant roles in maintaining Na^+/K^+ homeostasis in plants [18–20]. It was discovered that plants can also upregulate the expression of genes encoding antioxidases to enhance ROS degradation under salt stresses [13]. Additionally, the expression of many genes regulating the photosynthetic processes in plants has been found to be influenced significantly by salinity [21–23]. Nevertheless, the responses of genes involved in these processes in Kentucky bluegrass have not been well-documented.

To investigate the physiological response of Kentucky bluegrass to salt stress and preliminarily uncover the molecular basis underlying the maintenance of photosynthesis, as well as the alleviation of Na^+ toxicity and oxidative stress, we compared the growth and photosynthesis of a salt-tolerant cultivar, "Explorer", and a salt-sensitive cultivar, "Blue Best", under NaCl treatment; determined the contents of organic osmolytes, as well as Na^+ and K^+, in tissues; and analyzed the ROS contents and activities of antioxidases after salt treatment. We also monitored the expression changes in key genes associated with photosynthesis, Na^+/K^+ transport, and ROS degradation in both cultivars under NaCl treatment.

2. Results

2.1. *Effects of NaCl Treatment on the Growth and Photosynthesis of Two Kentucky Bluegrass Cultivars*

Under 400 mM NaCl treatment, the leaves of "Explorer" remained green, while the leaves of "Blue Best" were visually wilted and yellow (Figure 1A). In "Explorer" and "Blue Best", the salt treatment significantly reduced the tiller number, total DW, and leaf RWC, while it significantly increased the leaf RMP and root activity (except for the root activity in "Blue Best") (Figure 1B–F). Notably, under NaCl treatment, the leaf RMP in "Explorer" was significantly lower than that in "Blue Best", whilst the leaf RWC in the former was notably higher than that in the latter (Figure 1D,E).

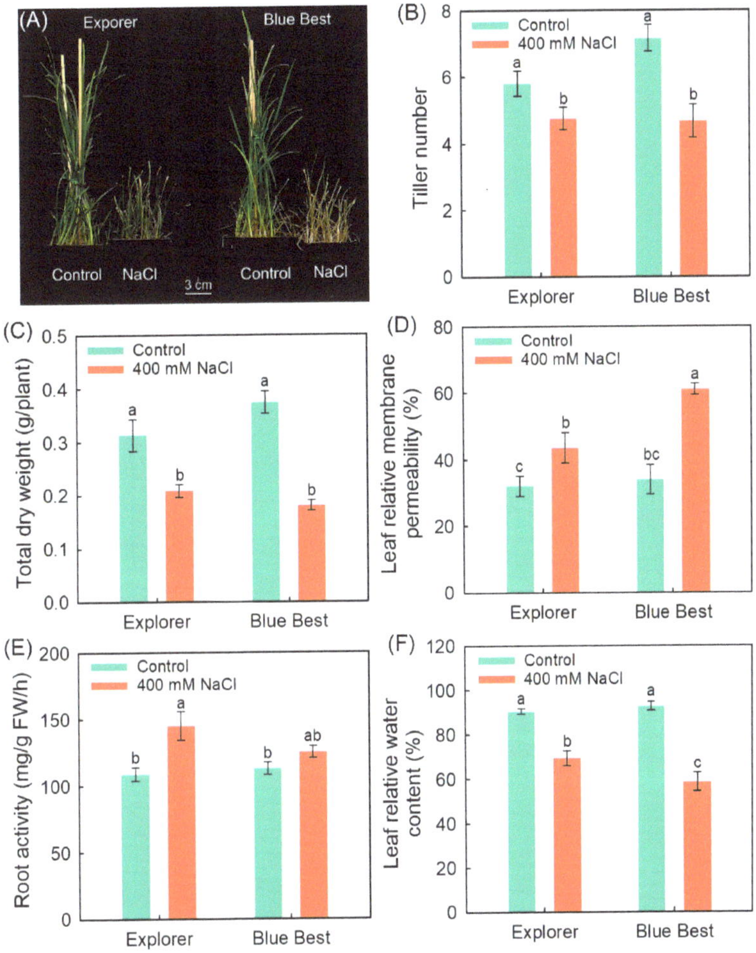

Figure 1. Effects of 400 mM NaCl treatment on growth-related parameters of Kentucky bluegrass cultivars "Explorer" and "Blue Best". (**A**) Growth photograph, (**B**) tiller number, (**C**) total dry weight, (**D**) leaf relative membrane permeability, (**E**) root activity, and (**F**) leaf relative water content. Data are means (±SDs), $n = 6$. Different letters indicate significant differences as determined using Tukey's HSD test ($p < 0.05$).

As shown in Figure 2, under the control condition, most photosynthesis-related parameters in "Blue Best", including chlorophyll b content, Pn, Gs, and Tr, were significantly higher than those in "Explorer", indicating that the former exhibits a stronger photosynthetic capacity than the latter when grown under normal conditions. The 400 mM NaCl treatment significantly decreased the chlorophyll contents, Pn, Gs, and Tr in both cultivars compared to the control (Figure 2). The WUE in "Explorer" under salt treatment was significantly elevated, while this parameter in "Blue Best" under salt treatment was sustained at the control level (Figure 2F). Moreover, under NaCl treatment, "Explorer" displayed obviously higher chlorophyll contents, Pn, Gs, Tr, and WUE compared to "Blue Best" (Figure 2).

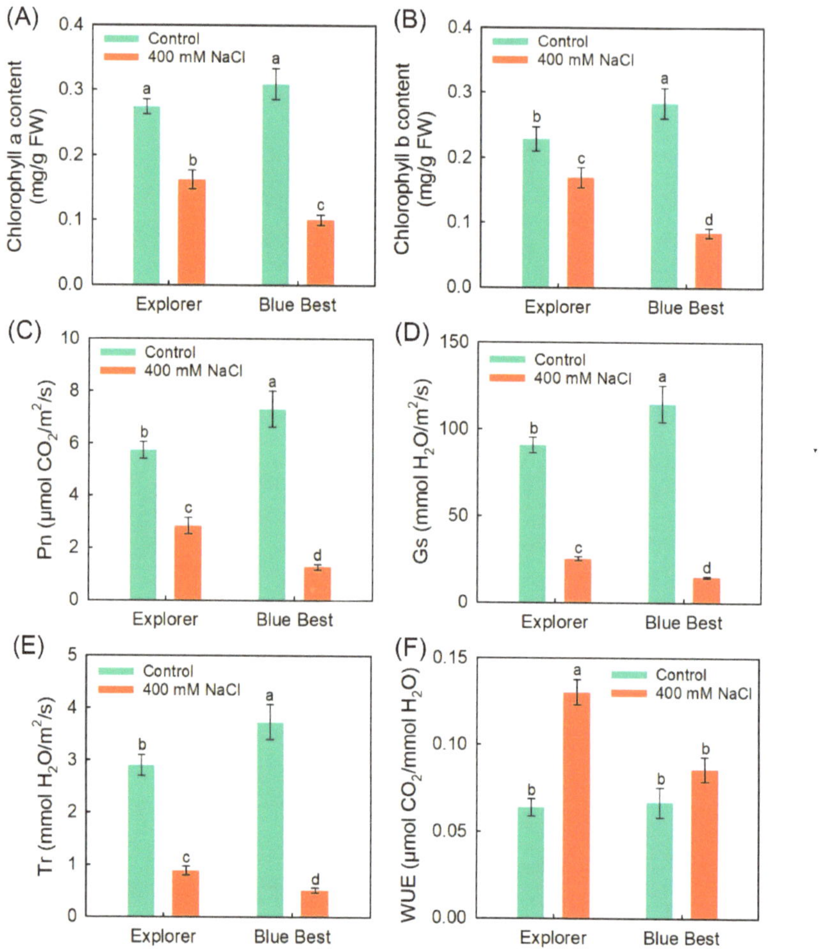

Figure 2. Effects of 400 mM NaCl treatment on photosynthesis-related parameters of Kentucky bluegrass cultivars "Explorer" and "Blue Best". (**A**) Chlorophyll a content, (**B**) chlorophyll b content, (**C**) net photosynthesis rate (Pn), (**D**) stomatal conductance (Gs), (**E**) transpirational rate (Tr), and (**F**) water-use efficiency (WUE). Data are means (±SDs), $n = 6$. Different letters indicate significant differences as determined using Tukey's HSD test ($p < 0.05$).

2.2. Effects of NaCl Treatment on Tissue Na^+ and K^+ Accumulation in Two Kentucky Bluegrass Cultivars

The tissue Na^+ contents of both cultivars under NaCl treatment were drastically elevated; furthermore, "Explorer" showed a lower leaf Na^+ content than "Blue Best" (Figure 3A,B). NaCl treatment did not affect the leaf K^+ content in "Explorer", but it significantly decreased this parameter in "Blue Best" (Figure 3C,D). Additionally, the K^+/Na^+ ratio in the leaves of "Explorer" was significantly higher than that in "Blue Best" under salt treatment (Figure 3E,F).

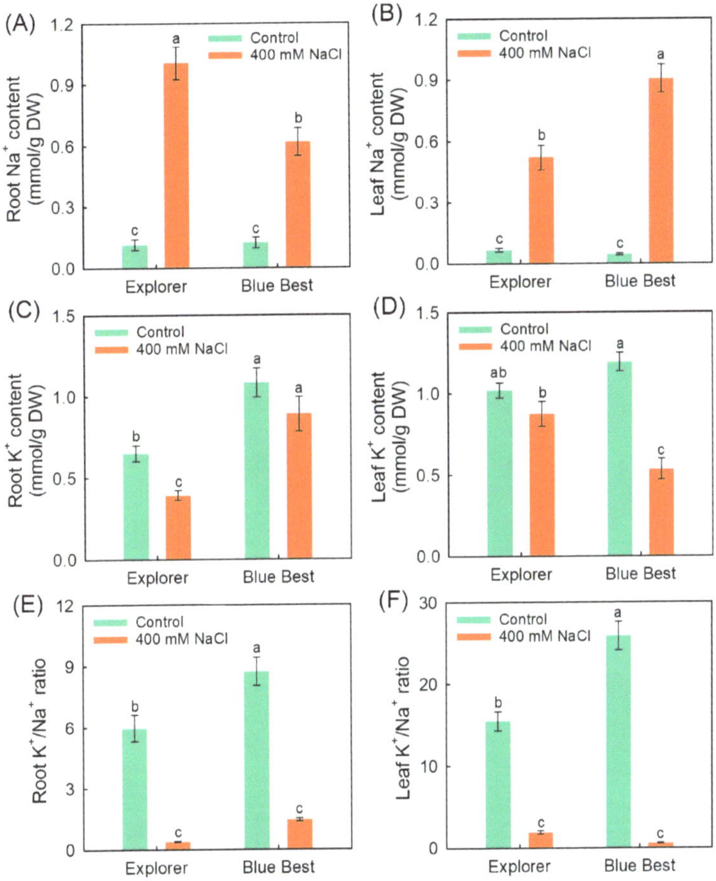

Figure 3. Effects of 400 mM NaCl treatment on tissue K^+ and Na^+ contents as well as K^+/Na^+ ratio of Kentucky bluegrass cultivars "Explorer" and "Blue Best". (**A**) Root Na^+ content, (**B**) leaf Na^+ content, (**C**) root K^+ content, (**D**) leaf K^+ content, (**E**) root K^+/Na^+ ratio, and (**F**) leaf K^+/Na^+ ratio. Data are means (±SDs), $n = 6$. Different letters indicate significant differences as determined using Tukey's HSD test ($p < 0.05$).

2.3. Effects of NaCl Treatment on Contents of Organic Osmolytes in the Leaves of Two Kentucky Bluegrass Cultivars

NaCl treatment significantly elevated the free proline and betaine contents in both cultivars, while it did not affect the leaf soluble sugar contents (Figure 4). Under NaCl treatment, no difference in soluble sugar contents between the two cultivars was observed, but the free proline and betaine contents in "Explorer" were significantly higher than those in "Blue Best" (Figure 4).

2.4. Effects of NaCl Treatment on ROS Contents and Activities of Antioxidases in the Leaves of Two Kentucky Bluegrass Cultivars

The 400 mM NaCl treatment did not affect the O_2^- content in the leaves of "Explorer"; moreover, the H_2O_2 content in "Explorer" was significantly lower than that in "Blue Best" (Figure 5A,B). As shown in Figure 5C,D, the NaCl treatment significantly increased the SOD and CAT activities in "Explorer" but had no effect on their activities in "Blue Best". Under the NaCl treatment, the APX activity of both cultivars was increased, but no significant difference was observed between "Explorer" and "Blue Best" (Figure 5E). The POD activity in "Explorer" and "Blue Best" under the NaCl treatment was also significantly increased, while the increase in "Blue Best" was smaller than that in "Explorer" (Figure 5F).

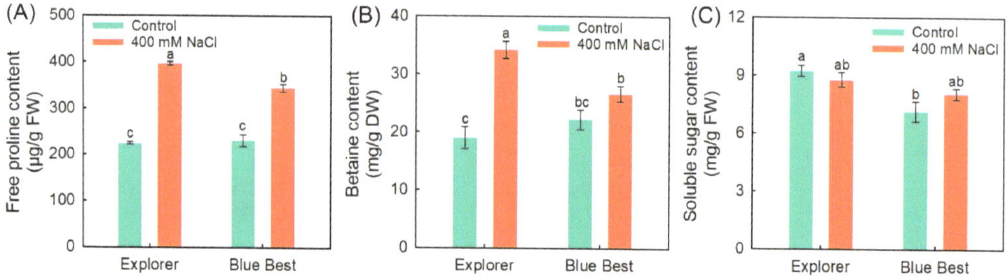

Figure 4. Effects of 400 mM NaCl on leaf organic osmolyte contents of Kentucky bluegrass cultivars "Explorer" and "Blue Best". (**A**) Free proline content, (**B**) betaine content, and (**C**) soluble sugar content. Data are means (±SDs), $n = 6$. Different letters indicate significant differences as determined using Tukey's HSD test ($p < 0.05$).

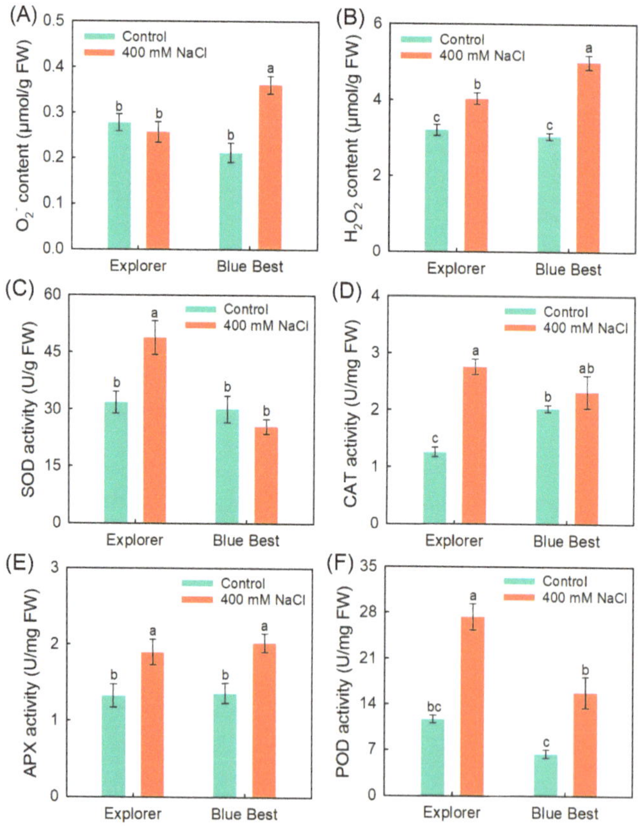

Figure 5. Effects of 400 mM NaCl on contents of O_2^- and H_2O_2 as well as activities of antioxidases in leaves of Kentucky bluegrass cultivars "Explorer" and "Blue Best". (**A**) O_2^- content; (**B**) H_2O_2 content; (**C**) superoxide dismutase (SOD) activity; (**D**) catalase (CAT) activity; (**E**) ascorbate peroxidase (APX) activity; (**F**) peroxidase (POD) activity. Data are means (±SDs), $n = 6$. Different letters indicate significant differences as determined using Tukey's HSD test ($p < 0.05$).

2.5. Effects of NaCl Treatment on the Expression of Key Genes Associated with Photosynthesis in the Leaves of Two Kentucky Bluegrass Cultivars

We identified nine genes associated with photosynthesis—*HEMA1* (responsible for chlorophyll biosynthesis [24]), *psbA*, and *psbD* (which encode subunits in photosystem II [25]), *rbcL*, *rbcS*, *PRK*, *PGK*, *FBPase*, and *SBPase* (all involved in the Calvin cycle [26])—in the leaves of both Kentucky bluegrass cultivars. The expression of *PpHEMA1*, *PppsbA*, and *PppsbD* in "Explorer" was induced or maintained at high levels after salt treatment; in contrast, these genes in "Blue Best" under salt treatment were substantially downregulated (Figure 6 and Table S2). *PprbcL*, *PprbcS*, *PpPRK*, *PpPGK*, and *PpFBPase* in "Explorer" were immediately upregulated after short-term treatment (Figure 6 and Table S2), whilst these genes (except for *PpPRK*) were downregulated after 6 h of treatment, and some members (such as *PprbcS*, *PpPRK*, *PpPGK*, and *PpFBPase*) showed increasing trends when the treatment time was prolonged to 48 h (Figure 6 and Table S2). In addition, the expression of *PpSBPase* after salt treatment was downregulated in both cultivars (Figure 6 and Table S2).

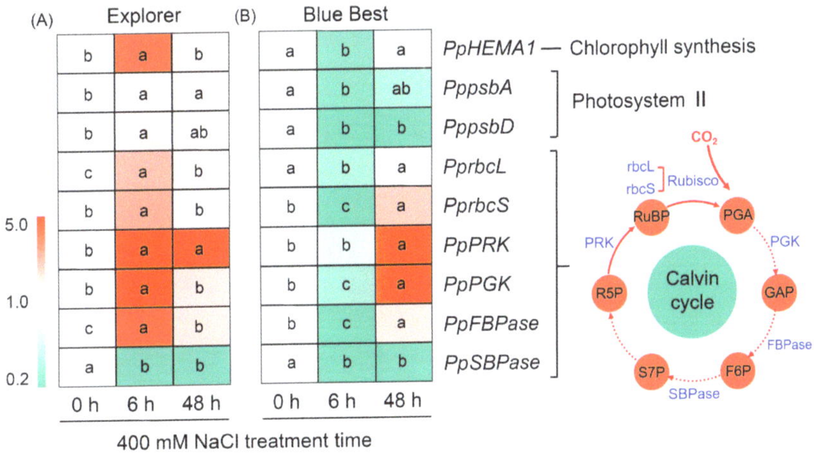

Figure 6. The expression changes of genes related to photosynthesis processes of Kentucky bluegrass cultivars "Explorer" (**A**) and "Blue Best" (**B**) after 400 mM NaCl treatment. Different letters in blocks indicate significant differences as determined using Tukey's HSD test ($p < 0.05$). HEMA: glutamyl-tRNA reductase, psbA: photosystem II protein D1, psbD: photosystem II protein D2, rbcL: ribulose 1,5-bisphosphate carboxylase/oxygenase large subunit, rbcS: ribulose 1,5-bisphosphate carboxylase/oxygenase small subunit, PRK: phosphoribulokinase, PGK: phosphoglycerate kinase, FBPase: fructose-1,6-bisphosphatase, SBPase: sedoheptulose-1,7-bisphosphatase, RuBP: ribulose-1,5-bisphosphate, PGA: glycerate-3-phosphate, GAP: glyceraldehyde-3-phosphate, F6P: fructose-6-phosphate, S7P: sedoheptulose-7-phosphate, R5P: ribulose-5-phosphate.

2.6. Effects of NaCl Treatment on the Expression of Key Genes Related to Na^+ and K^+ Transport in the Roots of Two Kentucky Bluegrass Cultivars

It has been documented that NHX1 mediates the sequestration of Na^+ into vacuoles and SOS1 mediates the efflux of Na^+ from roots [27,28]. *PpNHX1* in "Explorer" and "Blue Best" showed an increasing trend, while *PpSOS1* in both cultivars exhibited a declining trend after salt treatment (Figure 7B,C and Table S3). HKTs are thought to govern the uptake of Na^+ and/or K^+ into plant cells, among which HKT1;5 is involved in root xylem Na^+ unloading, and HKT2;1 mediates Na^+ absorption by roots [29,30]. The expression of *PpHKT1;5* was greatly induced in "Explorer" after short-term treatment but suppressed in "Blue Best" after long-term treatment; the expression of *PpHKT2;1* in "Explorer" was constantly downregulated after salt treatment, while it was upregulated in "Blue Best" after long-term NaCl treatment (Figure 7B,C and Table S3).

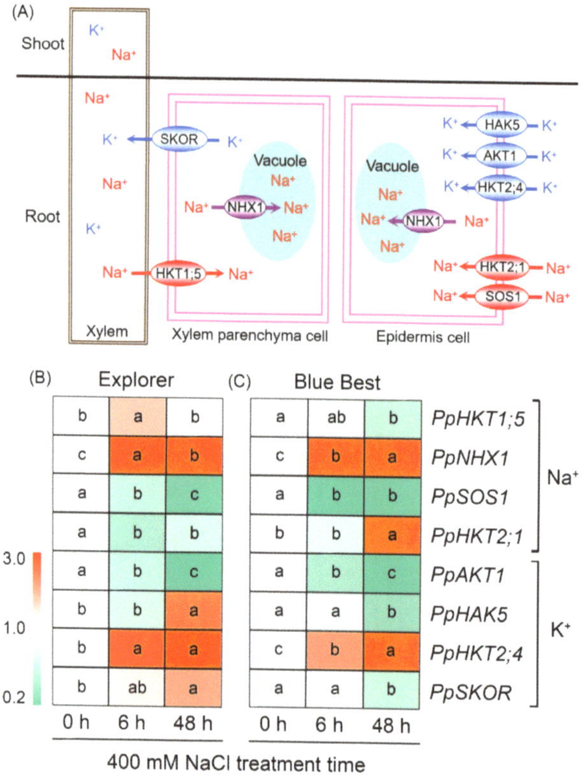

Figure 7. The diagram of proteins involved in Na^+/K^+ transport in plants (**A**). The HAK5, AKT1, and HKT2;4 mediate the uptake of K^+ into the root epidermis cells; the HKT2;1 and SOS1 mediate the uptake of Na^+ into the root epidermis cells; the NHX1 mediates the transport of Na^+ into vacuoles; the SKOR mediates the efflux of K^+ from root xylem parenchyma cells; the HKT1;5 mediates the unloading of Na^+ from root xylem into parenchyma cells. The expression changes of genes related to Na^+ and K^+ transport in Kentucky bluegrass cultivars "Explorer" (**B**) and "Blue Best" (**C**) after 400 mM NaCl treatment. Different letters in blocks indicate significant differences as determined using Tukey's HSD test ($p < 0.05$). HKTs and HAK5: high-affinity K^+ transporters, NHX1: tonoplast Na^+/H^+ antiporter, SOS1: plasma membrane Na^+/H^+ antiporter, AKT1: inwardly rectifying K^+ channel, SKOR: stelar K^+ outwardly rectifying channel.

HKT2;4 is thought to mediate the uptake of K^+ into plant cells [31]. The expression of *PpHKT2;4* was highly inducible by NaCl treatment in both Kentucky bluegrass cultivars (Figure 7B,C and Table S3). Shaker-type potassium channels (including AKT1) and potassium transporters (including HAK5) also mediate K^+ uptake [32]. As shown in Figure 7 and Table S3, the expression of *PpAKT1* in both cultivars and *PpHAKT5* in "Explorer" was constantly downregulated after NaCl treatment, while *PpHAKT5* in "Blue Best" was inducible by long-term treatment. SKOR mediates the translocation of K^+ into shoots [33]. Interestingly, *PpSKOR* was substantially downregulated in "Blue Best" but remained stable or was significantly upregulated in "Explorer" after salt treatment (Figure 7B,C and Table S3).

2.7. Effects of NaCl Treatment on the Expression of Key Genes Involved in ROS Degradation in the Leaves of Two Kentucky Bluegrass Cultivars

The ROS-scavenging system in plants constitutes the SOD pathway, GPX pathway, CAT pathway, Prdx/Trx pathway, and AsA-GSH cycle [23]. The expression of *PpSOD* and

PpCAT (involved in the SOD pathway and CAT pathway, respectively) in "Explorer" was upregulated after salt treatment, but in "Blue Best" they were downregulated after short-term treatment (Figure 8 and Table S4). Interestingly, after 6 h of treatment, the expression of *PpAPX*, *PpMDAR*, *PpDHAR*, *PpGr*, and *PpGST* in "Explorer" was significantly upregulated, while most of these genes in "Blue Best" were downregulated (Figure 8 and Table S4). It was found that the expression of *PpGPX* and *PpPOD* (involved in the GPX pathway) in "Explorer" was substantially upregulated (over 3-fold) after NaCl treatment, while the expression of *PpGLP* (also involved in the GPX pathway) in both cultivars was significantly downregulated after salt treatment. Furthermore, *PpTrx* and *PpPrdx* (both involved in the Prdx/Trx pathway) in both cultivars were also substantially downregulated under salt treatment (Figure 8 and Table S4).

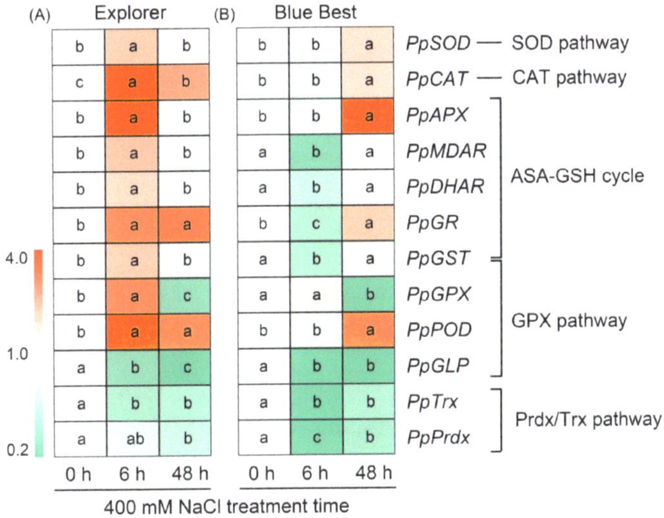

Figure 8. The expression changes of genes related to biosynthesis of key enzymes in the ROS-scavenging system of Kentucky bluegrass cultivars "Explorer" (**A**) and "Blue Best" (**B**) after 400 mM NaCl treatment. Different letters in blocks indicate significant differences as determined using Tukey's HSD test ($p < 0.05$). SOD: superoxide dismutase, CAT: catalase, APX: ascorbate peroxidase, MDAR: monodehydroascorbate reductase, DHAR: dehydroascorbate reductase, GR: glutathione reductase, GST: glutathione S-transferase, GPX: glutathione peroxidase, POD: peroxidases, GLP: germin-like protein, Trx: thioredoxin, Prdx: peroxiredoxin.

3. Discussion

3.1. Maintaining a High Photosynthetic Performance by Upregulating the Expression of Photosynthesis-Related Genes Is Important for Salt Tolerance in Kentucky Bluegrass

Although Kentucky bluegrass is a typical glycophyte that is more sensitive to salinity than many other turfgrasses, this species is thought to have high plasticity in terms of salt adaptation [3,34]. We previously found that the Kentucky bluegrass cultivar "Explorer" is a relatively salt-tolerant species and "Blue Best" is a sensitive one [35]. Here, "Explorer" also exhibited a much stronger salt tolerance than "Blue Best", as evidenced by the fact that the shoot growth and leaf hydration of the former were significantly better than those of the latter under 400 mM NaCl treatment for 15 d (Figure 1). Thus, the elucidation of the prominent traits possessed by "Explorer" relative to "Blue Best" under salt stress would provide a theoretical basis for improving the salt tolerance of Kentucky bluegrass using genetic approaches.

The net photosynthesis rate (Pn) is used to represent the photosynthetic capacity of plants, and it has been found that the Pn in most glycophytes (and even some halophytes) is inhibited greatly under saline conditions [36–38]. Although the Pn in the two Kentucky bluegrass cultivars significantly decreased under salt treatment, the decrease in "Explorer" was much smaller than that in "Blue Best" (Figure 2C). Given that "Explorer" exhibits a stronger salt tolerance than "Blue Best" (Figure 1), the ability to sustain a relatively high photosynthetic capacity is essential for Kentucky bluegrass to adapt to salinity. The stomatal pores constitute the only means for CO_2 to enter the photosynthesis organs [39]. Notably, the stomatal conductance in "Explorer" under salt treatment was significantly higher than that in "Blue Best" (Figure 2D), suggesting that the maintenance of stomatal opening to ensure CO_2 acquisition is also closely related to the salt tolerance of Kentucky bluegrass.

The biosynthesis process of chlorophyll is hindered in many plants under salt stress [5,40]. In this study, it was observable that the leaves of "Explorer" remained green but most tillers of "Blue Best" turned yellow under the 400 mM NaCl treatment (Figure 1A). Correspondingly, the chlorophyll a and b contents in "Blue Best" sharply declined after the salt treatment (Figure 2A,B). HEMA is a rate-limiting enzyme that catalyzes chlorophyll biosynthesis, and *HEMA1* in rice (*Oryza sativa*) is substantially upregulated under salt treatments [24,40]. In this study, *PpHEMA1* under NaCl treatment showed upregulated expression in "Explorer" but downregulated expression in "Blue Best" (Figure 6). These results suggest that "Explorer" should possess a stronger ability to maintain chlorophyll biosynthesis than "Blue Best".

The expression of genes encoding components in PSII, such as *psbA*, *psbD*, and *psbQ* in sweet sorghum and black locust, has been found to substantially decline after salt treatments [22,25]. In this study, *PppsbA* and *PppsbD* showed opposing expression patterns in "Explorer" (significantly upregulated) and "Blue Best" (significantly downregulated) after NaCl treatment (Figure 6), indicating that the upregulation of *psbA* and *psbD* should also play vital roles in maintaining photosynthetic capacity in Kentucky bluegrass under saline conditions. We also analyzed the expression of key genes in the Calvin cycle and found that, except for *PpSBPase*, the expression of other determined genes in "Explorer"—including *PprbcL*, *PprbcS*, *PpPRK*, *PpPGK*, and *PpFBPase*—was immediately upregulated after short-term salt treatment and maintained at high levels when the treatment time was prolonged to 48 h; in contrast, most of these genes in "Blue Best" showed suppressed expression under salt treatment (Figure 6). Therefore, the rapid upregulation of genes in the Calvin cycle could be another reason why "Explorer" displays relatively high salt adaptation.

3.2. Restricting Na^+ Overaccumulation and Maintaining K^+ Homeostasis in Leaves Play Essential Roles in Kentucky Bluegrass's Adaptation to Saline Environments

There are diverse strategies for plants to cope with Na^+ toxicity; for example, plants with succulent leaves can accumulate Na^+ into developed vacuoles, and some halophytes can export Na^+ through salt-secreting organs on the leaf surface [41,42]. However, for most plant species within the Poaceae family, restricting the transport of Na^+ into leaves is more important for alleviating Na^+ toxicity [30,43]. In this study, "Explorer" under NaCl treatment showed a significantly lower leaf Na^+ content compared to "Blue Best" (Figure 3A,B). Furthermore, the Na^+ content in the roots of "Explorer" was approximately 2-fold higher than that in the leaves, while this parameter in the roots of "Blue Best" (approximately 0.6 mmol/g DW) was lower than that in the leaves (approximately 0.9 mmol/g DW) under NaCl treatment (Figure 2A,B). Therefore, the restriction of Na^+ overaccumulation in leaves is critical for the salt tolerance of Kentucky bluegrass.

The root xylem's unloading of Na^+ is particularly important for restricting Na^+ accumulation in shoots, and HKT1;5 has been proven to mediate this process [44,45]. *PpHKT1;5* was significantly upregulated in the roots of "Explorer", while in "Blue Best" it was downregulated, after salt treatment (Figure 7B,C). Therefore, the enhanced expression of *PpHKT1;5* may be closely associated with the restriction of Na^+ overaccumulation in the leaves of Kentucky bluegrass under salt stress. In addition, HKT2;1 has been discovered to mediate roots' Na^+ absorption, and both *TaHKT2;1* in wheat (*Triticum aestivum*) and *OsHKT2;1* in rice are upregulated under salt stress [29,46]. Interestingly, *PpHKT2;1* in "Explorer" exhibited substantially decreased expression after NaCl treatment, while this gene in "Blue Best" was upregulated after treatment for 48 h (Figure 7B,C). Therefore, the downregulation of *HKT2;1* may also be essential for alleviating Na^+ toxicity in Kentucky bluegrass under salt stress.

Na^+ in saline soils would disrupt the K^+ homeostasis in most plants [41]. Interestingly, the leaf K^+ content under NaCl treatment in "Explorer" remained at a relatively high level, while this parameter in "Blue Best" sharply declined after salt treatment (Figure 3C,D). A K^+/Na^+ ratio < 1 in plant tissues is considered to exert significant disturbances on the metabolism in plants [1]. We found that the leaf K^+/Na^+ ratio in "Explorer" was approximately 2, while it was less than 0.6 in "Blue Best", under salt stress (Figure 3F). AKT1 and HAK5 constitute important pathways for K^+ absorption in roots, and *AKT1* has been reported to be upregulated in some xerophytic species under salt stress [19,32]. In contrast, the expression of *PpAKT1* in both Kentucky bluegrass cultivars was significantly downregulated after NaCl treatment (Figure 7B,C). Notably, *PpHAK5* in the roots of "Blue Best" was also downregulated after NaCl treatment, while this gene in the roots of "Explorer" was significantly upregulated under salt treatment (Figure 7B,C). Therefore, the increased expression of *HAK5* might be indispensable for the acquisition of K^+ by Kentucky bluegrass under saline conditions. Additionally, *PpSKOR* (responsible for the transport of K^+ into shoots) in "Blue Best" remained stable or even significantly declined after NaCl treatment, whilst this gene in "Explorer" displayed significantly upregulated expression after NaCl treatment (Figure 7B,C), which may be a primary reason why this cultivar could maintain K^+ homeostasis in leaves under salt treatment (Figure 3D). All of these results suggest that improving K^+ uptake and translocation ability would be a promising way to enhance the salt tolerance of Kentucky bluegrass.

3.3. Enhancing the Ability to Degrade ROS Is Crucial for Kentucky Bluegrass to Cope with Salt Stress

ROS are continuously produced during the metabolic processes of plants; however, their production is elevated under various stress conditions, further exerting oxidative stress on the plants [11]. The ROS content in terms of O_2^- and H_2O_2 in "Explorer" under NaCl treatment was much lower than that in "Blue Best" (Figure 5A,B). Correspondingly, the leaf RMP in "Explorer" was substantially lower than that in "Blue Best" after salt treatment (Figure 1D). These results suggest that "Explorer" has a stronger ability to degrade ROS than "Blue Best" under salt stress. The ROS-scavenging system, consisting of many antioxidases, is the most effective way to relieve the damage to cell membranes and organelles caused by ROS [23]. The activities of antioxidases (especially SOD and POD) in "Explorer" substantially increased after NaCl treatment (Figure 5), indicating that the enhanced activities of antioxidases could be particularly important for alleviating oxidative stress in Kentucky bluegrass.

The increased biosynthesis of antioxidases could also enhance the capacity to degrade ROS in plants [23]. Our results showed that the genes encoding key antioxidases involved the SOD, CAT, and GPX pathways, and the ASA-GSH cycle of ROS-scavenging systems in the leaves of "Explorer", were immediately upregulated after NaCl treatment, while the expression of most genes in this system was stable or downregulated in the leaves of "Blue Best" (Figure 8), suggesting that the rapid upregulation of genes encoding antioxidases could also contribute to the adaptation of Kentucky bluegrass to salt stress.

3.4. The Increased Accumulation of Organic Osmolytes Plays Essential Roles in the Osmotic Adjustment of Kentucky Bluegrass

K^+ is the most important inorganic osmolyte in plants [47]. However, it was observed that the K^+ contents in the tissues of the Kentucky bluegrass cultivars did not increase under NaCl treatment (Figure 3D). In contrast, the contents of organic osmolytes in the leaves of the Kentucky bluegrass cultivars were substantially elevated after NaCl treatment

(Figure 5). The leaf hydration status in "Explorer" under NaCl treatment was substantially better than that in "Blue Best" (Figure 1F). Correspondingly, the contents of free proline and betaine in "Explorer" were obviously higher than those in "Blue Best" (Figure 5). Therefore, the increased biosynthesis of proline and betaine may be important for Kentucky bluegrass to maintain hydration status under saline conditions.

In addition to serving as an organic osmolyte, proline in plants is also a non-enzymatic antioxidant; meanwhile, it can increase the activities of antioxidases, such as CAT [48]. Considering that "Explorer" displayed a much stronger ability to degrade ROS than "Blue Best", the increased accumulation of proline in leaves may also be important for alleviating oxidative stress under saline conditions. Taken together, the large-scale biosynthesis of organic osmolytes, such as free proline and betaine, may play an essential role in the salt tolerance of Kentucky bluegrass.

4. Materials and Methods

4.1. Plant Materials and Growth Conditions

Seeds of the Kentucky bluegrass cultivars "Explorer" and "Blue Best" were first placed in Petri dishes for germination, and then the seedlings were transferred into 0.4 L plastic pots containing silica sand (8 seedlings/pot) and irrigated with Hoagland solution [43]. All seedlings were constantly cultured in a growth chamber with 16 h/8 h light/dark at 25/20 °C, 60% relative humidity, and ~600 $\mu mol \cdot m^{-2} \cdot s^{-1}$ of illumination intensity.

After 2 months, uniform seedlings of both cultivars were selected. Our previous study evaluated the salt tolerance of 30 Kentucky bluegrass cultivars under 400 mM NaCl treatment, and we found that detrimental effects of 400 mM NaCl treatment on the growth of "Blue Best" were more severe than those on "Explorer" [35]. Therefore, to analyze the physiological responses of these two cultivars to salt stress, the seedlings were exposed to Hoagland solution containing 400 mM NaCl, and seedlings that we continued to irrigate with Hoagland solution were assigned as the control. All solutions were renewed every 2 d. After 15 d, all seedlings in one pot were harvested and assigned as one replicate. Six replicates were used for all measurements (n = 6).

To monitor the expression of genes related to photosynthesis, Na^+ and K^+ transport, and ROS degradation in Kentucky bluegrass after salt treatment, uniform seedlings were exposed to 400 mM NaCl treatment solution for 0 (control), 6, and 48 h. Three replicates were used for the gene expression analysis (n = 3).

4.2. Determination of Growth-Related Parameters

The tiller numbers of the seedlings were counted first, and then the whole plant was placed in an oven at 80 °C for 3 d, after which the total dry weight (DW) was measured. Finally, the leaves' relative water content (RWC) was calculated [5].

For determination of the leaves' relative membrane permeability (RMP), all leaves of the seedlings were collected, immersed in deionized water, vacuum-filtered 3 times, and shaken at 25 °C. After 2 h, the electrolyte leakage was determined with a conductivity meter. After incubation at 100 °C for 2 h, the total electrolytes of the leaves were determined. The RMP was calculated as described by Cui et al. [5].

All roots of the seedlings were collected, immersed in 0.4% TTC solution at 37 °C in the dark, and then sealed in methyl alcohol for 3 h. Finally, the absorbance was recorded at 485 nm using a UV spectrophotometer (UV-2102C, Unico Instrument Co., Ltd., Shanghai, China) to calculate the root activity [49].

4.3. Determination of Photosynthesis-Related Parameters

The net photosynthesis rate (Pn), stomatal conductance (Gs), transpiration rate (Tr), and water-use efficiency (WUE) were measured in the growth chamber between 3 h and 5.5 h after the start of photoperiod using the LI-6800 Portable Photosynthesis System [5].

Fresh leaf samples weighing approximately 1 g were collected and then crushed with a mixed solution of 80% acetone and 95% ethanol. Then, the absorbances at 645 nm and 663 nm were determined to calculate the chlorophyll (Chl) a and Chl b contents [50].

4.4. Determination of Tissue Na^+ and K^+ Contents

Thoroughly dried root and leaf samples weighing approximately 0.1 g were immersed in 100 mM glacial acetic acid at 90 °C. After filtering, the solutions were collected to determine the Na^+ and K^+ contents using a flame spectrophotometer (Model 410 Flame; Sherwood Scientific, Ltd., Cambridge, UK) [51]. The tissue K^+/Na^+ ratio was calculated as K^+ content/Na^+ content.

4.5. Determination of the Contents of Free Proline, Betaine, and Soluble Sugars in Leaves

Fresh leaf samples weighing 0.5 g were homogenized with 3% sulfosalicylic acid at 100 °C. The supernatant was reacted with the mixed solution of 2.5% acid–ninhydrin and glacial acetic acid at 100 °C for 1 h. The free proline was leached with toluene, and then the absorbance at 520 nm was measured to calculate the free proline content [52].

The betaine content in 0.1 g dried leaf samples was determined using a Reinecke Salt Kit (Comin Biotechnology, Co., Ltd., Suzhou, China) [53].

Fresh leaf samples weighing 0.5 g were homogenized with 80% ethanol in a boiling water bath for 1 h, and then the soluble sugar content was determined via the classic anthrone colorimetric method [54].

4.6. Determination of the Contents of O_2^- and H_2O_2 in Leaves and the Activities of SOD, POD, CAT, and APX

The O_2^- content in fresh leaf samples was determined with an O_2^- content detection kit (Solarbio, Beijing, China), and the H_2O_2 content in fresh leaf samples was determined with a H_2O_2 content detection kit (Jiangcheng, Nanjing, China) [55].

The activities of antioxidases, including superoxide dismutase (SOD), peroxidase (POD), catalase (CAT), and ascorbate peroxidase (APX), were determined using the enzyme activity assay kits (Solarbio, Beijing, China) [56].

4.7. Real-Time Quantitative PCR Analysis

The leaves and roots were transiently frozen, and then the total RNA was extracted with the TRIzol reagent (Tiangen, Beijing, China). The gene-specific primer pairs with product lengths of 180–220 bp were designed using Premier Primer 5.0 (Table S1). After removal of genomic DNA, RNA samples were converted into cDNA [57]. The cDNA samples were used as templates for qRT-PCR using a Real-Time PCR Thermocycler (QuantStudio7 Flex, Thermo Fisher Scientific, Waltham, MA, USA) [58].

4.8. Statistical Analysis

The data were subjected to one-way analysis of variance (Tukey's HSD, $p < 0.05$) using SPSS 19.0 (IMB Corp, Armonk, NY, USA). All histograms were drawn with SigmaPlot 14.0, and all heatmaps were drawn using the Multiple Experiment Viewer 4.9.0 (J. Craig Venter Institute, La Jolla, CA, USA).

5. Conclusions

In conclusion, this study obtained the following findings: (I) the increased expression of genes associated with chlorophyll biosynthesis, components in PSII, and the Calvin cycle contributes to the maintenance of photosynthesis of Kentucky bluegrass under salt stress; (II) the restriction of Na^+ overaccumulation in leaves to maintain K^+ homeostasis by upregulating the expression of genes involved in Na^+/K^+ transport (such as *HKT1;5, HAK5,* and *SKOR*) plays a key role in Kentucky bluegrass's adaptation to saline environments; (III) the enhanced activities of antioxidases, such as SOD and POD, along with the increased expression of genes involved in the biosynthesis of enzymes in the ROS-scavenging system, are crucial for ROS degradation in Kentucky bluegrass under saline conditions; and (IV) the significant accumulation of free proline and betaine should be important for maintaining the hydration status of Kentucky bluegrass under salt stress.

Supplementary Materials: The following supporting information can be downloaded at: https://www.mdpi.com/article/10.3390/plants13152107/s1, Table S1: The primers for qRT-PCR analysis; Table S2: The relative expression level of genes related to photosynthesis in Kentucky bluegrass cultivars "Explorer" and "Blue Best" under 400 mM NaCl for 0 h (control), 6 h, or 48 h; Table S3: The relative expression level of genes related to Na^+ and K^+ transport in roots of Kentucky bluegrass cultivars "Explorer" and "Blue Best" under 400 mM NaCl for 0 h (control), 6 h, or 48 h; Table S4: The relative expression level of genes related to the ROS-scavenging system in leaves of Kentucky bluegrass cultivars "Explorer" and "Blue Best" under 400 mM NaCl for 0 h (control), 6 h, or 48 h.

Author Contributions: Conceptualization, H.G. and Y.-N.C.; methodology, R.W. and S.-J.Y.; software, R.W. and C.L.; validation, R.W., S.-J.Y. and C.L.; formal analysis, R.W.; investigation, R.W. and S.-J.Y.; data curation, R.W., S.-J.Y. and C.L.; writing—original draft preparation, R.W.; writing—review and editing, H.G. and Y.-N.C.; visualization, R.W. and S.-J.Y.; supervision, H.G. and Y.-N.C.; funding acquisition, H.G. and Y.-N.C. All authors have read and agreed to the published version of the manuscript.

Funding: This research was financially supported by the National Natural Science Foundation of China (32101253 and 32301505), the Key Research and Development Program of Shaanxi (2023-YBNY-046), the Natural Science Basic Research Program of Shaanxi (2023-JC-QN-0255), and open funds of the State Key Laboratory of Plant Environmental Resilience (SKLPERKF2404).

Data Availability Statement: The data that support the findings of this study are available from the corresponding author upon reasonable request.

Conflicts of Interest: The authors declare no conflict of interest.

References

1. Munns, R.; Tester, M. Mechanisms of salinity tolerance. *Annu. Rev. Plant Biol.* **2008**, *59*, 651–681. [CrossRef]
2. Xu, R.; Fujiyama, H. Comparison of ionic concentration, organic solute accumulation and osmotic adaptation in Kentucky bluegrass and tall fescue under NaCl stress. *Soil Sci. Plant Nutr.* **2013**, *59*, 168–179. [CrossRef]
3. Dunn, J.H.; Ervin, E.H.; Fresenburg, B.S. Turf performance of mixtures and blends of tall fescue, Kentucky bluegrass, and perennial ryegrass. *HortScience* **2002**, *37*, 214–217. [CrossRef]
4. Zuo, Z.F.; Li, Y.; Mi, X.F.; Li, Y.L.; Zhai, C.Y.; Yang, G.F.; Wang, Z.Y.; Zhang, K. Physiological and lipidomic response of exogenous choline chloride alleviating salt stress injury in Kentucky bluegrass (*Poa pratensis*). *Front. Plant Sci.* **2023**, *14*, 1269286. [CrossRef]
5. Cui, Y.N.; Li, X.T.; Yuan, J.Z.; Wang, F.Z.; Guo, H.; Xia, Z.R.; Wang, S.M.; Ma, Q. Chloride is beneficial for growth of the xerophyte *Pugionium cornutum* by enhancing osmotic adjustment capacity under salt and drought stresses. *J. Exp. Bot.* **2020**, *71*, 4215–4231. [CrossRef]
6. Chen, H.; Jiang, J.G. Osmotic responses of *Dunaliella* to the changes of salinity. *J. Cell. Physiol.* **2009**, *219*, 251–258. [CrossRef]
7. Kaur, G.; Asthir, B. Proline: A key player in plant abiotic stress tolerance. *Biol. Plant.* **2015**, *59*, 609–619. [CrossRef]
8. Zhu, J.K. Abiotic stress signaling and responses in plants. *Cell* **2016**, *167*, 313–324. [CrossRef]
9. James, R.A.; Davenport, R.J.; Munns, R. Physiological characterization of two genes for Na^+ exclusion in durum wheat, *Nax1* and *Nax2*. *Plant Physiol.* **2016**, *142*, 1537–1547. [CrossRef]
10. Munns, R.; James, R.A.; Xu, B.; Athman, A.; Conn, S.J.; Jordans, C.; Byrt, C.S.; Hare, R.A.; Tyerman, S.D.; Tester, M.; et al. Wheat grain yield on saline soils is improved by an ancestral Na^+ transporter gene. *Nat. Biotechnol.* **2012**, *30*, 360–364. [CrossRef]
11. Akyol, T.Y.; Yilmaz, O.; Uzilday, B.; Uzilday, R.O.; Turkan, I. Plant response to salinity: An analysis of ROS formation, signaling, and antioxidant defense. *Turk. J. Bot.* **2020**, *44*, 1–13.
12. Garg, N.; Manchanda, G. ROS generation in plants: Boon or bane? *Plant Biosyst.* **2009**, *143*, 81–96. [CrossRef]
13. Gill, S.S.; Tuteja, N. Reactive oxygen species and antioxidant machinery in abiotic stress tolerance in crop plants. *Plant Physiol. Biochem.* **2010**, *48*, 909–930. [CrossRef]
14. Dennison, K.L.; Robertson, W.R.; Lewis, B.D.; Hirsch, R.E.; Sussman, M.R.; Spalding, E.P. Functions of AKT1 and AKT2 potassium channels determined by studies of single and double mutants of Arabidopsis. *Plant Physiol.* **2001**, *127*, 1012–1019. [CrossRef]
15. Ivashikina, N.; Becker, D.; Ache, P.; Meyerhoff, O.; Felle, H.H.; Hedrich, R. K^+ channel profile and electrical properties of Arabidopsis root hairs. *FEBS Lett.* **2001**, *508*, 463–469. [CrossRef]
16. Bao, A.K.; Du, B.Q.; Touil, L.; Kang, P.; Wang, Q.L.; Wang, S.M. Co-expression of tonoplast Cation/H^+ antiporter and H^+-pyrophosphatase from xerophyte *Zygophyllum xanthoxylum* improves alfalfa plant growth under salinity, drought and field conditions. *Plant Biotechnol. J.* **2016**, *14*, 964–975. [CrossRef]

17. Wang, W.Y.; Liu, Y.Q.; Duan, H.R.; Yin, X.X.; Cui, Y.N.; Chai, W.W.; Song, X.; Flowers, T.J.; Wang, S.M. SsHKT1;1 is coordinated with SsSOS1 and SsNHX1 to regulate Na^+ homeostasis in *Suaeda salsa* under saline conditions. *Plant Soil.* **2020**, *449*, 117–131. [CrossRef]
18. Maathuis, F.J.M. The role of monovalent cation transporters in plant responses to salinity. *J. Exp. Bot.* **2006**, *57*, 1137–1147. [CrossRef]
19. Ma, Q.; Hu, J.; Zhou, X.R.; Yuan, H.J.; Kumar, T.; Luan, S.; Wang, S.M. ZxAKT1 is essential for K^+ uptake and K^+/Na^+ homeostasis in the succulent xerophyte *Zygophyllum xanthoxylum*. *Plant J.* **2017**, *90*, 48–60. [CrossRef]
20. Zhang, M.; Liang, X.; Wang, L.; Cao, Y.; Song, W.; Shi, J.; Lai, J.; Jiang, C. A HAK family Na^+ transporter confers natural variation of salt tolerance in maize. *Nat. Plants* **2019**, *5*, 1297–1308. [CrossRef]
21. Nouri, M.Z.; Moumeni, A.; Komatsu, S. Abiotic stresses: Insight into gene regulation and protein expression in photosynthetic pathways of plants. *Int. J. Mol. Sci.* **2015**, *16*, 20392–20416. [CrossRef]
22. Sui, N.; Yang, Z.; Liu, M.; Wang, B. Identification and transcriptomic profiling of genes involved in increasing sugar content during salt stress in sweet sorghum leaves. *BMC Genom.* **2015**, *16*, 534. [CrossRef] [PubMed]
23. Ma, Q.; Bao, A.K.; Chai, W.W.; Wang, W.Y.; Zhang, J.L.; Li, Y.X.; Wang, S.M. Transcriptomic analysis of the succulent xerophyte *Zygophyllum xanthoxylum* in response to salt treatment and osmotic stress. *Plant Soil* **2016**, *402*, 343–361. [CrossRef]
24. McCormac, A.C.; Fischer, A.; Kumar, A.M.; Söll, D.; Terry, M.J. Regulation of *HEMA1* expression by phytochrome and a plastid signal during de-etiolation in *Arabidopsis thaliana*. *Plant J.* **2001**, *25*, 549–561. [CrossRef]
25. Chen, J.; Zhang, H.; Zhang, X.; Tang, M. Arbuscular mycorrhizal symbiosis alleviates salt stress in black locust through improved photosynthesis, water status, and K^+/Na^+ homeostasis. *Front. Plant Sci.* **2017**, *8*, 1739. [CrossRef] [PubMed]
26. Chen, J.H.; Tang, M.; Jin, X.Q.; Li, H.; Chen, L.S.; Wang, Q.L.; Sun, A.Z.; Yi, Y.; Guo, F.Q. Regulation of Calvin-Benson cycle enzymes under high temperature stress. *aBIOTECH* **2022**, *3*, 65–77. [CrossRef]
27. Shi, H.Z.; Quintero, F.J.; Pardo, J.M.; Zhu, J.K. The putative plasma membrane Na^+/H^+ antiporter SOS1 controls long-distance Na^+ transport in plants. *Plant Cell* **2002**, *14*, 465–477. [CrossRef]
28. Apse, M.P.; Blumwald, E. Na^+ transport in plants. *FEBS Lett.* **2007**, *581*, 2247–2254. [CrossRef]
29. Laurie, S.; Feeney, K.A.; Maathuis, F.J.M.; Heard, P.J.; Brown, S.J.; Leigh, R.A. A role for HKT1 in sodium uptake by wheat roots. *Plant J.* **2002**, *32*, 139–149. [CrossRef]
30. Zhang, M.; Cao, Y.; Wang, Z.; Wang, Z.Q.; Shi, J.; Liang, X.; Song, W.; Chen, Q.; Lai, J.; Jiang, C. A retrotransposon in an HKT1 family sodium transporter causes variation of leaf Na^+ exclusion and salt tolerance in maize. *New Phytol.* **2018**, *217*, 1161–1176. [CrossRef]
31. Horie, T.; Brodsky, D.E.; Costa, A.; Kaneko, T.; Schiavo, F.L.; Katsuhara, M.; Schroeder, J.I. K^+ transport by the OsHKT2;4 transporter from rice with atypical Na^+ transport properties and competition in permeation of K^+ over Mg^{2+} and Ca^{2+} Ions. *Plant Physiol.* **2009**, *156*, 1493–1507. [CrossRef] [PubMed]
32. Rubio, F.; Nieves-Cordones, M.; Aleman, F.; Martinez, V. Relative contribution of AtHAK5 and AtAKT1 to K^+ uptake in the high-affinity range of concentrations. *Physiol. Plant.* **2008**, *134*, 598–608. [CrossRef] [PubMed]
33. Liu, K.; Li, L.; Luan, S. Intracellular K^+ sensing of SKOR, a shaker-type K^+ channel from Arabidopsis. *Plant J.* **2006**, *46*, 260–268. [CrossRef] [PubMed]
34. Johnson, R.C.; Johnston, W.J.; Golob, C.T. Residue management, seed production, crop development, and turf quality in diverse Kentucky bluegrass germplasm. *Crop Sci.* **2003**, *43*, 1091–1099. [CrossRef]
35. Xia, H.; Cao, Z.; Yu, M.; Li, Z.; Nie, C.; Zhang, P.; Guo, H. Tolerance of 30 Kentucky bluegrass varieties to NaCl stress during the seedling stage. *Pratacultural Sci.* **2023**, *12*, 3124–3137.
36. Guo, H.; Zhang, L.; Cui, Y.N.; Wang, S.M.; Bao, A.K. Identification of candidate genes related to salt tolerance of the secretohalophyte *Atriplex canescens* by transcriptomic analysis. *BMC Plant Biol.* **2019**, *19*, 213. [CrossRef] [PubMed]
37. Bose, J.; Munns, R.; Shabala, S.; Gilliham, M.; Pogson, B.; Tyerman, S. Chloroplast function and ion regulation in plants growing on saline soils: Lessons from halophytes. *J. Exp. Bot.* **2017**, *68*, 3129–3143. [CrossRef] [PubMed]
38. Hedrich, R.; Shabala, S. Stomata in a saline world. *Curr. Opin. Plant Biol.* **2018**, *46*, 87–95. [CrossRef] [PubMed]
39. Franks, P.J.; Farquhar, G.D. The mechanical diversity of stomata and its significance in gas-exchange control. *Plant Physiol.* **2007**, *143*, 78–87. [CrossRef]
40. Turan, S.; Tripathy, B.C. Salt-stress induced modulation of chlorophyll biosynthesis during de-etiolation of rice seedlings. *Physiol. Plant.* **2015**, *153*, 477–491. [CrossRef]
41. Ma, Q.; Yue, L.J.; Zhang, J.L.; Wu, G.Q.; Bao, A.K.; Wang, S.M. Sodium chloride improves photosynthesis and water status in the succulent xerophyte *Zygophyllum xanthoxylum*. *Tree Physiol.* **2012**, *32*, 4–13. [CrossRef] [PubMed]
42. Yang, Z.; Zhang, Z.; Qiao, Z.; Guo, X.; Wen, Y.; Zhou, Y.; Yao, C.; Fan, H.; Wang, B.; Han, G. The RING zinc finger protein LbRZF1 promotes salt gland development and salt tolerance in *Limonium bicolor*. *J. Integr. Plant Biol.* **2024**, *66*, 787–809. [CrossRef] [PubMed]
43. Guo, H.; Nie, C.Y.; Li, Z.; Kang, J.; Wang, X.L.; Cui, Y.N. Physiological and transcriptional analyses provide insight into maintaining ion homeostasis of sweet sorghum under salt stress. *Int. J. Mol. Sci.* **2023**, *24*, 11045. [CrossRef] [PubMed]
44. Olías, R.; Eljakaoui, Z.; Li, J.; Morales, P.A.D.; Marín-Manzano, M.C.; Pardo, J.M.; Belver, A. The plasma membrane Na^+/H^+ antiporter SOS1 is essential for salt tolerance in tomato and affects the partitioning of Na^+ between plant organs. *Plant Cell Environ.* **2009**, *32*, 904–916. [CrossRef] [PubMed]

45. Horie, T.; Hauser, F.; Schroeder, J.I. HKT transporter-mediated salinity resistance mechanisms in Arabidopsis and monocot crop plants. *Trends Plant Sci.* **2009**, *14*, 660–668. [CrossRef] [PubMed]
46. Horie, T.; Costa, A.; Kim, T.H.; Han, M.J.; Horie, R.; Leung, H.Y.; Miyao, A.; Hirochika, H.; An, G.; Schroeder, J.I. Rice OsHKT2;1 transporter mediates large Na^+ influx component into K^+-starved roots for growth. *Embo J.* **2007**, *26*, 3003–3014. [CrossRef] [PubMed]
47. Shabala, S.; Shabala, L. Ion transport and osmotic adjustment in plants and bacteria. *Biomol. Concepts* **2011**, *2*, 407–419. [CrossRef]
48. Hoque, M.A.; Okuma, E.; Banu, M.N.A.; Nakamura, Y.; Shimoishi, Y.; Murata, Y. Exogenous proline mitigates the detrimental effects of salt stress more than exogenous betaine by increasing antioxidant enzyme activities. *J. Plant Physiol.* **2007**, *164*, 553–561. [CrossRef]
49. Zhang, X.; Huang, G.; Bian, X.; Zhao, Q. Effects of root interaction and nitrogen fertilization on the chlorophyll content, root activity, photosynthetic characteristics of intercropped soybean and microbial quantity in the rhizosphere. *Plant Soil Environ.* **2013**, *59*, 80–88. [CrossRef]
50. Arnon, D.I. Copper enzymes in isolated chloroplasts. Polyphenoloxidase in *Beta vulgaris*. *Plant Physiol.* **1949**, *24*, 1–15. [CrossRef]
51. Wang, S.M.; Zhang, J.L.; Flowers, T.J. Low-affinity Na^+ uptake in the halophyte *Suaeda maritima*. *Plant Physiol.* **2007**, *145*, 559–571. [CrossRef] [PubMed]
52. Bates, L.S.; Waldren, R.P.; Teare, I.D. Rapid determination of free proline for water-stress studies. *Plant Soil* **1973**, *39*, 205–207. [CrossRef]
53. Pan, Y.Q.; Guo, H.; Wang, S.M.; Zhao, B.; Zhang, J.L.; Ma, Q.; Yin, H.J.; Bao, A.K. The photosynthesis, Na^+/K^+ homeostasis and osmotic adjustment of *Atriplex canescens* in response to salinity. *Front. Plant Sci.* **2016**, *7*, 848. [CrossRef] [PubMed]
54. Pe'er, S.; Cohen, Y. Sugar accumulation in tobacco plants systemically protected against blue mold (*Peronospora tabacina*). *Phytoparasitica* **1987**, *15*, 307–315. [CrossRef]
55. Yu, P.; Song, X.; Zhang, W.; Yao, Y.; Ren, J.; Wang, L.; Liu, W.; Meng, Z.; Meng, X. Analysis of ginseng rusty root symptoms transcriptome and its pathogenesis directed by reactive oxygen species theory. *Plant Direct* **2024**, *8*, e586. [CrossRef] [PubMed]
56. Liu, Y.; Wang, W.; Ren, G.; Cao, Y.; Di, J.; Wang, Y.; Zhang, L. Selenium treatment regulated the accumulation of reactive oxygen species and the expressions of related genes in postharvest broccoli. *Agronomy* **2024**, *14*, 1047. [CrossRef]
57. Cui, Y.N.; Lin, Z.R.; Cai, M.M.; Liu, R.W.; Wang, S.M.; Ma, Q. PcCLCg is involved in the accumulation of Cl^- in shoots for osmotic adjustment and salinity resistance in the Cl^--tolerant xerophyte *Pugionium cornutum*. *Plant Soil* **2023**, *487*, 283–298. [CrossRef]
58. Duan, H.R.; Ma, Q.; Zhang, J.L.; Hu, J.; Bao, A.K.; Wei, L.; Wang, Q.; Luan, S.; Wang, S.M. The inward-rectifying K^+ channel SsAKT1 is a candidate involved in K^+ uptake in the halophyte *Suaeda salsa* under saline condition. *Plant Soil* **2015**, *395*, 173–187. [CrossRef]

plants

MDPI

Article

Effects of High Salinity and Water Stress on Wetland Grasses from the Spanish Mediterranean Coast

Adrián Sapiña-Solano [1], Monica Boscaiu [2], Francisco Collado [3], Oscar Vicente [1,*] and Mario X. Ruiz-González [1,*]

1 Institute for Conservation and Improvement of Valencian Agrodiversity (COMAV), Universitat Politècnica de València, Camino de Vera s/n, 46022 Valencia, Spain; adsaso@doctor.upv.es
2 Mediterranean Agroforestry Institute (IAM), Universitat Politècnica de València, Camino de Vera s/n, 46022 Valencia, Spain; mobosnea@eaf.upv.es
3 Servici Devesa-Albufera, Vivers Municipals de El Saler, CV-500, km 8.5, 46012 Valencia, Spain; fjcollado@valencia.es
* Correspondence: ovicente@upvnet.upv.es (O.V.); maruigon@upvnet.upv.es (M.X.R.-G.)

Abstract: The impacts of climate change are reaching unprecedented levels, heightening the risk of species loss and ecosystem service degradation. Wetlands, highly threatened ecosystems, serve vital ecological functions by capturing carbon, filtering water, and harbouring diverse wildlife. Coastal wetlands encounter many challenges, such as increased drought periods and escalating salinity levels, severely impacting plant biodiversity. Assessing how plants respond to various environmental stress factors is imperative for devising successful conservation strategies. In the present study, we examined three representative grass species found in various habitats within the Albufera Natural Park, close to the city of Valencia on the Spanish Mediterranean coast: *Imperata cylindrica*, *Phragmites australis*, and *Saccharum ravennae*. High salinity and water stress conditions were induced by subjecting the plants to irrigation with solutions containing 200, 400, 600, and 800 mM NaCl or withholding irrigation altogether to mimic coastal flooding and drought scenarios. The treatments were maintained until noticeable wilting of the plants occurred, at which point a range of stress biomarkers were determined, including photosynthetic pigments, ions, osmolytes, oxidative stress markers, and antioxidant metabolites, as well as antioxidant enzyme activities. *Saccharum ravennae* displayed the highest sensitivity to salt stress, whereas *I. cylindrica* appeared to be the most tolerant. The primary salinity tolerance mechanism observed in *I. cylindrica* and *P. australis* was a blockage of ion transport from the root zone to the aerial part, together with the salt-induced accumulation of proline and soluble sugars to high concentrations in the former. No significant effects of the water deficit treatment on the growth or biochemical parameters were observed for any of the analysed species. These findings offer valuable information for the effective management and conservation of coastal wetlands facing the challenges posed by climate change.

Keywords: salt stress; water deficit; *Imperata cylindrica*; *Phragmites australis*; *Saccharum ravennae*; biotic interactions; mycorrhizae; ion content; osmolytes; photosynthetic pigments

check for updates

Citation: Sapiña-Solano, A.; Boscaiu, M.; Collado, F.; Vicente, O.; Ruiz-González, M.X. Effects of High Salinity and Water Stress on Wetland Grasses from the Spanish Mediterranean Coast. *Plants* **2024**, *13*, 1939. https://doi.org/10.3390/plants13141939

Academic Editor: Emilia Apostolova

Received: 1 June 2024
Revised: 28 June 2024
Accepted: 11 July 2024
Published: 15 July 2024

1. Introduction

One of the threats associated with climate change is sea level rise, with the consequent risk of inundating critical ecosystems such as coastal wetlands and dune systems [1–3]. Coastal wetlands provide essential ecosystem services [4–6]; however, they are heavily threatened by pollution [7,8], invasive species [9], the overexploitation of biological resources [10], human activities such as urbanisation or agriculture [11], and, especially, the effects of climate change [12,13].

Amongst the latter effects, drought and soil salinisation are the abiotic factors that have a major impact on plants and soil organisms [14]. Mediterranean wetlands are both biodiversity hotspots and fragile environments [15]. These areas harbour numerous

2. Results

2.1. Effect of High Salinity and Water Stress Treatments on Plant Biomass

The first distinct effect that the application of the highest-salinity treatment (800 mM NaCl) had on the plants was the time it took for them to show signs of wilting. This strongly relates to the accumulation of salt in the pots during the treatments, reflected in the progressive increase in the substrate electrical conductivity (Table 1). The *Imperata cylindrica* plants resisted for 14 days, *P. australis* for 11 days, and *S. ravennae* was the most sensitive to salt stress, exhibiting drastic wilting symptoms after seven days. The high salt and water stress treatments had different significant overall effects on the biomass of the selected plant species except for the root water contents (Table 2). The treatments had significant general effects on the leaf and root fresh weight, the production of new leaves and stems, and the root/shoot ratio in *I. cylindrica*, on the number of leaves and stems produced in *P. australis*, and on the leaf fresh weight, leaves and stems produced, leaf water content, and root/shoot ratio in *S. ravennae* (Table 2).

Table 1. Substrate electrical conductivity (EC). The electrical conductivity was measured weekly using a WET sensor (Delta Devices, Cambridge, England), with results reported in milli Siemens per centimetre (mS/cm). The plants were monitored until they exhibited wilting symptoms. Wilting occurred after 14 days for *Imperata cylindrica*, 11 days for *Phragmites australis*, and 7 days for *Saccharum ravennae*. Different lowercase letters indicate significant differences between treatments for each day and species, and different uppercase letters indicate significant differences across different days for each treatment and species.

Species	***I. cylindrica***			***P. australis***			***S. ravennae***	
Day	**0**	**7**	**14**	**0**	**7**	**11**	**0**	**7**
Treatment								
Control	2.0 ± 0.2 ^{A}	1.3 ± 0.2 ^{bA}	1.8 ± 0.4 ^{bcA}	2.0 ± 0.1 ^{abB}	1.1 ± 0.1 ^{abA}	1.3 ± 0.1 ^{bcA}	1.5 ± 0.2 ^{bA}	1.1 ± 0.1 ^{cB}
0.2 M NaCl	1.8 ± 0.1 ^{B}	2.7 ± 0.1 ^{abB}	7.0 ± 0.3 ^{abcA}	2.0 ± 0.1 ^{abA}	3.5 ± 0.6 ^{abA}	5.3 ± 1.4 ^{abcA}	1.4 ± 0.2 ^{bB}	3.8 ± 0.3 ^{abcA}
0.4 M NaCl	1.8 ± 0.1 ^{B}	3.8 ± 0.5 ^{abB}	30.3 ± 5.2 ^{abA}	2.3 ± 0.1 ^{aB}	5.4 ± 0.8 ^{abAB}	9.9 ± 2.4 ^{abcA}	2.3 ± 0.2 ^{abB}	9.1 ± 1.2 ^{abcA}
0.6 M NaCl	2.3 ± 0.1 ^{B}	4.1 ± 3.0 ^{abB}	40.2 ± 14.8 ^{abA}	2.4 ± 0.1 ^{aB}	7.7 ± 2.6 ^{aAB}	42.4 ± 16.2 ^{abA}	2.3 ± 0.1 ^{abB}	18.7 ± 4.5 ^{abA}
0.8 M NaCl	2.2 ± 0.1 ^{B}	8.8 ± 1.4 ^{aB}	53.8 ± 0.0 ^{aA}	2.2 ± 0.1 ^{abB}	20.8 ± 14.3 ^{aAB}	46.7 ± 0.0 ^{aA}	2 ± 0.1 ^{abB}	46.9 ± 6.9 ^{aA}

Table 2. Kruskal–Wallis analysis for overall main treatment effects on plant biomass, biochemical response, and mycorrhizal colonisation. Thirty plants of each species (five plants per treatment) were subjected to four salt concentrations (200, 400, 600, and 800 mM NaCl) and water stress (withholding of irrigation) on pots containing fresh soil from the Albufera Natural Park. Control plants were watered twice weekly with 2 L of tap water (EC = 1.051 mS/cm). Treatment lengths were 14 days for *I. cylindrica*, 11 days for *P. australis*, and seven days for *S. ravennae*. Values in bold denote statistically significant differences.

Species	***I. cylindrica***			***P. australis***			***S. ravennae***		
Variable	***N***	**$\chi2$**	***p*-Value**	***N***	**$\chi2$**	***p*-Value**	***N***	**$\chi2$**	***p*-Value**
BIOMASS									
Leaf fresh weight	30	21.263	**0.001**	26	4.055	0.542	28	21.012	**0.001**
Root fresh weight	30	11.147	**0.049**	27	8.230	0.144	30	8.055	0.153
Leaves produced	30	23.810	**<0.001**	25	15.749	**0.008**	30	23.882	**<0.001**
Stems produced	30	13.340	**0.020**	29	16.393	**0.006**	30	12.425	**0.029**
Leaf water content	26	8.809	0.117	27	9.756	0.082	29	16.531	**0.005**
Root water content	30	8.310	0.140	27	6.140	0.293	29	7.853	0.165
Root/shoot ratio	26	19.464	**0.002**	26	7.007	0.220	26	11.073	**0.050**

Table 2. *Cont.*

Species		*I. cylindrica*			*P. australis*			*S. ravennae*	
Variable	*N*	$\chi2$	*p*-Value	*N*	$\chi2$	*p*-Value	*N*	$\chi2$	*p*-Value
PIGMENTS									
Chlorophyll *a*	26	15.166	**0.010**	19	4.947	0.422	26	14.850	**0.011**
Chlorophyll *b*	26	11.669	**0.040**	19	4.211	0.519	26	14.786	**0.011**
Carotenoids	26	15.329	**0.009**	18	6.763	0.239	27	10.552	0.061
Chlorophylls/carotenoids	25	14.720	**0.012**	18	2.986	0.702	27	6.943	0.225
IONS									
Na^+ in roots	30	25.377	**<0.001**	27	19.976	**0.001**	29	24.061	**<0.001**
Na^+ in leaves	28	22.154	**<0.001**	25	3.574	0.612	23	8.441	0.134
Cl^- in roots	30	25.036	**<0.001**	27	19.084	**0.002**	30	20.102	**0.001**
Cl^- in leaves	28	19.482	**0.002**	27	12.680	**0.027**	24	17.460	**0.004**
K^+ in roots	30	11.921	**0.036**	26	7.805	0.167	29	12.723	**0.026**
K^+ in leaves	29	8.685	0.122	27	4.814	0.439	24	7.098	0.213
Ca^{2+} in roots	30	24.272	**<0.001**	27	17.121	**0.004**	30	20.055	**0.001**
Ca^{2+} in leaves	29	23.600	**<0.001**	27	7.042	0.218	24	18.885	**0.002**
OSMOLYTES									
Proline	26	15.440	**0.009**	20	6.972	0.223	25	12.883	**0.024**
Total soluble sugars	24	14.186	**0.014**	17	1.686	0.891	25	7.751	0.170
Glycine betaine	24	3.477	0.627	21	2.288	0.808	27	5.362	0.373
OXIDATIVE STRESS MARKERS									
Malondialdehyde	23	11.982	**0.035**	19	2.411	0.790	26	9.390	0.094
Hydrogen peroxide	24	3.802	0.578	18	3.973	0.553	27	7.410	0.192
NON-ENZYMATIC ANTIOXIDANTS									
Total phenolic compounds	25	11.346	**0.045**	19	0.180	0.999	26	7.716	0.173
Total flavonoids	25	11.527	**0.042**	19	0.816	0.976	25	6.295	0.279
TOTAL SOLUBLE PROTEINS									
Total soluble proteins	23	4.145	0.529	19	4.589	0.468	28	18.015	**0.003**
ANTIOXIDANT ENZYMES									
Superoxide dismutase	22	4.040	0.544	19	4.054	0.542	24	8.798	0.117
Catalase	22	3.519	0.621	21	5.747	0.332	28	3.596	0.609
Ascorbate peroxidase	23	10.954	0.052	20	2.654	0.753	26	3.817	0.576
Glutathione reductase	23	5.892	0.317	21	2.152	0.828	27	11.836	**0.037**
MYCORRHIZAE ANALYSIS									
Arbuscule abundance	24	4.012	0.548	13	3.899	0.564	26	3.133	0.679
Vesicle abundance	23	4.769	0.445	13	8.189	0.146	26	4.637	0.462
Hyphae abundance	22	3.961	0.555	13	4.292	0.508	26	2.526	0.773

Salt stress had negative effects on the leaf fresh weight, the production of leaves and stems, and increased the root/shoot ratio in *I. cylindrica* (Figure 1A,C,D,G), whereas, in *P. australis*, salt stress only reduced the production of leaves and stems (Figure 1C,D). In *S. ravennae*, salt stress strongly negatively affected the leaf fresh weight, leaf and stem production, and leaf water content (Figure 1A,C–E). The water stress treatment, however, did not produce significant differences in any biomass marker compared to the control plants.

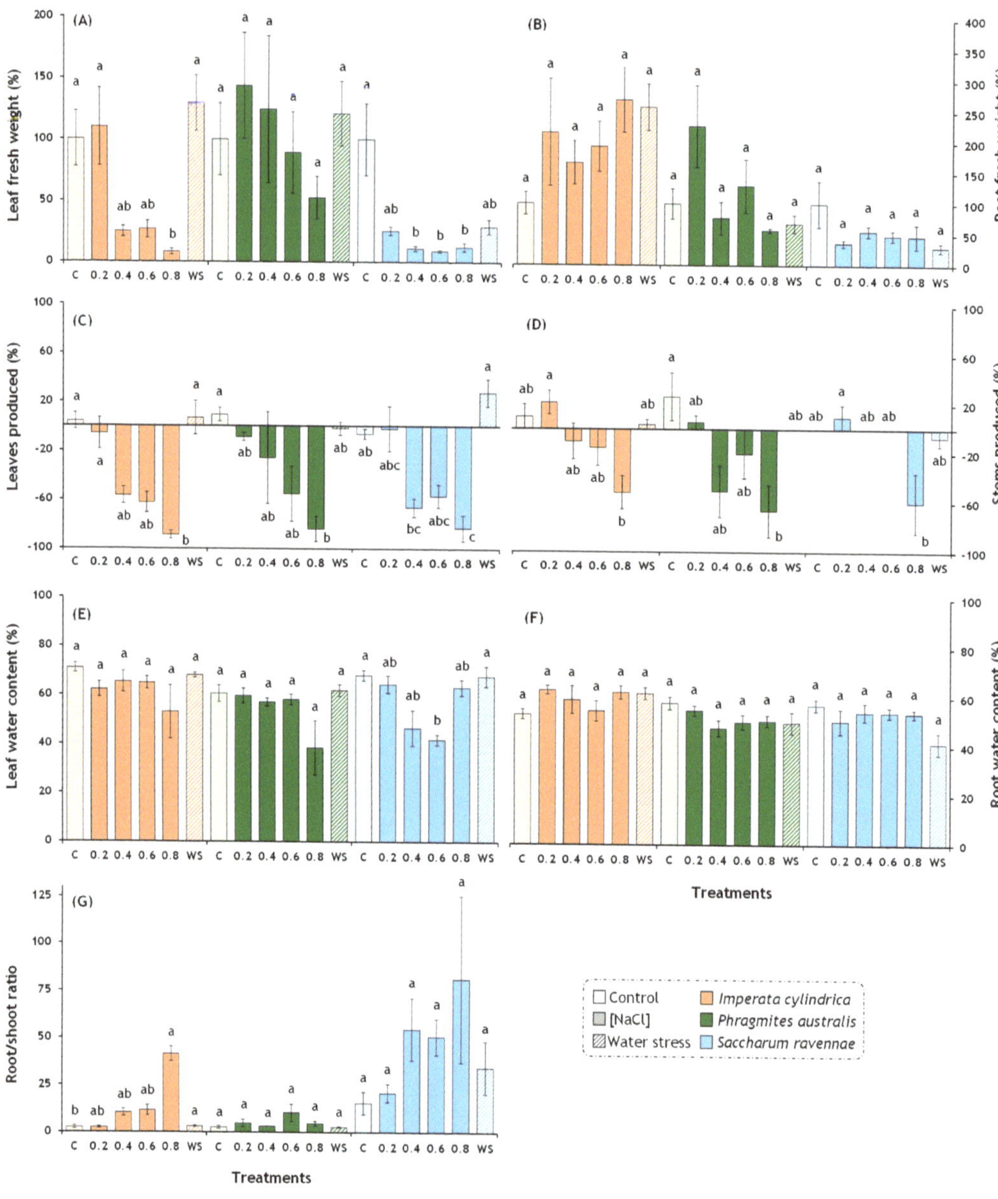

Figure 1. Effects of salt and water stress treatments on plant biomass. Thirty plants of each species (five plants per treatment) were exposed to four salt concentrations (0.2, 0.4, 0.6, and 0.8 M NaCl), water stress (WS, withholding of irrigation) on pots containing fresh soil from the Albufera Natural Park, and control plants (C). Control plants were watered twice weekly with 2 L of tap water (EC = 1.051 mS/cm). Treatment lengths were 14 days for *I. cylindrica*, 11 days for *P. australis*, and seven days for *S. ravennae*. (**A**) Leaf fresh weight; (**B**) root fresh weight; (**C**) leaves produced or lost during the treatment; (**D**) stems produced or lost during the treatment; (**E**) leaf water content; (**F**) root water content; (**G**) root/stem ratio. Bars represent mean ± SE. Letters denote post hoc significant pairwise differences between treatments corrected after Bonferroni. Statistical analyses were performed independently for each species.

2.2. Photosynthetic Pigments

Significant overall effects of the treatments were found for *I. cylindrica* in all the photosynthetic pigment variables (chlorophylls *a* and *b*, carotenoids, and the chlorophyll/carotenoid ratio) and for *S. ravennae* in chlorophylls *a* and *b* (Table 2). Salt stress significantly decreased the concentration of chlorophylls *a* and *b* in *S. ravennae*, reducing the Chl *a* content up to 84% and the Chl *b* content up to 75% (Figure 2A,B). A similar pattern (non-significant) was observed in *I. cylindrica* for Chl *a*, carotenoids, and the chlorophyll/carotenoids ratio (Figure 2A,C,D), as well as in *P. australis* for Chl *a* and *b* (Figure 2A,B).

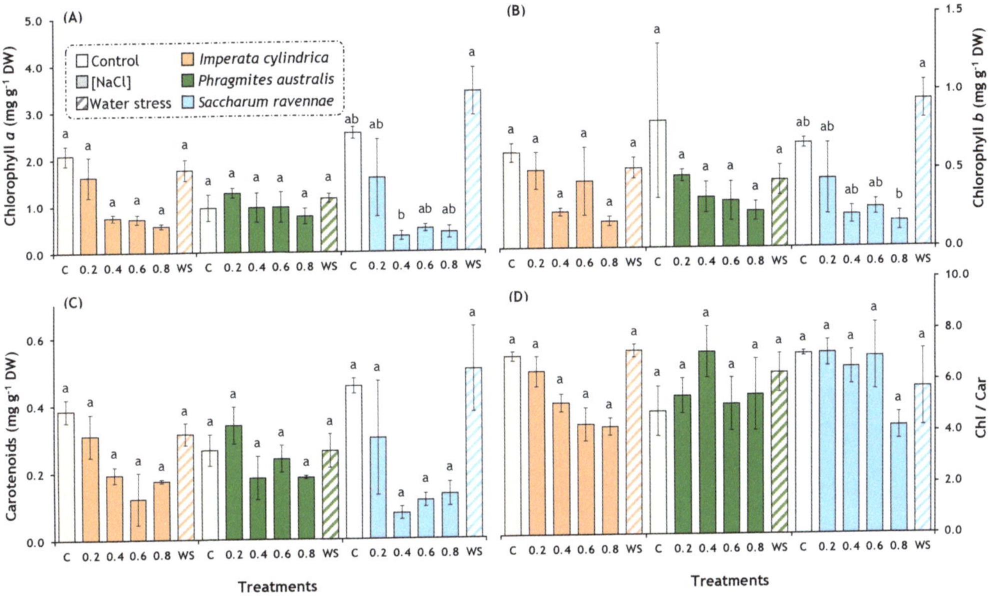

Figure 2. Effects of salt and water stress treatment on photosynthetic pigments of *Imperata cylindrica*, *Phragmites australis*, and *Saccharum ravennae*. Thirty plants of each species (five plants per treatment) were exposed to four salt concentrations (0.2, 0.4, 0.6, and 0.8 M NaCl), water stress (WS, withholding of irrigation) on pots containing fresh soil from the Albufera Natural Park, and control plants (C). Control plants were watered twice weekly with 2 L of tap water (EC = 1.051 mS/cm). Treatment lengths were 14 days for *I. cylindrica*, 11 days for *P. australis*, and 7 days for *S. ravennae*. (**A**) Chlorophyll *a*; (**B**) chlorophyll *b*; (**C**) carotenoids; (**D**) total chlorophylls/carotenoids ratio. Bars represent mean ± SE. Letters denote post hoc significant pairwise differences between treatments corrected after Bonferroni. Statistical analyses were performed independently for each species.

2.3. Ion Contents

The high salt and water stress treatments produced significant overall effects on the ion contents in the roots and leaves (Table 2). In *I. cylindrica*, the treatments had overall effects on the Na^+ concentration in the roots and leaves, the Cl^- concentration in the roots and leaves, the K^+ concentration in the roots, and the Ca^{2+} concentration in the roots and leaves (Table 2). Similarly, in *P. australis*, the treatment affected the Na^+ concentration in the roots, the Cl^- concentration in the roots and leaves, and the Ca^{2+} concentration in the roots. Finally, in *S. ravennae*, the treatment affected the Na^+ concentration in the roots, the Cl^- concentration in the roots and leaves, the K^+ concentration in the roots, and the Ca^{2+} concentration in the roots and leaves (Table 2). In *I. cylindrica*, salt stress increased the concentration of Na^+ up to 20-fold in the roots and 5.5-fold in the leaves; Cl^- increased up to 11.2-fold in the roots and 4.1-fold in the leaves; and Ca^{2+} increased up to 6.8-fold in the roots and 8.4-fold in the leaves but decreased the concentration of K^+ up to 50% in the roots (Figure 3A–E,G,H). Similarly, in *P. australis*, salt stress increased the concentration of Na^+ up to 13.6-fold in the roots and up to 1.3-fold in the leaves; Cl^- increased up to 6.6-fold in the roots and 1.8-fold in the leaves; and Ca^{2+} increased up to 4-fold in the roots and 2.8-fold in the leaves (Figure 3A,C,D,G). Likewise, in *S. ravennae*, salt stress increased the concentration of Na^+ up to 28.5-fold in the roots and 9.3-fold in the leaves; Cl^- and Ca^{2+} increased up to 5.8-fold in the roots and up to 10.2-fold in the leaves, and up to 6-fold in the roots and up to 5.8-fold in the leaves, respectively (Figure 3A,C,D,G,H). Water stress did not produce significant differences compared to the control plants.

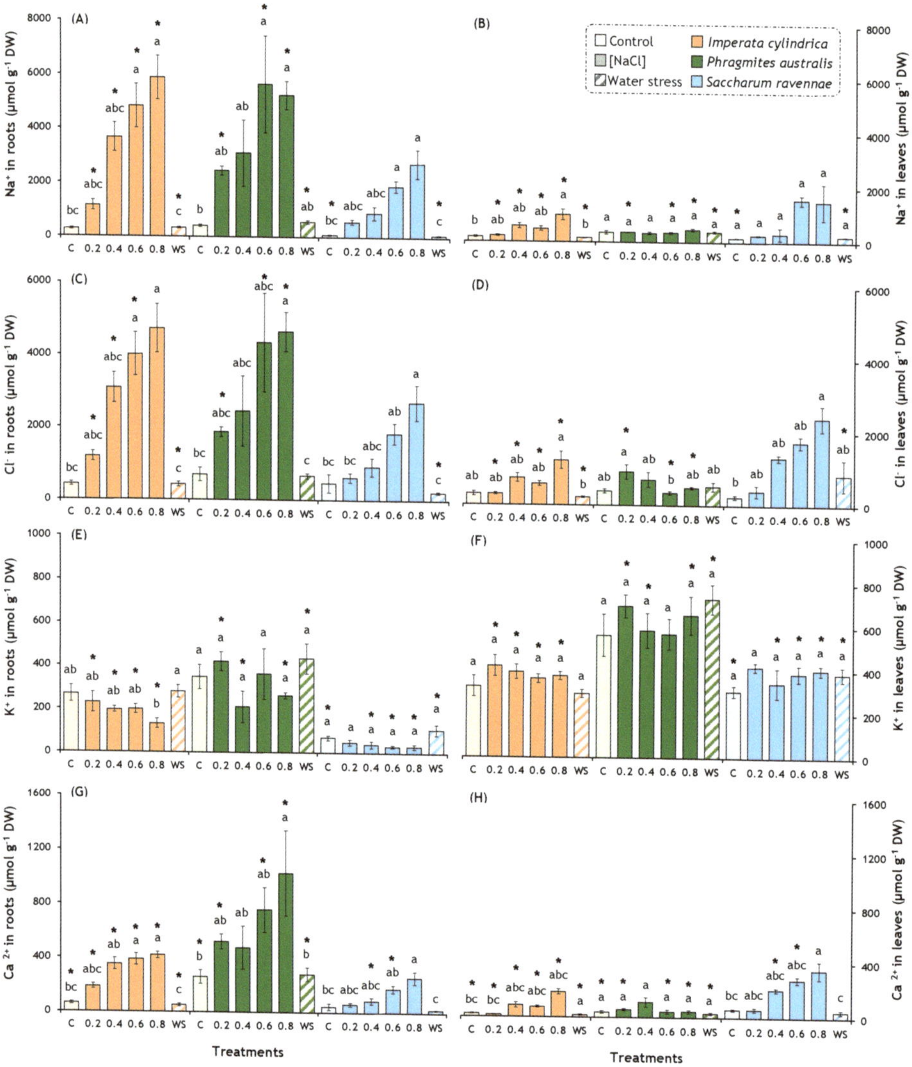

Figure 3. Effects of salt and water stress treatments on ions contents in *Imperata cylindrica, Phragmites australis*, and *Saccharum ravennae* roots and leaves. Thirty plants of each species (five plants per

treatment) were exposed to four salt concentrations (0.2, 0.4, 0.6, and 0.8 M NaCl), water stress (WS, withholding of irrigation) on pots containing fresh soil from the Albufera Natural Park, and control plants (C). Control plants were watered twice weekly with 2 L of tap water (EC = 1.051 mS/cm). Treatment lengths were 14 days for *I. cylindrica*, 11 days for *P. australis*, and 7 days for *S. ravennae*. (**A**,**B**) Na^+ ions in roots and leaves; (**C**,**D**) Cl^- ions in roots and leaves; (**E**,**F**); K^+ ions in roots and leaves; (**G**,**H**) Ca^{2+} ions in roots and leaves. Bars represent mean ± SE. Letters denote post hoc significant pairwise differences between treatments corrected after Bonferroni. Asterisks indicate significant differences between roots and leaves for the same treatment and species. Statistical analyses were performed independently for each species.

The concentration of Na^+ was significantly higher in the roots than in the leaves for the salt and water stress treatments in *I. cylindrica*. The same trend was observed in *P. australis* except for the 400 mM NaCl treatment (Figure 3A,B). Similar findings were detected for Cl^- and Ca^{2+} (Figure 3C,D,G,H), although the *S. ravennae* plants treated with 400 and 600 mM NaCl showed higher concentrations of Ca^{2+} in the leaves than in the roots. The concentration of K^+ was significantly lower in the roots than in the leaves for the three species (Figure 3E,F).

2.4. Osmolytes

The high salt and water stress treatments had significant overall effects on the proline and total soluble sugar concentrations but not on glycine betaine (Table 2). In *I. cylindrica*, the treatments had overall effects on the proline content and total sugar content, whereas, in *S. ravennae*, the treatment only affected the proline content. No significant post hoc pairwise comparisons were detected within each species for the treatments. However, a positive trend in the proline content, total sugar content, and glycine betaine content (non-significant) was observed with increasing salt concentrations in *I. cylindrica*. The same trend (non-significant) was observed for the proline content with increasing salt concentrations in *S. ravennae* and *P. australis* (Figure 4A–C).

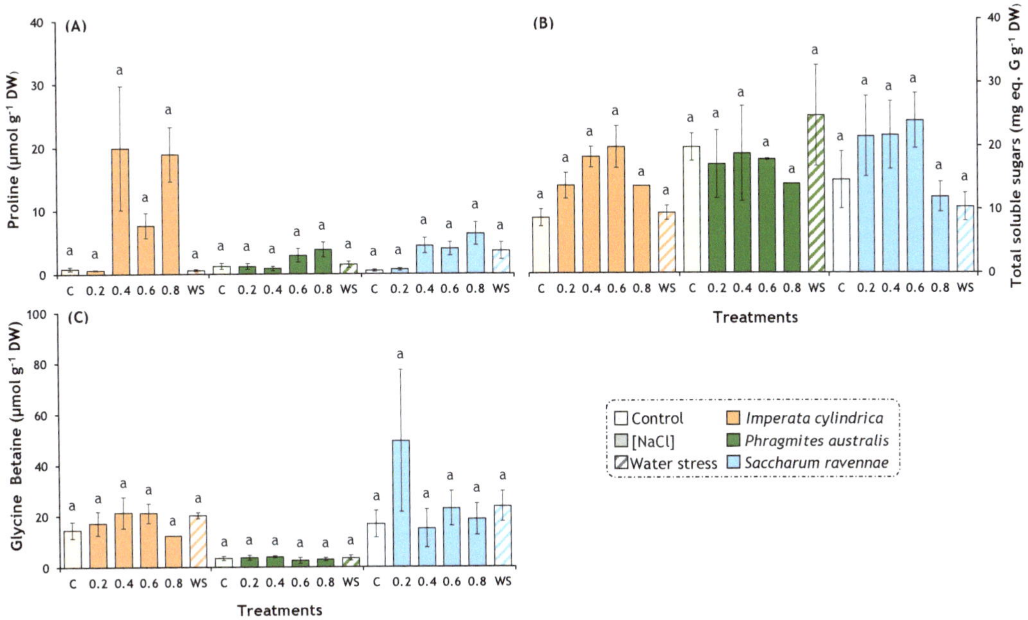

Figure 4. Effects of salt and water stress treatments on osmolyte contents in *Imperata cylindrica*, *Phragmites australis*, and *Saccharum ravennae*. Thirty plants of each species (five plants per treatment)

were exposed to four salt concentrations (0.2, 0.4, 0.6, and 0.8 M NaCl), water stress (WS, withholding of irrigation) on pots containing fresh soil from Albufera Natural Park, and control plants (C). Control plants were watered twice weekly with 2 L of tap water (EC = 1.051 mS/cm). Treatment lengths were 14 days for *I. cylindrica*, 11 days for *P. australis*, and 7 days for *S. ravennae*. (**A**) Proline; (**B**) total soluble sugars; (**C**) glycine betaine. Bars represent mean ± SE. Letters denote post hoc significant pairwise differences between treatments corrected after Bonferroni. Statistical analyses were performed independently for each species.

2.5. Oxidative Stress Biomarkers and Antioxidant Compounds

The high salt and water stress treatments only produced significant overall effects on the malondialdehyde content, total phenolic compounds, and total flavonoids in *I. cylindrica* and on the glutathione reductase (GR) activity in *S. ravennae* (Table 2). However, the post hoc pairwise analyses were not statistically significant (Table 3).

Table 3. Summary of oxidative stress markers, antioxidant compounds, soluble proteins, and antioxidant enzyme activities from *Imperata cylindrica*, *Phragmites australis*, and *Saccharum ravennae*. Thirty plants of each species (five plants per treatment) were exposed to 200, 400, 600, and 800 mM NaCl and water stress (WS) on pots containing fresh soil from the Albufera Natural Park. Control plants were watered twice weekly with 2 L of tap water (EC = 1.051 mS/cm). Treatment lengths were 14 days for *I. cylindrica*, 11 days for *P. australis*, and 7 days for *S. ravennae*. Letters represent post hoc pairwise statistically significant differences corrected after Bonferroni between treatments. Mean ± SE. values are shown. Abbreviations: dry weight (DW), catechin (C), gallic acid (GA), superoxide dismutase (SOD), catalase (CAT), ascorbate peroxidase (APX), and glutathione reductase (GR).

Parameter	Treatment	N	*I. cylindrica*	N	*P. australis*	N	*S. ravennae*
Malondialdehyde (nmol g^{-1} DW)	Control	5	17.6 ± 9.0^{a}	4	96.2 ± 12.3^{a}	4	31.9 ± 11.8^{a}
	200 mM	5	17.4 ± 9.8^{a}	5	78.7 ± 18.5^{a}	4	0.0 ± 0.0^{a}
	400 mM	5	0.0 ± 0.0^{a}	3	78.3 ± 23.1^{a}	5	35.9 ± 21.0^{a}
	600 mM	3	0.0 ± 0.0^{a}	2	94.8 ± 13.8^{a}	5	41.5 ± 27.1^{a}
	800 mM	1	6.3 ± 0.0^{a}	2	65.3 ± 5.5^{a}	4	0.7 ± 0.7^{a}
	WS	4	31.7 ± 11.0^{a}	3	95.5 ± 30.2^{a}	4	0.0 ± 0.0^{a}
Hydrogen peroxide (μmol H_2O_2 g^{-1} DW)	Control	5	4.6 ± 0.5^{a}	3	0.6 ± 0.1^{a}	5	5.0 ± 1.4^{a}
	200 mM	5	3.8 ± 0.5^{a}	5	0.7 ± 0.1^{a}	4	6.0 ± 1.4^{a}
	400 mM	5	5.4 ± 0.8^{a}	2	0.3 ± 0.2^{a}	5	2.2 ± 0.7^{a}
	600 mM	4	3.7 ± 1.0^{a}	2	0.4 ± 0.4^{a}	5	2.9 ± 0.5^{a}
	800 mM	1	3.3 ± 0.0^{a}	2	0.2 ± 0.2^{a}	4	4.0 ± 2.1^{a}
	WS	4	4.7 ± 0.5^{a}	4	0.4 ± 0.2^{a}	4	4.2 ± 0.5^{a}
Total phenolic compounds (mg eq. GA g^{-1} DW)	Control	5	2.2 ± 0.3^{a}	4	6.6 ± 2.3^{a}	4	2.2 ± 0.1^{a}
	200 mM	5	2.1 ± 0.2^{a}	5	4.5 ± 1.3^{a}	4	2.8 ± 0.5^{a}
	400 mM	5	2.7 ± 0.4^{a}	2	5.8 ± 2.7^{a}	5	1.3 ± 0.4^{a}
	600 mM	4	3.2 ± 0.3^{a}	2	4.5 ± 0.4^{a}	5	2.0 ± 0.3^{a}
	800 mM	1	1.3 ± 0.0^{a}	2	4.3 ± 0.1^{a}	4	1.5 ± 0.6^{a}
	WS	5	1.7 ± 0.1^{a}	4	4.4 ± 0.3^{a}	4	1.6 ± 0.3^{a}
Total flavonoids (mg eq. C g^{-1} DW)	Control	5	4.3 ± 0.6^{ab}	4	2.4 ± 0.8^{a}	4	3.3 ± 0.3^{a}
	200 mM	5	4.9 ± 0.7^{ab}	4	1.7 ± 0.7^{a}	4	5.4 ± 1.1^{a}
	400 mM	5	5.9 ± 1.1^{ab}	3	1.5 ± 0.5^{a}	5	2.3 ± 0.8^{a}
	600 mM	4	6.5 ± 0.5a	2	1.6 ± 0.3^{a}	5	3.7 ± 0.4^{a}
	800 mM	1	2.8 ± 0.0^{ab}	2	1.5 ± 0.1^{a}	3	3.0 ± 2.0^{a}
	WS	5	3.4 ± 0.3^{b}	4	1.7 ± 0.3^{a}	4	2.6 ± 0.8^{a}

Table 3. *Cont.*

Parameter	Treatment	N	*I. cylindrica*	N	*P. australis*	N	*S. ravennae*
Total soluble proteins (mg protein g^{-1} DW)	Control	5	2.2 ± 0.3 [a]	4	3.71 ± 0.4 [a]	5	2.5 ± 0.6 [a b]
	200 mM	5	2.1 ± 0.4 [a]	4	3.03 ± 0.3 [a]	4	4.1 ± 0.5 [a b]
	400 mM	4	2.7 ± 0.4 [a]	3	2.8 ± 0.2 [a]	5	1.4 ± 0.4 [b]
	600 mM	4	2.7 ± 0.4 [a]	2	3.1 ± 0.4 [a]	5	2.5 ± 0.3 [b]
	800 mM	1	2.0 ± 0.0 [a]	2	2.7 ± 0.3 [a]	4	1.3 ± 0.4 [a b]
	WS	4	1.9 ± 0.3 [a]	4	2.9 ± 0.3 [a]	5	4.2 ± 0.4a
SOD activity (U g^{-1} protein)	Control	5	388.7 ± 27.1 [a]	4	137.2 ± 49.8 [a]	4	145.2 ± 46.4 [a]
	200 mM	5	429.1 ± 51.1 [a]	5	287.3 ± 105.2 [a]	4	433.1 ± 79.0 [a]
	400 mM	4	415.0 ± 53.2 [a]	3	280.7 ± 151.6 [a]	4	185.0 ± 44.6 [a]
	600 mM	4	484.2 ± 80.2 [a]	2	421.3 ± 213.5 [a]	5	544.4 ± 153.4 [a]
	800 mM	1	198.7 ± 0.0 [a]	2	365.4 ± 4.6 [a]	3	241.1 ± 166.1 [a]
	WS	3	447.1 ± 57.7 [a]	3	330.0 ± 54.0 [a]	4	275.5 ± 103.0 [a]
CAT activity (U g^{-1} protein)	Control	4	3.5 ± 1.3 [a]	5	6.0 ± 4.0 [a]	5	6.06 ± 4.4 [a]
	200 mM	5	17.6 ± 6.7 [a]	5	13.6 ± 7.1 [a]	4	12.32 ± 5.5 [a]
	400 mM	4	10.2 ± 2.4 [a]	3	35.0 ± 33.0 [a]	5	35.1 ± 23.4 [a]
	600 mM	4	13.0 ± 7.0 [a]	2	28.1 ± 8.5 [a]	5	4.5 ± 2.7 [a]
	800 mM	1	3.9 ± 0.0 [a]	2	90.2 ± 72.4 [a]	4	25.2 ± 18.2 [a]
	WS	4	12.4 ± 6.3 [a]	4	22.8 ± 14.3 [a]	5	5.6 ± 3.8 [a]
APX activity (U g^{-1} protein)	Control	5	293.5 ± 47.9 [a]	4	1287.7 ± 264.8 [a]	5	322.0 ± 112.9 [a]
	200 mM	5	454.7 ± 68.4 [a]	5	1436.4 ± 398.4 [a]	4	526.5 ± 76.6 [a]
	400 mM	4	391.4 ± 107.7 [a]	3	1000.3 ± 214.9 [a]	5	531.9 ± 167.4 [a]
	600 mM	4	665.2 ± 142.2 [a]	2	1660.6 ± 667.6 [a]	4	536.6 ± 28.7 [a]
	800 mM	1	413.6 ± 0.0 [a]	2	1413.5 ± 112.0 [a]	3	283.9 ± 174. 2 [a]
	WS	4	706.4 ± 144.8 [a]	4	1040.7 ± 369.6 [a]	5	637.5 ± 132.9 [a]
GR activity (U g^{-1} protein)	Control	5	27.3 ± 5.3 [a]	5	1.2 ± 1.1 [a]	4	13.2 ± 6.9 [a]
	200 mM	5	22.6 ± 6.3 [a]	5	0.2 ± 0.2 [a]	4	15. 0 ± 6.5 [a]
	400 mM	4	12.6 ± 7.7 [a]	3	21.6 ± 21.1 [a]	5	28.1 ± 9.9 [a]
	600 mM	4	27.8 ± 13.6 [a]	2	5.4 ± 5.4 [a]	5	17.9 ± 9.3 [a]
	800 mM	1	35.5 ± 0.0 [a]	2	0.0 ± 0.0 [a]	4	21.5 ± 5.7 [a]
	WS	4	41.7 ± 7.8 [a]	4	0.3 ± 0.3 [a]	5	4.3 ± 3.5 [a]

2.6. Quantification of Mycorrhiza

Saccharum ravennae showed the highest percentage of mycorrhizal colonisation compared to *I. cylindrica* and *P. australis* (Figure 5B). Thus, the presence of arbuscules, vesicles, and hyphae on the roots was higher in *S. ravennae* compared to *I. cylindrica* and *S. ravennae* (all p-values < 0.001).

2.7. Principal Component and Correlation Analyses

The PCA and the heat map (Figure 6) show the impact of the salt and water stress treatments on *I. cylindrica*, *P. australis*, and *S. ravennae* with respect to the plant biomass and stress response biomarkers (Figure 6). Seven principal components accounted for 67% of the total variance. The PC2 nicely separates the three species, although *P. australis* and *S. ravennae* clearly separate into two distinct groups (Figure 6A). PC1 separates the control and water-stressed from the salt-stressed plants (Figure 6A), as shown in the above results. The heatmap enabled detecting some interactions (Figure 6B): the pigment concentrations were higher in the control and water stress than in the salt-treated plants. *Saccharum ravennae* was the species that accumulated the most Na^+, Cl^-, and Ca^{2+} in the leaves. *Phragmites australis* showed a group of variables with values higher than those in the other species: K^+ in the roots and leaves, malondialdehyde, ascorbate peroxidase activity, and total phenolic compounds (Figure 6B).

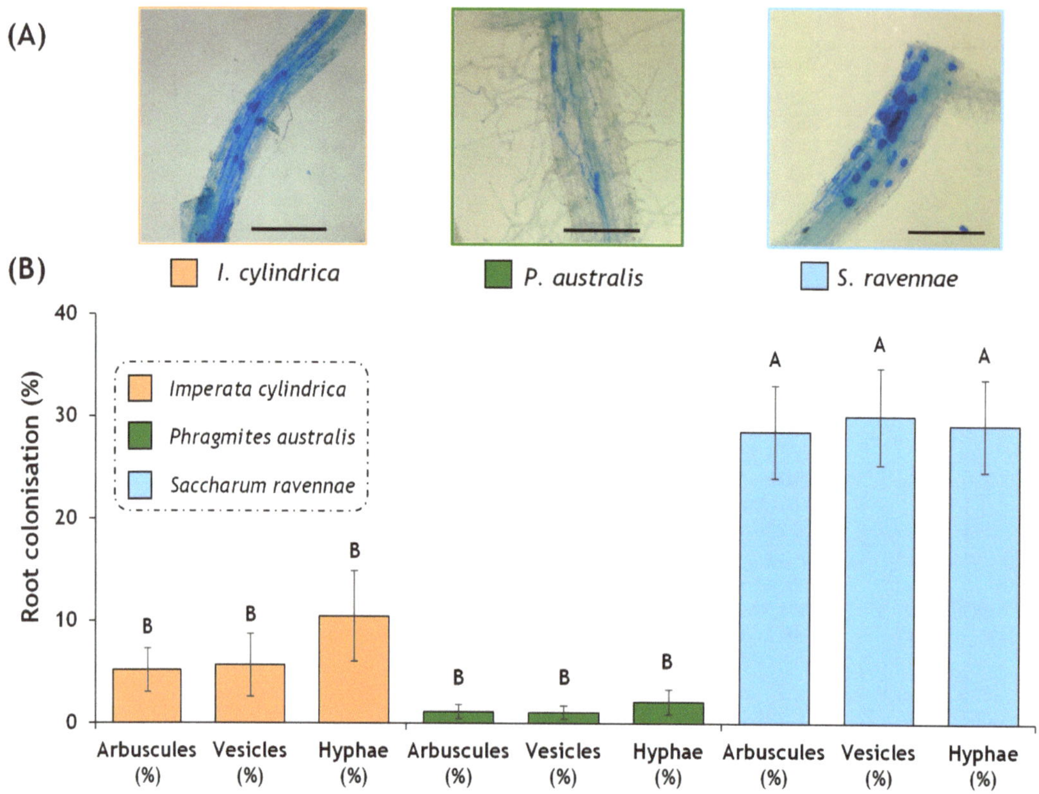

Figure 5. Percentage of mycorrhiza colonisation (arbuscules, vesicles, and hyphae) in *Imperata cylindrica*, *Phragmites australis*, and *Saccharum ravennae* roots. (**A**) Example of mycorrhization status in *I. cylindrica*, *P. australis*, and *S. ravennae*. Roots were stained with acidified Shaeffer blue ink. The scale corresponds to 500 µm. (**B**) Percentage of root colonisation in *I. cylindrica*, *P. australis*, and *S. ravennae* roots. Bars represent mean ± SE. Capital letters denote post hoc significant pairwise differences between species for each mycorrhizae structure corrected after Bonferroni.

In *I. cylindrica*, the partial correlation analysis detected a positive interaction between the leaf fresh weight and the chlorophyll/carotenoid ratio. In addition, chlorophyll *a* and *b*, root K^+, and MDA positively interacted with the number of leaves produced. The Na^+ content in the roots and leaves, the Ca^{2+} content in the roots and leaves, and the Cl^- content in the roots demonstrated positive interactions with the root/shoot ratio. On the other hand, there was a negative interaction between chlorophyll *a*, carotenoids, and chlorophyll/carotenoids with the root/shoot ratio. The Na^+, Cl^-, and Ca^{2+} contents in the roots and leaves also demonstrated negative interactions with the leaf fresh weight and the number of leaves produced (Figure 7).

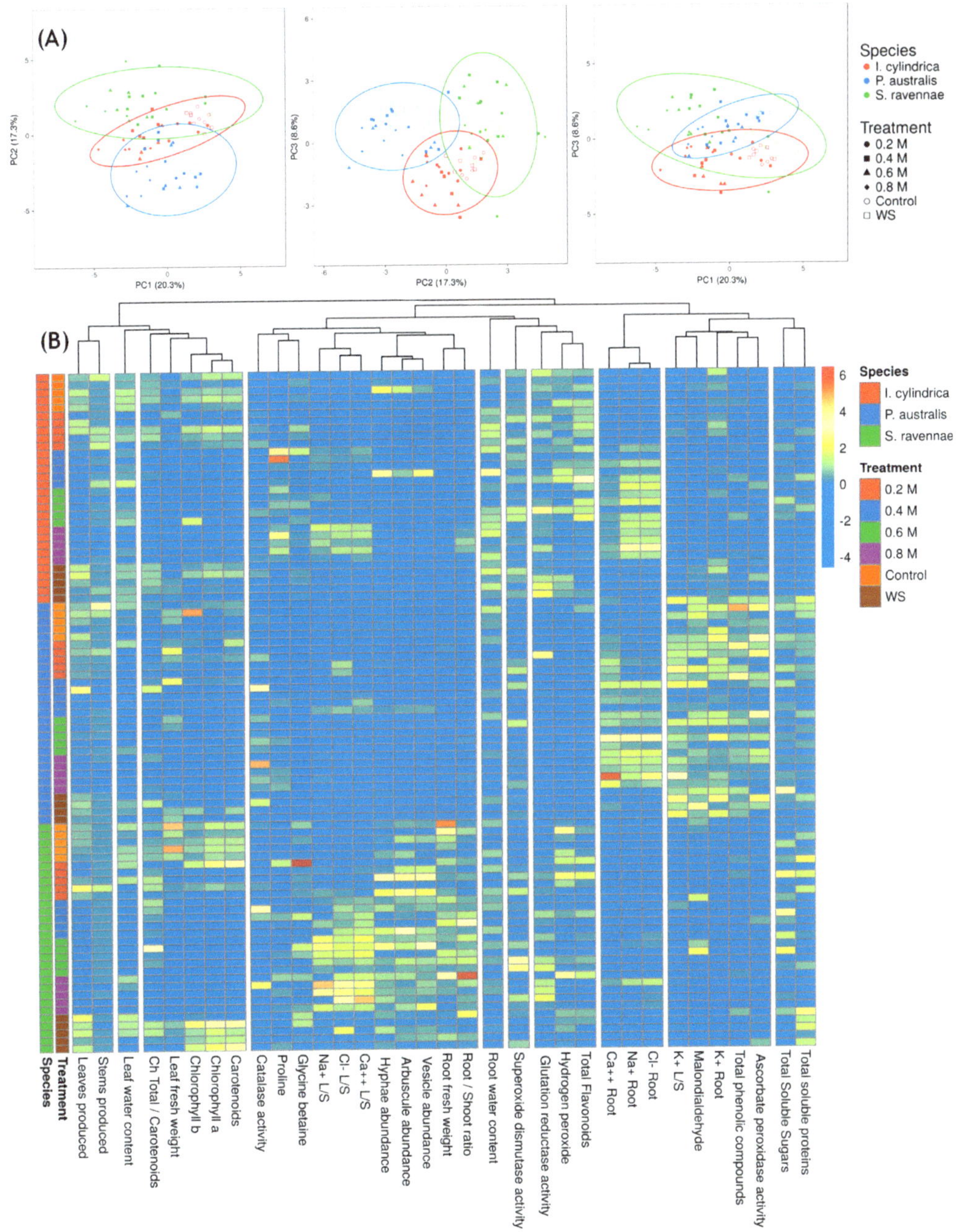

Figure 6. PCA analysis and heatmap for species and treatment. (**A**) Contribution of the three main components to data variance by treatment. (**B**) Manhattan distance and complete linkage heatmap of treatment effects on *I. cylindrica*, *P. australis*, and *S. ravennae*. The seven groups of variables explained 67% of the variance.

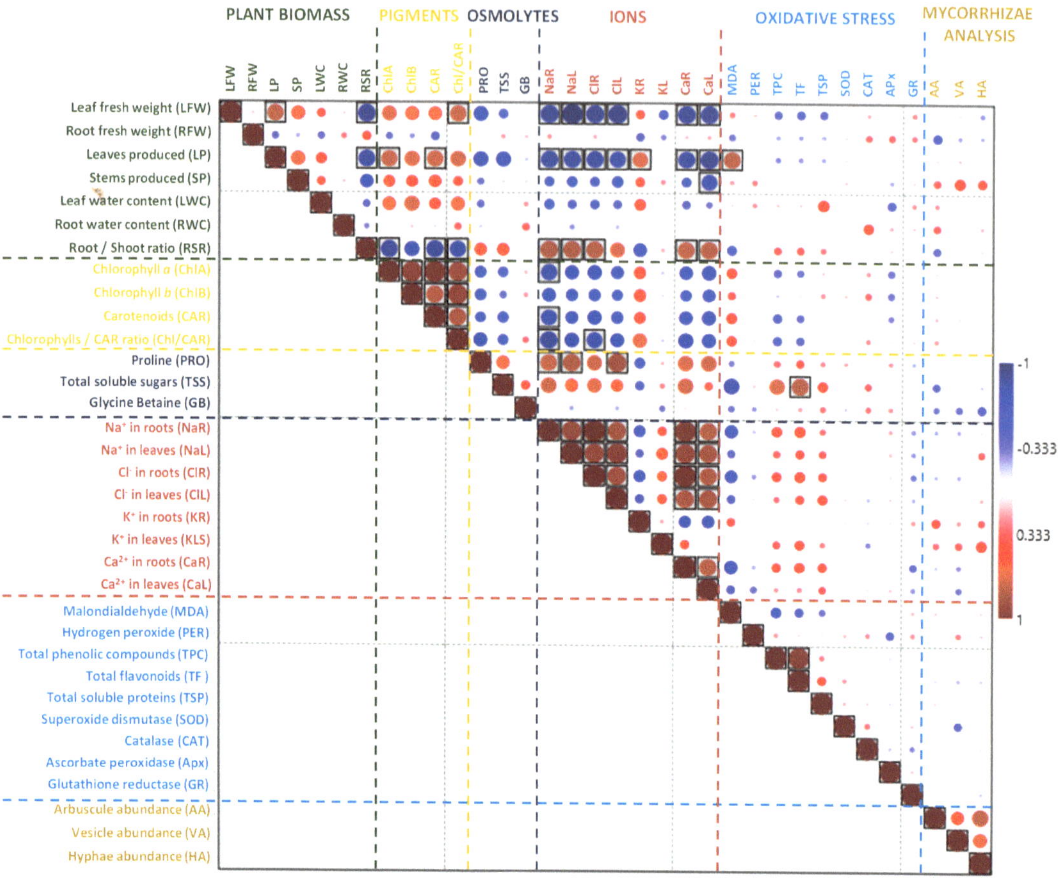

Figure 7. Spearman's rho correlation between biomass and stress markers in *I. cylindrica*. Red fillings denote positive and blue fillings negative correlations. Significant correlations after Bonferroni correction are shown in boxes (p-value < 0.05).

In *P. australis*, no positive or negative correlation was detected between the biomass parameters and the other variables analysed (Figure 8). In *S. ravennae*, the partial correlation analysis detected a positive interaction between the amount of soluble protein and the number of leaves produced. The analysis detected a negative correlation between the Na^+ content in the roots and the Ca^{2+} content in the leaves and the leaf fresh weight, and there was also a negative interaction between the Na^+ content in the roots, the Cl^- content in the roots, and the Ca^{2+} content in the roots and leaves and the number of leaves produced (Figure 9).

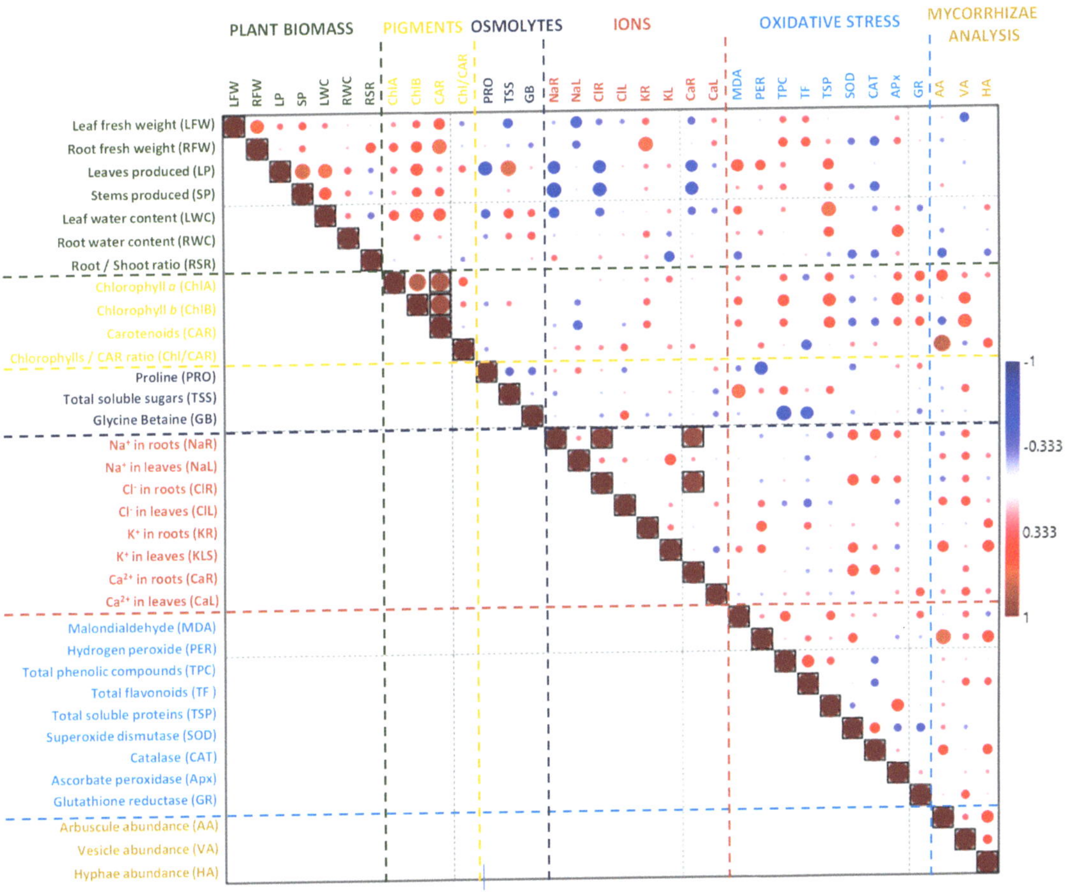

Figure 8. Spearman's rho correlation between biomass and stress markers in *P. australis*. Red fillings denote positive and blue fillings negative correlations. Significant correlations after Bonferroni correction are shown in boxes (p-value < 0.05).

Overall, the control, mild salt stress, and water stress conditions had little or no effect on the plants, highlighting that high salt stress is the main challenge for these species. We found large variability for some characteristics when plants of different species were exposed to salt stress above 400 mM NaCl, which may be due to the genetic differences of the individual plants, in addition to the low sample size.

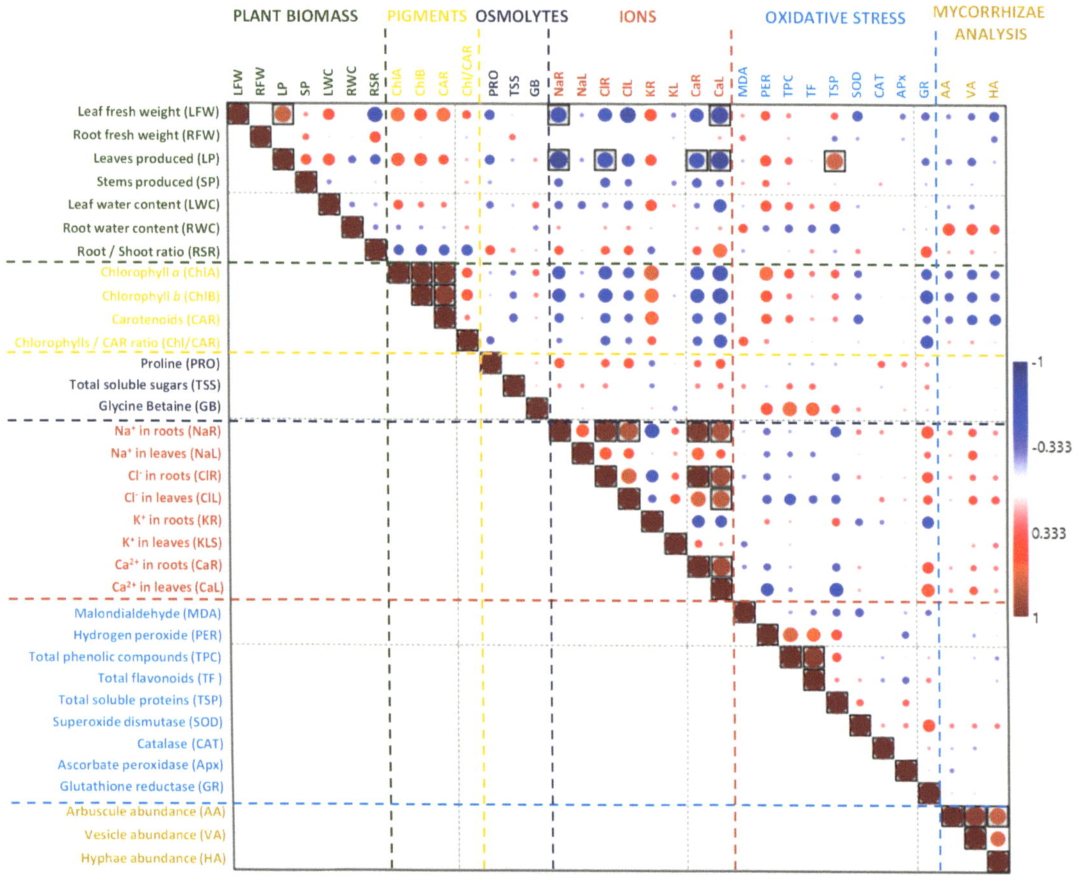

Figure 9. Spearman's rho correlation between biomass and stress markers in *S. ravennae*. Red fillings denote positive and blue fillings negative correlations. Significant correlations after Bonferroni correction are shown in boxes (p-value < 0.05).

3. Discussion

Our results confirm the higher resistance of *I. cylindrica* and *P. australis* to salinity, compared to *S. ravennae*, which seems to be mainly based on constitutive defence mechanisms such as blocking ion transport from the roots to the aerial part of the plants. In addition, the possible activation of K^+ transport into the leaves under high-salinity conditions could contribute to their increased tolerance.

Imperata cylindrica showed other tolerance mechanisms, such as the accumulation of osmolytes (proline and soluble sugars) to maintain osmotic balance. It also increased the accumulation of antioxidant metabolites (total phenolic compounds and total flavonoids), which probably helped to counteract the deleterious effects of the salt-induced oxidative stress. Therefore, *I. cylindrica* might be the most suitable species for slowing down the erosion of soils exposed to salinisation.

Despite activating antioxidant enzymes, such as GR, and having a higher degree of mycorrhizal colonisation, *S. ravennae* was the least salt-tolerant species as it was the first to show wilt symptoms.

When facing increased salt concentrations, *I. cylindrica*, *P. australis*, and *S. ravennae* reduced their fresh weights and lost leaves and stems. This response is due to a trade-off between allocating plant resources for biomass accumulation and activating the stress tolerance mechanisms [52,61]. Plant biomass production depends on the accumulation of carbon products through photosynthesis. However, the presence of high salinity levels can negatively affect photosynthesis [27,61–64], leading to a reduction in plant biomass. There is evidence of the effect of increased salt concentration on plant biomass and a reduction in the number of leaves and stems in these species [27,28,40]. In addition, under low salt and water stress, *S. ravennae* exhibits greater variability in leaf wilting compared to control (healthy) conditions or high salt concentrations (wilted).

High salinity and a water deficit in the soil generate osmotic stress in the plants, leading to leaf dehydration [52]. However, *I. cylindrica* and *P. australis* seem to be tolerant to this osmotic effect because no significant water loss was detected in the presence of high NaCl concentrations, or even after a period without irrigation. This ability to retain water suggests efficient adaptive mechanisms in these species to counteract the osmotic stress associated with salinity or a lack of water [28,65]. However, *S. ravennae* experienced a substantial decrease in the aerial water content, indicating greater sensitivity to the osmotic stress conditions and a less efficient response to regulate the water balance under high salinity.

Our experiments detected a decrease in the chlorophyll *a* and *b* levels with increasing salt concentrations in all three species. This decrease in the chlorophyll content is caused by the inhibition of the enzymes involved in chlorophyll biosynthesis, such as Rubisco and PEP carboxylase, and the activation of chlorophyllase, which breaks down chlorophyll [66,67]. The reduction in the chlorophyll *a* and *b* content was more noticeable in *S. ravennae* than in *I. cylindrica* and *P. australis*, which appear to have more efficient mechanisms to preserve the chlorophyll integrity, crucial for photosynthesis, under salt stress conditions. The more pronounced response in *S. ravennae* suggests a greater susceptibility of this species to salinity, manifested in a more marked reduction in the photosynthetic pigments. Research by Hameed et al. [64] supports our findings by showing a decrease in chlorophylls *a* and *b* and carotenoids due to an increasing salt concentration in *I. cylindrica*. Furthermore, similar results were obtained by Gorai et al. [27] in *P. australis*.

Imperata cylindrica and *P. australis* exhibited a blockage of ion transport from the root zone to the aerial part, suggesting an effective strategy to mitigate the adverse effects of salt stress. Similar results have been observed in *P. australis* [26–28,41]. This was expected because a common response to salt stress in many plant species, contributing to tolerance, is based on blocking ion transport. Indeed, glycophytes and monocotyledonous halophytes cope with high salinity mainly by limiting the uptake and transport of Na^+ and Cl^- to the leaves [68–70]. However, this blockade of ion transport was not observed in *S. ravennae*, which could indicate a less efficient response to exposure to high salinity. Our experiments also detected a significant decrease in the K^+ levels in the roots of *I. cylindrica* as the Na^+ concentrations increased. However, the K^+ levels in the aerial part remained high in all three species. This result suggests the activation of K^+ transport to the leaves, which could partly compensate for Na^+ accumulation and, thus, contribute to salt tolerance.

In general, the accumulation of Na^+ is accompanied by a reduction in the K^+ levels in plants as both cations compete for the same binding sites on proteins, including membrane channels and K^+ transporters [68–70]. Given the fundamental role of K^+ in photosynthesis, osmoregulation, turgor generation, membrane potential regulation, and essential cellular functions such as protein synthesis and enzyme activity, it seems clear that the adverse effects of salt stress are due, at least in part, to decreased cytosolic K^+ concentrations, especially in photosynthetic tissues [71]. Therefore, maintaining a relatively high cytosolic K^+ level (mainly in leaf cells) is another fundamental tolerance mechanism described in some halophytes [72].

Imperata cylindrica displayed a significant increase in the proline levels and a trend to increase the concentrations of the total soluble sugars and glycine betaine in response to the rising salt concentrations. These findings are consistent with those of Hameed et al. [64]. The accumulation of proline, soluble sugars, and glycine betaine in *I. cylindrica* suggests the activation of inherent stress response mechanisms in this species. When these molecules accumulate to a sufficiently high level, they help in cellular osmotic adjustments, act as ROS scavengers, and influence gene expression regulation [73–75], ultimately favouring stress tolerance [76,77]. Although a significant increase in the concentrations of proline and soluble sugars in *P. australis* was not observed in the present study, there is evidence of an increase in these molecules under salt stress conditions [27,28].

Only *I. cylindrica* showed an increase in the total phenol and flavonoid contents with increasing salt concentrations. The accumulation of these compounds may reduce the oxidative damage caused by abiotic stress [78,79]. In addition, flavonoids seem to constitute a secondary ROS scavenging system, activated when plants face severe stress conditions and the primary antioxidant defence systems (i.e., antioxidant enzymes) are weakened [80–83]. These results support the idea that salinity-tolerant species, such as *I. cylindrica*, use efficient mechanisms, such as accumulating potent non-enzymatic antioxidants, in response to salt stress.

Regarding the four antioxidant enzymes tested (SOD, CAT, AP, and GR), a significant overall increase in specific activity was only observed in *S. ravennae* for glutathione reductase activity. The GR enzyme plays a crucial role in the recovery and maintenance of cellular redox balance in stressed plants by reducing oxidised glutathione (GSSG) to GSH using NADPH as a cofactor [84]. However, the GR activation appears to be insufficient to counteract or limit the adverse effects of the salinity-induced oxidative stress in *S. ravennae*, given the higher sensitivity of this species.

All three species were naturally colonised by mycorrhizal fungi. Such symbiotic associations benefit plants because, amongst other reasons, they facilitate the uptake of essential nutrients. Mycorrhizal fungi have also been associated with enhancing stress tolerance in plants by strengthening the enzymatic and non-enzymatic antioxidant defence systems [85,86], affecting lipid peroxidation [87], and phytohormone synthesis [88].

Saccharum ravennae was observed to have a higher percentage of arbuscules, vesicles, and hyphae compared to *I. cylindrica* and *P. australis*. As discussed above, the mycorrhizal formation can improve the plants' ability to take up nutrients. Given the higher susceptibility of this species, *S. ravennae* may have more effective strategies for using mycorrhizal symbiosis to overcome the nutritional limitations in a saline environment.

Rising sea levels, as a consequence of climate change, could cause a drastic increase in soil salinity and, therefore, changes in biodiversity, ecology, and ecosystem balance in areas close to the coast. Species that do not possess efficient salt stress tolerance mechanisms, such as *S. ravennae*, might be displaced or driven to local extinction, along with many other endemic species that provide essential ecosystem services. Then, the opportunity arises for tolerant species to prosper and even facilitate the establishment of invasive species. The main consequence is the drastic alteration of the population structure of native species. Thus, careful observation of sea flooding events in coastal areas where *S. ravennae* prospers should become an activity for monitoring the natural park. Another severe consequence is when the local salt-tolerant species, such as *I. cylindrica* or *P. australis*, cannot compete with the invader species, e.g., *Spartina patens*. The latter can reduce the presence and coverage of the native species, altering the original vegetation structure. Moreover, it has been reported that *S. patens* can transform the original ecological community to the point where it becomes unrecognisable, especially in marshes and beaches associated with estuaries, lagoons, and delta areas, such as the Albufera Natural Park [89]. It is noteworthy that *P. australis* behaves similarly to *S. patens* on the US coast, where *S. patens* is native [90]. Additionally, the changes in soil conditions caused by high salinity and drought might also affect the "unseen communities", that is, the microorganisms that establish mutualistic relationships with plants, such as mycorrhizal fungi. Mycorrhizal communities play a vital role in maintaining plant health and many wetland ecosystem services [60]. Assessing the presence of mycorrhizae in the three host species adds extra value to the ecosystem and strengthens the need for increased investment in research in wild plant–microbe interactions. Similarly, more research on the effects of soil salinisation or drought on these microorganisms and their interaction with plants in unique ecosystems, such as the Albufera Natural Park, is needed to better understand the consequences of climate change and to develop appropriate conservation strategies.

4. Materials and Methods

4.1. Plant Material, Location, and Cultivation

Plant material was provided by the Technical Management Office of the Albufera Natural Park, located in the Devesa de El Saler (Valencia Province, Spain; 39°21′28.91 N, 0°19′32.56 W). Thirty rhizomes of each selected species, *Imperata cylindrica*, *Phragmites australis*, and *Saccharum ravennae*, were collected from different locations in the park (Figure 10A). Plants had similar sizes and characteristics, but there was no age control because we wanted to investigate the effects of high salinity and water stress on the natural populations. The rhizomes were placed in individual 1 L pots and kept in a nursery within the natural park for three weeks. Then, the pots were transferred to the greenhouse at the Polytechnic University of Valencia, where they were acclimatised for two months before starting the treatments, with irrigation every five days on the lower part of the pots to simulate water uptake by the root system. The plants were kept under controlled conditions throughout the experiment, with a long day photoperiod of natural light (16 h of light and 8 h of darkness), relative humidity between 50% and 80%, and temperatures of 30 °C during the day and 20 °C at night.

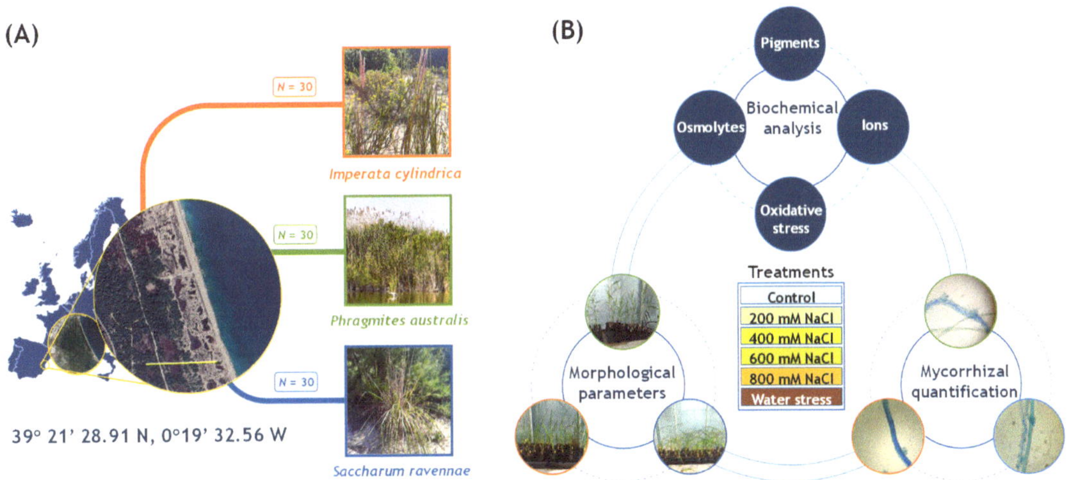

Figure 10. Plant sampling, treatments, and analysis. (**A**) Thirty plants from each species were randomly collected from different locations in the park. The scale corresponds to 500 m. (**B**) Plants were subjected to salt and water stress treatments after the acclimation period: 200 mM, 400 mM, 600 mM, and 800 mM NaCl, and complete withholding of irrigation (N = 5 plants per species and treatment). Control plants were watered twice weekly with 2 L of tap water (EC = 1.051 mS/cm). Plants were harvested as soon as they showed wilting symptoms. Wilt symptoms occurred after 14 days in *I. cylindrica*, 11 days in *P. australis*, and 7 days in *S. ravennae*. Then, plant morphological parameters, root mycorrhization, photosynthetic pigments, concentration of ions, osmolytes, and different aspects of the response against oxidative stress were analysed.

After acclimation, the plants were randomly distributed in trays for each treatment. Five plant replicates were allocated to each treatment: four salt concentrations, water stress, and the control. Each tray was supplied with 2 L of salt solutions or water. The treatment was applied twice a week (on Tuesday and Friday). The plants subjected to salt treatments were irrigated with 200, 400, 600, or 800 mM NaCl solutions in tap water, respectively.

Control plants were irrigated with tap water, and plants under water stress treatment were not irrigated at all. The electrical conductivity of the substrate in the pots was measured weekly with a WET sensor (Delta Devices, Cambridge, England). The treatments were applied until the plants started to show severe wilt symptoms, at which time the experiment was stopped, and morphological and biochemical analyses of each plant were carried out.

Stress conditions affected each species differently; thus, *I. cylindrica*, *P. australis*, and *S. ravennae* were harvested after 14, 11, and 7 days, respectively. Once the plants were harvested, they were transported to the laboratory to be measured and weighed in their aerial parts and roots. A weighed fraction of the aerial part was flash-frozen in liquid nitrogen and stored at −75 °C for subsequent biochemical analysis. The remaining fraction of the aerial part was reweighed (FW) and dried in an oven at 65 °C. For quantification of potential mycorrhizae structures, root fragments were cut to lengths around 0.5 cm and stored in 70% alcohol. The remaining fragments were dried in an oven at 65 °C. Dried leaves and root samples were used to measure ion contents.

4.2. Growth Parameters

The following parameters related to plant biomass were recorded: the increase in the number of live leaves (NL) and live stems (NT) produced during the treatment and the total fresh weight (FW) of the aerial part (FWT) and roots of each plant (FWR).

Once the aerial part and root samples reached a constant weight in the oven, they were reweighed to obtain the dry weight (DW) of the aerial part (DWT) and roots (DWR) of each plant. The water content percentage was calculated according to the following formula: $WC\% = [(FW - DW)/FW] \times 100$

4.3. Biochemical Analysis of Stress Markers

Different components at the molecular level potentially involved in plant responses to abiotic stress were analysed to understand the possible tolerance mechanisms activated in *I. cylindrica*, *P. australis*, and *S. ravennae*. These components included photosynthetic pigments (chlorophyll *a* and *b*, and carotenoids), three types of osmolytes (proline, total soluble sugars, and glycine betaine), ion contents (Na^+, Cl^-, K^+, and Ca^{2+}), biomarkers of oxidative stress (malondialdehyde and hydrogen peroxide), as well as representative antioxidant metabolites (total phenolic compounds and flavonoids) and antioxidant enzyme activities (superoxide dismutase, catalase, ascorbate peroxidase, and glutathione reductase).

4.4. Photosynthetic Pigments

Chlorophylls *a* and *b* and total carotenoid contents were quantified by spectrophotometric techniques according to Lichtenthaler and Wellburn [91]. Between 0.05 and 0.1 g of fresh plant material was ground and mixed with 1 mL of 80% (v/v) acetone. The samples were then shaken for 24 h in the dark. After centrifugation at 13,400× g for 15 min at 4 °C, the supernatant was diluted 10-fold with 80% (v/v) acetone and its absorbance was measured at 470, 646, and 663 nm. The concentration of each group of compounds was calculated according to the equations described by Lichtenthaler and Wellburn [91]. Photosynthetic pigment contents were expressed in mg g^{-1} DW.

4.5. Ion Contents

Concentrations of sodium (Na^+), chloride (Cl^-), potassium (K^+), and calcium (Ca^{2+}) were measured in leaves and roots, according to Weimberg [92]. Between 0.05 and 0.1 g of dried plant material was ground, and 2 mL of Mili-Q water was added. Samples were incubated at 95 °C in a heating block for one hour and kept shaking for 24 h in the dark. After vortexing for 5–10 s, the samples were centrifuged at 13,400× g for 10 min at room temperature. The supernatant was collected and diluted 15-fold. Sodium, potassium, and calcium ions were quantified using a PFP7 flame photometer (Jenway Inc., Burlington, VT, USA), whereas chlorides were measured on a MKII 926 chloride analyser (Sherwood, Inc., Cambridge, UK).

4.6. Osmolytes

Proline (Pro) content was quantified according to Bates et al. [93], as adapted by Vicente et al. [94]. Extracts were prepared by grinding 0.05–0.1 g of fresh leaf material in 1 mL of 3% (*w*/*v*) aqueous sulphosalicylic acid. The extracts were then centrifuged at 16,100× *g* for 15 min at 25 °C, and the supernatant was collected and mixed with acid ninhydrin and glacial acetic acid. Samples were incubated in a 98 °C water bath for 1 h, cooled on ice for 10 min, and extracted with toluene. The absorbance of the organic phase was measured at 520 nm. Proline content was expressed in μmol g^{-1} DW based on a calibration curve obtained from known proline concentrations.

Total soluble sugars (TSSs) were quantified according to Dubois et al. [95]. Extracts were prepared by grinding 0.05–0.1 g of fresh leaf material with 2 mL of 80% (*v*/*v*) methanol. Samples were mixed in a shaker for 24 h and centrifuged at 15,700× *g* for 10 min. The supernatant was collected and mixed with concentrated sulphuric acid and 5% phenol. After 20 min at room temperature, the absorbance of the samples was measured at 490 nm. TSS content was expressed as glucose equivalents (mg eq. glucose g^{-1} DW) according to a calibration curve prepared with known glucose concentrations.

Glycine betaine (GB) was quantified according to Grieve and Grattan [96]. Extracts were prepared by grinding 0.05–0.1 g of fresh leaf material with 1.5 mL of distilled water and kept in agitation for 24 h. After centrifuging the samples, 400 μL of the supernatant was mixed with 400 μL of 2 N H_2SO_4, and 50 μL of potassium triiodide solution was added to 125 μL of the mixture. The samples were incubated for 16 h at 4 °C in the dark. After carefully removing the supernatant, the precipitate was dissolved in 1.4 mL of 1,2-dichloroethane. After 2.5 h under cold and dark conditions, the absorbance was measured at 365 nm. The glycine betaine content was expressed in μmol g^{-1} DW based on a calibration curve obtained from known glycine betaine concentrations.

4.7. Oxidative Stress Biomarkers and Antioxidant Compounds

Malondialdehyde (MDA) concentration was measured to assess the oxidative stress level of the samples, whereas total phenolic compounds (TPCs) and total flavonoid (TF) contents were measured as representative antioxidant metabolites. MDA, TPC, and TF were quantified in the same methanol extracts used for TSS measurements. MDA was determined according to Hodges et al. [97] with some modifications [98]. Two methanol extracts were used per plant; to the first one, 0.5% thiobarbituric acid (TBA) prepared in 20% trichloroacetic acid (TCA) was added, whereas only 20% TCA was added to the second extract. The samples were incubated at 95 °C for 15 min and then cooled on ice for five minutes to stop the reaction. The extracts were centrifuged at 13,400× *g* for 10 min at 4 °C, and the absorbance of the supernatant was measured at 440, 532, and 600 nm. Samples containing only TCA acted as blanks for the corresponding samples with TBA + TCA. The concentration of MDA was calculated according to the equations proposed by Taulavuori et al. [98]. MDA content was expressed as nmol g^{-1} DW.

Hydrogen peroxide levels were assayed following the protocol of Alexieva et al. [99]. The leaf samples (0.05 g) were homogenised in 0.5 mL of 0.1% (*w/v*) trichloroacetic acid (TCA). After centrifugation at 15,700× *g* for 15 min at 4 °C, 0.5 mL of the supernatant was mixed with 0.5 mL of 10 mM Tris (pH 7) and 1 mL of 1 M KI. The reaction was run for one hour in the dark at room temperature. The absorbance was measured at 390 nm. Hydrogen peroxide content was expressed as μmol H_2O_2 g^{-1} DW based on a calibration curve from known hydrogen peroxide concentrations.

Total phenolic compounds (TPCs) content was measured according to the protocol of Blainski et al. [100]. Methanol extracts were mixed with the Folin–Ciocalteu reagent and 15% (*w/v*) sodium carbonate. Subsequently, the samples were incubated at room temperature under dark conditions for 90 min, and the absorbance was measured at 765 nm. In parallel, a calibration curve was obtained with known concentrations of gallic acid (GA), which was used as the standard. The TPC content was expressed as GA equivalents (mg eq. GA g^{-1} DW).

Total flavonoid (TF) content was quantified according to Zhishen et al. [101]. Methanol extracts were mixed with 5% (*w*/*v*) sodium nitrite, followed by 10% (*w*/*v*) aluminium chloride and 1 M sodium hydroxide. After the reaction, the absorbance was measured at 510 nm. The TF content was expressed as catechin (C) equivalents (mg eq. C g^{-1} DW) based on a calibration curve obtained with known catechin concentrations.

4.8. Antioxidant Enzyme Activities

Crude protein extracts were prepared according to Gil et al. [102], and proteins were quantified according to Bradford [103] using the commercial BioRad reagent and bovine serum albumin (BSA) as standard. Superoxide dismutase (SOD) activity was measured according to Beyer and Fridovich [104], considering one activity unit as the amount of enzyme causing a 50% inhibition of nitro blue tetrazolium (NBT) photoreduction, using riboflavin as a source of superoxide radicals, and measuring the absorbance at 560 nm. Catalase (CAT) activity was determined according to Aebi [105] by the decrease in absorbance at 240 nm due to the consumption of H_2O_2 added to the extracts. Ascorbate peroxidase (APX) activity was determined according to Nakano and Asada [106] by the decrease in absorbance at 290 nm due to ascorbate oxidation. Glutathione reductase (GR) activity was determined according to Connell and Mullet [107] by the decrease in absorbance at 340 nm due to the oxidation of the NADPH cofactor. For the three latter enzymes, one unit of enzyme activity was defined as the amount of enzyme required to oxidise one nmol of substrate per minute at 25 °C. The enzymes' specific activities were expressed as units per gram of protein. All spectrophotometric measurements were performed on a UV-3100PC spectrophotometer (VWR International, LLC., Ragnor, PA, USA).

4.9. Root Staining and Quantification of Mycorrhiza

Root samples were preserved in 70% alcohol before acidified blue Shaeffer ink staining [108]. Root fragments were washed with distilled water and cut into 0.5 cm long fragments with a scalpel. Then, the samples were incubated with 10% potassium hydroxide for 40 min in a water bath at 100 °C and washed with distilled water three times. Further, 10% Sheaffer blue dye was added until covering the roots [108], and the samples were re-incubated at 100 °C for 3 min. To remove excess dye from the roots, first, they were rinsed with 1% acetic acid and then left overnight in fresh 1% acetic acid. The next day, the acetic acid was replaced with 50% glycerol, prepared on slides with a drop of 50% lactoglycerol, and evaluated under the microscope. Arbuscules, vesicles, and hyphae were quantified according to the protocol of Trouvelot et al. [109]. Three to seven stained root fragments were evaluated under the microscope and classified according to the range of classes indicated in Figure 5. The abundance of arbuscules (% A), vesicles (% V), and hyphae (% H) in the root system was assessed (mycorrhiza manual at https://www2.dijon.inrae.fr/mychintec/Protocole/protoframe.html, accessed on 13 February 2024).

4.10. Statistical Analysis

Data were analysed using IBM SPSS Statistics for Windows, version 25 (IBM Corp., Armonk, NY, USA). Since some variables did not meet the variance analysis assumptions, the data were analysed using the non-parametric Kruskal–Wallis test. Pairwise post hoc comparisons were adjusted after Bonferroni. We used the ClustVis webtool [110] for a PCA analysis and the visualisation of the first three PCA components. Because the variables had different intensity ranges and units, unit variance scaling was applied as the variance normalisation method during the pre-processing. Then, we applied the NIPALSs (Nonlinear Iterative Partial Least Squares) PCA method. In addition, a clustered heatmap was produced with the three species under the six treatments with Manhattan distance and complete linkage methods. Finally, we conducted a Spearman's rho correlation for each species and used Bonferroni correction on significant correlations. Spearman's rho correlations were calculated with PAST 4.09 [111].

5. Conclusions

This study addressed the impact of high salinity and water stress on three wild plant species from the Albufera Natural Park on the Spanish Mediterranean coast, identifying *Saccharum ravennae* as the most sensitive and *Imperata cylindrica* as the most resilient to salt stress. The salt tolerance of the latter species is primarily based on an efficient mechanism of inhibition of Na^+ and Cl^- transport from the roots to the aerial part of the plants (a mechanism shared by *Phragmites australis*) and the salt-induced accumulation of proline and soluble sugars in the photosynthetic tissues; K^+ retention in the leaves in the presence of increasing Na^+ concentrations also seemed to contribute to salt tolerance. On the other hand, the three species showed relatively high resistance to water deficit stress. All three species had roots infected with mycorrhizae, regardless of their niche of origin within the park, highlighting the importance of protecting microbial soil diversity, alongside plant and animal diversity, under the current climate change scenario.

Author Contributions: Conceptualization, M.B., O.V. and M.X.R.-G.; methodology, M.B. and O.V.; validation, M.X.R.-G., A.S.-S. and O.V.; formal analysis, A.S.-S. and M.X.R.-G.; investigation, A.S.-S.; resources, F.C. and O.V.; data curation, A.S.-S. and M.X.R.-G.; writing—original draft preparation, A.S.-S. and M.X.R.-G.; writing—review and editing, M.B. O.V. and M.X.R.-G.; visualization, A.S.-S. and M.X.R.-G.; supervision, M.B., O.V. and M.X.R.-G.; project administration, O.V.; funding acquisition, O.V. All authors have read and agreed to the published version of the manuscript.

Funding: This study was supported by a María Zambrano distinguished researcher contract to M.X.R.-G. funded by the Ministry of Universities (Government of Spain) and the Next Generation EU. A.S.-S. was supported by the "Programa Investigo", "Plan de Recuperación, Transformación y Resiliencia", funded by the European Union, Next Generation EU.

Data Availability Statement: Data are available upon request.

Acknowledgments: We would like to thank Antonio Bellido Nadal and Victoria Dobin for technical support.

Conflicts of Interest: The authors declare no conflicts of interest.

References

1. Herbert, E.R.; Boon, P.; Burgin, A.J.; Neubauer, S.C.; Franklin, R.B.; Ardón, M.; Hopfensperger, K.N.; Lamers, L.P.M.; Gell, P. A global perspective on wetland salinization: Ecological consequences of a growing threat to freshwater wetlands. *Ecosphere* **2015**, *6*, 1–43. [CrossRef]
2. White, E.; Kaplan, D. Restore or Retreat? Saltwater Intrusion and Water Management in Coastal Wetlands. *Ecosyst. Health Sustain.* **2017**, *3*, e01258. [CrossRef]
3. Perry, J.E.; Hershner, C.H. Temporal Changes in the Vegetation Pattern in a Tidal Freshwater Marsh. *Wetlands* **1999**, *19*, 90–99. [CrossRef]
4. Costanza, R.; d'Arge, R.; De Groot, R.; Farber, S.; Grasso, M.; Hannon, B.; Limburg, K.; Naeem, S.; O'Neill, R.V.; Paruelo, J.; et al. The Value of the World's Ecosystem Services and Natural Capital. *Nature* **1997**, *387*, 253–260. [CrossRef]
5. Barbier, E.B.; Hacker, S.D.; Kennedy, C.; Koch, E.W.; Stier, A.C.; Silliman, B.R. The Value of Estuarine and Coastal Ecosystem Services. *Ecol. Monogr.* **2011**, *81*, 169–193. [CrossRef]
6. Rawat, U.S.; Agarwal, N.K. Biodiversity: Concept, Threats and Conservation. *Environ. Conserv. J.* **2015**, *16*, 19–28. [CrossRef]
7. Barker, J.R.; Tingey, D.T. *Air Pollution Effects on Biodiversity*, 3rd ed.; Springer Science & Business Media: New York, NY, USA, 1992; pp. 154–196. [CrossRef]
8. Li, C.; Wang, H.; Liao, X.; Xiao, R.; Liu, K.; Bai, J.; Li, B.; He, Q. Heavy Metal Pollution in Coastal Wetlands: A Systematic Review of Studies Globally Over the Past Three Decades. *J. Hazard. Mater.* **2022**, *424*, 127312. [CrossRef]
9. Charles, H.; Dukes, J.S. Impacts of invasive species on ecosystem services. In *Biological Invasions*; Nentwig, W., Ed.; Springer: Berlin/Heidelberg, Germany, 2007; pp. 217–237. [CrossRef]
10. Koh, L.P.; Kettle, C.J.; Sheil, D.; Lee, T.M.; Giam, X.; Gibson, L.G.; Clement, G.R. Biodiversity state and trends in Southeast Asia. In *Encyclopedia of Biodiversity*, 2nd ed.; Levin, S., Ed.; Academic Press: Amsterdam, The Netherlands, 2013; pp. 509–527. [CrossRef]
11. Gedan, K.B.; Silliman, B.R.; Bertness, M.D. Centuries of Human-Driven Change in Salt Marsh Ecosystems. *Annu. Rev. Mar. Sci.* **2009**, *1*, 117–141. [CrossRef]
12. Harley, C.D.G.; Hughes, A.R.; Hultgren, K.M.; Miner, B.G.; Sorte, C.J.B.; Thornber, C.S.; Rodriguez, L.F.; Tomanek, L.; Williams, S.L. The Impacts of Climate Change in Coastal Marine Systems. *Ecol. Lett.* **2006**, *9*, 228–241. [CrossRef]

13. He, Q.; Silliman, B.R. Climate change, human impacts, and coastal ecosystems in the Anthropocene. *Curr. Biol.* **2019**, *29*, R1021–R1035. [CrossRef]
14. Boyer, J.S. Plant Productivity and Environment. *Science* **1982**, *218*, 443–448. [CrossRef]
15. Davis, S.D.; Heywood, V. *Centres of Plant Diversity: A Guide and Strategy for Their Conservation*, 3rd ed.; IUCN: Cambridge, UK, 1994.
16. Aedo, C.; Medina, L.; Fernández-Albert, M. Species Richness and Endemicity in the Spanish Vascular Flora. *Nord. J. Bot.* **2013**, *31*, 478–488. [CrossRef]
17. Sutter, L.A.; Chambers, R.M.; Perry III, J.E. Seawater Intrusion Mediates Species Transition in Low Salinity, Tidal Marsh Vegetation. *Aquat. Bot.* **2015**, *122*, 32–39. [CrossRef]
18. Adams, D.A. Factors Influencing Vascular Plant Zonation in North Carolina Salt Marshes. *Ecology* **1963**, *44*, 445–456. [CrossRef]
19. Watt, S.C.L.; García-Berthou, E.; Vilar, L. The Influence of Water Level and Salinity on Plant Assemblages of a Seasonally Flooded Mediterranean Wetland. *Plant Ecol.* **2007**, *189*, 71–85. [CrossRef]
20. Rogel, J.A.; Ariza, F.A.; Silla, R.O. Soil Salinity and Moisture Gradients and Plant Zonation in Mediterranean Salt Marshes of Southeast Spain. *Wetlands* **2000**, *20*, 357–372. [CrossRef]
21. Mateo, G.; Vizcaíno, A. *Flora y Vegetación del Parque Natural de l'Albufera*, 1st ed.; Monografías de Botánica Ibérica: Jaca, Spain, 2023; ISBN 9788412665666.
22. Smith, S.E.; Read, D.J. *Mycorrhizal Symbiosis*, 3rd ed.; Academic Press: London, UK, 2008; ISBN 9780080559346.
23. Berg, G.; Alavi, M.; Schmidt, C.S.; Zachow, C.; Egamberdieva, D.; Kamilova, F.; Lugtenberg, B.J. Biocontrol and Osmoprotection for Plants under Salinated Conditions. *Mol. Microb. Ecol. Rhizosphere* **2013**, *1*, 587–592. [CrossRef]
24. Egamberdieva, D.; Kucharova, Z.; Davranov, K.; Berg, G.; Makarova, N.; Azarova, T.; Chebotar, V.; Tikhonovich, I.; Kamilova, F.; Validov, S.Z.; et al. Bacteria Able to Control Foot and Root Rot and to Promote Growth of Cucumber in Salinated Soils. *Biol. Fertil. Soils* **2011**, *47*, 197–205. [CrossRef]
25. Egamberdieva, D.; Berg, G.; Lindström, K.; Räsänen, L.A. Alleviation of Salt Stress of Symbiotic *Galega officinalis* L. (Goat's Rue) by Co-inoculation of *Rhizobium* with Root-Colonizing *Pseudomonas*. *Plant Soil* **2013**, *369*, 453–465. [CrossRef]
26. Hameed, A.; Dilfuza, E.; Abd-Allah, E.F.; Hashem, A.; Kumar, A.; Ahmad, P. Salinity stress and arbuscular mycorrhizal symbiosis in plants. In *Use of Microbes for the Alleviation of Soil Stresses*; Miransari, M., Ed.; Springer: New York, NY, USA, 2014; Volume 3, pp. 139–159, ISBN 978-1-4614-9466-9.
27. Gorai, M.; Ennajeh, M.; Khemira, H.; Neffati, M. Combined effect of NaCl-salinity and hypoxia on growth, photosynthesis, water relations and solute accumulation in *Phragmites australis* plants. *Flora* **2010**, *205*, 462–470. [CrossRef]
28. Pagter, M.; Bragato, C.; Malagoli, M.; Brix, H. Osmotic and Ionic Effects of NaCl and Na_2SO_4 Salinity on *Phragmites australis*. *Aquat. Bot.* **2009**, *90*, 43–51. [CrossRef]
29. Cope, T.A. Flora of Pakistan, No. 143: Poaceae. In *Flora of Pakistan*, 2nd ed.; Nasir, E., Ali, S.I., Eds.; Pakistan Agricultural Research Council and University of Karachi: Karachi, Pakistan, 1982; Volume 3, pp. 154–196.
30. Global Invasive Species Database. Available online: http://www.iucngisd.org/gisd/species.php?sc=16 (accessed on 28 February 2024).
31. MacDonald, G.E. Cogongrass (*Imperata cylindrica*)-Biology, Ecology, and Management. *Crit. Rev. Plant Sci.* **2004**, *23*, 367–380. [CrossRef]
32. Santoso, D.; Adiningsih, S.; Mutert, E.; Fairhurst, T.; Van Noordwijk, M. Soil Fertility Management for Reclamation of *Imperata* grasslands by Smallholder Agroforestry. *Agrofor. Syst.* **1996**, *36*, 181–202. [CrossRef]
33. Trautwig, A.N.; Eckhardt, L.G.; Loewenstein, N.J.; Hoeksema, J.D.; Carter, E.A.; Nadel, R.L. Cogongrass (*Imperata cylindrica*) Affects Above and Belowground Processes in Commercial Loblolly Pine (*Pinus taeda*) Stands. *For. Sci.* **2017**, *63*, 10–16. [CrossRef]
34. Bryson, C.T.; Carter, R. Cogongrass, *Imperata cylindrica*, in the United States. *Weed Technol.* **1993**, *7*, 1005–1009. [CrossRef]
35. Kuusipalo, J.; Ådjers, G.; Jafarsidik, Y.; Otsamo, A.; Tuomela, K.; Vuokko, R. Restoration of Natural Vegetation in Degraded *Imperata cylindrica* Grassland: Understorey Development in Forest Plantations. *J. Veg. Sci.* **1995**, *6*, 205–210. [CrossRef]
36. Dozier, H.; Gaffney, J.F.; McDonald, S.K.; Johnson, E.R.; Shilling, D.G. Cogongrass in the United States: History, Ecology, Impacts, and Management. *Weed Technol.* **1998**, *12*, 737–743. [CrossRef]
37. Brewer, S. Declines in Plant Species Richness and Endemic Plant Species in Longleaf Pine Savannas Invaded by *Imperata cylindrica*. *Biol. Invasions* **2008**, *10*, 1257–1264. [CrossRef]
38. Lissner, J.; Schierup, H.H.; Comín, F.A.; Astorga, V. Effect of Climate on the Salt Tolerance of Two *Phragmites australis* Populations: I. Growth, Inorganic Solutes, Nitrogen Relations and Osmoregulation. *Aquat. Bot.* **1999**, *64*, 317–333. [CrossRef]
39. Lessmann, J.M.; Brix, H.; Bauer, V.; Clevering, O.A.; Comín, F.A. Effect of Climatic Gradients on the Photosynthetic Responses of Four *Phragmites australis* Populations. *Aquat. Bot.* **2001**, *69*, 109–126. [CrossRef]
40. Pagter, M.; Bragato, C.; Brix, H. Tolerance and Physiological Responses of *Phragmites australis* to Water Deficit. *Aquat. Bot.* **2005**, *81*, 285–299. [CrossRef]
41. Vasquez, E.A.; Glenn, E.P.; Guntenspergen, G.R.; Brown, J.J.; Nelson, S.G. Salt Tolerance and Osmotic Adjustment of Spartina alterniflora (Poaceae) and the Invasive M Haplotype of *Phragmites australis* (Poaceae) along a Salinity Gradient. *Am. J. Bot.* **2006**, *93*, 1784–1790. [CrossRef]
42. Kiviat, E. Ecosystem services of *Phragmites* in North America with emphasis on habitat functions. *AoB Plants* **2013**, *5*, plt008. [CrossRef]

43. Meadows, R.E.; Saltonstall, K. Distribution of Native and Introduced *Phragmites australis* in Freshwater and Oligohaline Tidal Marshes of the Delmarva Peninsula and Southern New Jersey. *J. Torrey Bot. Soc.* **2007**, *134*, 99–107. [CrossRef]
44. Saltonstall, K. Cryptic Invasion by a Non-Native Genotype of the Common Reed, *Phragmites australis*, into North America. *Proc. Natl. Acad. Sci. USA* **2002**, *99*, 2445–2449. [CrossRef]
45. Flora of China. Saccharum in Flora of China @ efloras.org. Available online: http://www.efloras.org/florataxon.aspx?flora_id=2&taxon_id=128968 (accessed on 20 February 2024).
46. CABI. Laodelphax Striatellus (Small Brown Planthopper). 2021. Available online: https://doi.org/10.1079/cabicompendium.10935 (accessed on 22 February 2024).
47. Steury, B.W. District of Columbia and Maryland. *Castanea* **2004**, *69*, 154–157. [CrossRef]
48. Janaki-Ammal, E.K. *Intergeneric Hybrids of Saccharum*; John Innes Horticultural Institution: Merton, UK, 1941.
49. Hattori, T.; Shiotsu, F.; Doi, T.; Morita, S. Suppression of Tillering in *Erianthus ravennae* (L.) Beauv. due to Drought Stress at Establishment. *Plant Prod. Sci.* **2010**, *13*, 252–255. [CrossRef]
50. Pandey, V.C.; Singh, A.K. Saccharum spp. Potential Role in Ecorestoration and Biomass Production. In *Phytoremediation Potential of Perennial Grasses*, 1st ed.; Elsevier: Amsterdam, The Netherlands, 2020; pp. 211–226. ISBN 9780128177334.
51. IPCC. *Impacts, Adaptation and Vulnerability. Contribution of Working Group II to the Sixth Assessment Report of the Intergovernmental Panel on Climate Change*; Pörtner, H.-O., Roberts, D.C., Tignor, M., Poloczanska, E.S., Mintenbeck, K., Alegría, A., Craig, M., Langsdorf, S., Löschke, S., Möller, V., et al., Eds.; Cambridge University Press: Cambridge, UK; Cambridge University Press: New York, NY, USA, 2022; p. 3056. [CrossRef]
52. Munns, R.; Tester, M. Mechanisms of Salinity Tolerance. *Annu. Rev. Plant Biol.* **2008**, *59*, 651–681. [CrossRef]
53. Savchenko, T.; Tikhonov, K. Oxidative Stress-Induced Alteration of Plant Central Metabolism. *Life* **2021**, *11*, 304. [CrossRef]
54. Singh, P.; Choudhary, K.K.; Chaudhary, N.; Gupta, S.; Sahu, M.; Tejaswini, B.; Sarkar, S. Salt Stress Resilience in Plants Mediated through Osmolyte Accumulation and Its Crosstalk Mechanism with Phytohormones. *Front. Plant Sci.* **2022**, *13*, 1006617. [CrossRef]
55. Flowers, T.J.; Munns, R.; Colmer, T.D. Sodium Chloride Toxicity and the Cellular Basis of Salt Tolerance in Halophytes. *Ann. Bot.* **2015**, *115*, 419–431. [CrossRef] [PubMed]
56. Al Hassan, M.; del Pilar López-Gresa, M.; Boscaiu, M.; Vicente, O. Stress Tolerance Mechanisms in Juncus: Responses to Salinity and Drought in Three Juncus Species Adapted to Different Natural Environments. *Funct. Plant Biol.* **2016**, *43*, 949–960. [CrossRef] [PubMed]
57. Al Hassan, M.; Chaura, J.; Donat-Torres, M.P.; Boscaiu, M.; Vicente, O. Antioxidant Responses under Salinity and Drought in Three Closely Related Wild Monocots with Different Ecological Optima. *AoB Plants* **2017**, *9*, plx009. [CrossRef] [PubMed]
58. Albertos, B.; San Miguel, E.; Draper, I.; Garilleti, R.; Lara, F.; Varela, J.M. Estado de Conservación de la Vegetación Dunar en las Costas de la Comunidad Valenciana. *Rev. Digit. Cedex* **2010**, *158*, 121–134. Available online: https://ingenieriacivil.cedex.es/index.php/ingenieria-civil/article/view/160 (accessed on 10 November 2023).
59. Carex Vivers. Available online: http://www.carex.cat/es/vivers-carex/catalogo/saccharum-ravennae-.aspx (accessed on 31 May 2024).
60. Huang, G.M.; Srivastava, A.K.; Zou, Y.N.; Wu, Q.S.; Kuča, K. Exploring arbuscular mycorrhizal symbiosis in wetland plants with a focus on human impacts. *Symbiosis* **2021**, *84*, 311–320. [CrossRef]
61. Zhu, J.K. Plant Salt Tolerance. *Trends Plant Sci.* **2001**, *6*, 66–71. [CrossRef] [PubMed]
62. Ashraf, M. Relationships between Leaf Gas Exchange Characteristics and Growth of Differently Adapted Populations of Blue Panicgrass (*Panicum antidotale* Retz.) under Salinity or Waterlogging. *Plant Sci.* **2003**, *165*, 69–75. [CrossRef]
63. Wang, H.L.; Hao, L.M.; Wen, J.Q.; Zhang, C.L.; Liang, H.G. Differential expression of photosynthesis-related genes of reed ecotypes in response to drought and saline habitats. *Photosynthetica* **1998**, *35*, 61–69. [CrossRef]
64. Hameed, M.; Ashraf, M.; Naz, N.; Nawaz, T.; Batool, R.; Fatima, S.; Ahmad, F. Physiological Adaptative Characteristics of *Imperata cylindrica* for Salinity Tolerance. *Biologia* **2014**, *69*, 1148–1156. [CrossRef]
65. Hameed, M.; Ashraf, M.; Naz, N. Anatomical and Physiological Characteristics Relating to Ionic Relations in Some Salt Tolerant Grasses from the Salt Range, Pakistan. *Acta Physiol. Plant.* **2011**, *33*, 1399–1409. [CrossRef]
66. Ashraf, M.; Harris, P.J. Photosynthesis under Stressful Environments: An Overview. *Photosynthetica* **2013**, *51*, 163–190. [CrossRef]
67. Cazzonelli, C.I. Carotenoids in Nature: Insights from Plants and Beyond. *Funct. Plant Biol.* **2011**, *38*, 833–847. [CrossRef] [PubMed]
68. Greenway, H.; Munns, R. Mechanisms of Salt Tolerance in Non-halophytes. *Annu. Rev. Plant Physiol.* **1980**, *31*, 149–190. [CrossRef]
69. Tester, M.; Davenport, R. Na^+ Tolerance and Na^+ Transport in Higher Plants. *Ann. Bot.* **2003**, *91*, 503–527. [CrossRef] [PubMed]
70. Flowers, T.J.; Hajibagheri, M.A.; Clipson, N.J.W. Halophytes. *Q. Rev. Biol.* **1986**, *61*, 313–337. [CrossRef]
71. Almeida, D.M.; Oliveira, M.M.; Saibo, N.J. Regulation of Na^+ and K^+ Homeostasis in Plants: Towards Improved Salt Stress Tolerance in Crop Plants. *Genet. Mol. Biol.* **2017**, *40*, 326–345. [CrossRef] [PubMed]
72. Volkov, V.; Wang, B.; Dominy, P.J.; Fricke, W.; Amtmann, A. Thellungiella halophila, a Salt-Tolerant Relative of Arabidopsis thaliana, Possesses Effective Mechanisms to Discriminate between Potassium and Sodium. *Plant Cell Environ.* **2004**, *27*, 342–353. [CrossRef]
73. Ashraf, M.; Harris, P.J.C. Potential Biochemical Indicators of Salinity Tolerance in Plants. *Plant Sci.* **2004**, *166*, 3–16. [CrossRef]
74. Slama, I.; Abdelly, C.; Bouchereau, A.; Flowers, T.; Savouré, A. Diversity, Distribution and Roles of Osmoprotective Compounds Accumulated in Halophytes under Abiotic Stress. *Ann. Bot.* **2015**, *115*, 433–447. [CrossRef]

75. Szabados, L.; Savouré, A. Proline: A Multifunctional Amino Acid. *Trends Plant Sci.* **2010**, *15*, 89–97. [CrossRef]
76. Ashraf, M.; Foolad, M.R. Roles of Glycine Betaine and Proline in Improving Plant Abiotic Stress Resistance. *Environ. Exp. Bot.* **2007**, *59*, 206–216. [CrossRef]
77. Verbruggen, N.; Hermans, C. Proline Accumulation in Plants: A Review. *Amino Acids* **2008**, *35*, 753–759. [CrossRef]
78. Bautista, I.; Boscaiu, M.; Lidón, A.; Llinares, J.V.; Lull, C.; Donat, M.P.; Vicente, O. Environmentally Induced Changes in Antioxidant Phenolic Compounds Levels in Wild Plants. *Acta Physiol. Plant.* **2016**, *38*, 9. [CrossRef]
79. Yan, K.; Zhao, S.; Bian, L.; Chen, X. Saline Stress Enhanced Accumulation of Leaf Phenolics in Honeysuckle (*Lonicera japonica* Thunb.) without Induction of Oxidative Stress. *Plant Physiol. Biochem.* **2017**, *112*, 326–334. [CrossRef]
80. Fini, A.; Brunetti, C.; Di Ferdinando, M.; Ferrini, F.; Tattini, M. Stress-induced Flavonoid Biosynthesis and the Antioxidant Machinery of Plants. *Plant Signal. Behav.* **2011**, *6*, 709–711. [CrossRef]
81. Pollastri, S.; Tattini, M. Flavonols: Old Compounds for Old Roles. *Ann. Bot.* **2011**, *108*, 1225–1233. [CrossRef]
82. Treutter, D. Significance of Flavonoids in Plant Resistance and Enhancement of Their Biosynthesis. *Plant Biol.* **2005**, *7*, 581–591. [CrossRef]
83. Winkel-Shirley, B. Biosynthesis of Flavonoids and Effects of Stress. *Curr. Opin. Plant Biol.* **2002**, *5*, 218–223. [CrossRef]
84. Saed-Moucheshi, A.; Shekoofa, A.; Pessarakli, M. Reactive Oxygen Species (ROS) Generation and Detoxifying in Plants. *J. Plant Nutr.* **2014**, *37*, 1573–1585. [CrossRef]
85. Ahmad, P.; Hashem, A.; Abd-Allah, E.F.; Alqarawi, A.A.; John, R.; Egamberdieva, D.; Gucel, S. Role of Trichoderma harzianum in Mitigating NaCl Stress in Indian Mustard (*Brassica juncea* L) through Antioxidative Defense System. *Front. Plant Sci.* **2015**, *6*, 868. [CrossRef] [PubMed]
86. Wu, Q.S.; Zou, Y.N.; Abd-Allah, E.F. Mycorrhizal Association and ROS in Plants. In *Oxidative Damage to Plants*, 1st ed.; Ahmad, P., Ed.; Academic Press: New York, NY, USA, 2014; pp. 453–475. ISBN 9780128004609.
87. Ef, A.A.; Abeer, H.; Alqarawi, A.A.; Hend, A.A. Alleviation of Adverse Impact of Cadmium Stress in Sunflower (*Helianthus annuus* L.) by Arbuscular Mycorrhizal Fungi. *Pak. J. Bot.* **2015**, *47*, 785–795.
88. Navarro, J.M.; Pérez-Tornero, O.; Morte, A. Alleviation of Salt Stress in Citrus Seedlings Inoculated with Arbuscular Mycorrhizal Fungi Depends on the Rootstock Salt Tolerance. *J. Plant Physiol.* **2014**, *171*, 76–85. [CrossRef]
89. Maun, M.A. *The Biology of Coastal Sand Dunes*; Oxford University Press: New York, NY, USA, 2009; ISBN 978-0198570363.
90. Martin, R.M. Effects of Warming on Invasive *Phragmites australis* and Native *Spartina patens* Seed Germination Rates and Implications for Response to Climate Change. *Northeast. Nat.* **2017**, *24*, 235–238. [CrossRef]
91. Lichtenthaler, H.K.; Wellburn, A.R. Determinations of Total Carotenoids and Chlorophylls *a* and *b* of Leaf Extracts in Different Solvents. *Biochem. Soc. Trans.* **1983**, *11*, 591–592. [CrossRef]
92. Weimberg, R. Solute Adjustments in Leaves of Two Species of Wheat at Two Different Stages of Growth in Response to Salinity. *Physiol. Plant.* **1987**, *70*, 381–388. [CrossRef]
93. Bates, L.S.; Waldren, R.A.; Teare, I.D. Rapid Determination of Free Proline for Water-Stress Studies. *Plant Soil* **1973**, *39*, 205–207. [CrossRef]
94. Vicente, O.; Boscaiu, M.; Naranjo, M.Á.; Estrelles, E.; Bellés, J.M.; Soriano, P. Responses to Salt Stress in the Halophyte *Plantago crassifolia* (Plantaginaceae). *J. Arid Environ.* **2004**, *58*, 463–481. [CrossRef]
95. Dubois, M.; Gilles, K.A.; Hamilton, J.K.; Rebers, P.T.; Smith, F. Colorimetric Method for Determination of Sugars and Related Substances. *Anal. Chem.* **1956**, *28*, 350–356. [CrossRef]
96. Grieve, C.M.; Grattan, S.R. Rapid Assay for Determination of Water Soluble Quaternary Ammonium Compounds. *Plant Soil* **1983**, *70*, 303–307. [CrossRef]
97. Hodges, D.M.; DeLong, J.M.; Forney, C.F.; Prange, R.K. Improving the Thiobarbituric Acid-Reactive-Substances Assay for Estimating Lipid Peroxidation in Plant Tissues Containing Anthocyanin and Other Interfering Compounds. *Planta* **1999**, *207*, 604–611. [CrossRef]
98. Taulavuori, E.; Hellström, E.K.; Taulavuori, K.; Laine, K. Comparison of Two Methods Used to Analyse Lipid Peroxidation from *Vaccinium myrtillus* (L.) during Snow Removal, Reacclimation and Cold Acclimation. *J. Exp. Bot.* **2001**, *52*, 2375–2380. [CrossRef]
99. Alexieva, V.; Sergiev, I.; Mapelli, S.; Karanov, E. The Effect of Drought and Ultraviolet Radiation on Growth and Stress Markers in Pea and Wheat. *Plant Cell Environ.* **2001**, *24*, 1337–1344. [CrossRef]
100. Blainski, A.; Lopes, G.C.; De Mello, J.C.P. Application and Analysis of the Folin Ciocalteu Method for the Determination of the Total Phenolic Content from *Limonium brasiliense* L. *Molecules* **2013**, *18*, 6852–6865. [CrossRef]
101. Zhishen, J.; Mengcheng, T.; Jianming, W. The Determination of Flavonoid Contents in Mulberry and Their Scavenging Effects on Superoxide Radicals. *Food Chem.* **1999**, *64*, 555–559. [CrossRef]
102. Gil, R.; Bautista, I.; Boscaiu, M.; Lidón, A.; Wankhade, S.; Sánchez, H.; Llinares, J.; Vicente, O. Responses of Five Mediterranean Halophytes to Seasonal Changes in Environmental Conditions. *AoB Plants* **2014**, *6*, plu049. [CrossRef]
103. Bradford, M.M. A Rapid and Sensitive Method for the Quantitation of Microgram Quantities of Protein Utilizing the Principle of Protein-Dye Binding. *Anal. Biochem.* **1976**, *72*, 248–254. [CrossRef]
104. Beyer, W.F., Jr.; Fridovich, I. Assaying for Superoxide Dismutase Activity: Some Large Consequences of Minor Changes in Conditions. *Anal. Biochem.* **1987**, *161*, 559–566. [CrossRef]
105. Aebi, H. Catalase in Vitro. In *Methods in Enzymology*; Academic Press: London, UK, 1984; Volume 105, pp. 121–126. [CrossRef]

106. Nakano, Y.; Asada, K. Hydrogen Peroxide Is Scavenged by Ascorbate-Specific Peroxidase in Spinach Chloroplasts. *Plant Cell Physiol.* **1981**, *22*, 867–880. [CrossRef]
107. Connell, J.P.; Mullet, J.E. Pea Chloroplast Glutathione Reductase: Purification and Characterization. *Plant Physiol.* **1986**, *82*, 351–356. [CrossRef]
108. Kowal, J.; Arrigoni, E.; Lane, S. Acidified Blue Ink-Staining Procedure for the Observation of Fungal Structures Inside Roots of Two Disparate Plant Lineages. *Bio-protocol* **2020**, *10*, e3786. [CrossRef]
109. Trouvelot, A. Measure du taux de mycorrhization d'un système radiculaire. Recherche de méthodes d'estimation ayant une signification fonctionnelle. In *Physiological and Genetic Aspects of Mycorrhizae*; Gianinazzi-Pearson, V., Gianinazzi, S., Eds.; INRA: Paris, France, 1986; pp. 217–221. ISBN 2-85340-774-8.
110. Metsalu, T.; Vilo, J. ClustVis: A web tool for visualizing clustering of multivariate data using Principal Component Analysis and heatmap. *Nucleic Acids Res.* **2015**, *43*, W566–W570. [CrossRef] [PubMed]
111. Hammer, Ø.; Harper, A.A.T.; Ryan, P.D. Past: Paleontological statistics software package for education and data analysis. *Palaeontol. Electron.* **2001**, *4*, 1–9. Available online: http://palaeo-electronica.org/2001_1/past/issue1_01.htm (accessed on 12 October 2023).

plants

MDPI

Article

Response of Tomato Plants, *Ailsa Craig* and Carotenoid Mutant *tangerine*, to Simultaneous Treatment by Low Light and Low Temperature

Antoaneta V. Popova [1,*], Martin Stefanov [1], Gergana Mihailova [2], Preslava Borisova [1] and Katya Georgieva [2]

[1] Institute of Biophysics and Biomedical Engineering, Bulgarian Academy of Sciences, Acad, G. Bonchev Str. Bl. 21, 1113 Sofia, Bulgaria; martin@bio21.bas.bg (M.S.); preslavab12345@gmail.com (P.B.)

[2] Institute of Plant Physiology and Genetics, Bulgarian Academy of Sciences, Acad, G. Bonchev Str. Bl. 21, 1113 Sofia, Bulgaria; gmihailova@bio21.bas.bg (G.M.); georgieva.katya.m@gmail.com (K.G.)

* Correspondence: popova@bio21.bas.bg

Abstract: Tomato (*Solanum lycopersicum* L.) plants, wild type *Ailsa Craig*, and carotenoid mutant *tangerine* that accumulates prolycopene instead of all-*trans*-lycopene were exposed to a combined treatment by low light and low temperature for 5 days. The ability of plants to recover from the stress after development for 3 days at control conditions was followed as well. The suffered oxidative stress was evaluated by the extent of pigment content, lipid peroxidation, membrane stability, and H_2O_2 generation. The level of MDA content under combined treatment in *tangerine* implies that the mutant demonstrates lower sensitivity to stress in comparison with *Ailsa Craig*. The oxidative protective strategy of plants was estimated by following the antioxidant and antiradical activity of phenolic metabolites, including anthocyanins, as well as the activities of antioxidant enzymes superoxide dismutase (SOD), ascorbate peroxidase (APX) and catalase (CAT). Presented results revealed that the oxidative stress was much stronger expressed after exposure of both types of plants to low light combined with low temperature compared to that after treatment with only low light. The most significant antioxidant protection was provided by phenolic substances, including anthocyanins. The lower sensitivity of *tangerine* plants to low light can be attributed to the higher activity of the antioxidant enzyme CAT.

Keywords: *Solanum lycopersicum*; carotenoid mutant *tangerine*; low light; low temperature; abiotic stress; antioxidants

check for updates

Citation: Popova, A.V.; Stefanov, M.; Mihailova, G.; Borisova, P.; Georgieva, K. Response of Tomato Plants, *Ailsa Craig* and Carotenoid Mutant *tangerine*, to Simultaneous Treatment by Low Light and Low Temperature. *Plants* **2024**, *13*, 1929. https://doi.org/10.3390/plants13141929

Academic Editor: Liming Yang

Received: 27 June 2024
Revised: 4 July 2024
Accepted: 10 July 2024
Published: 12 July 2024

1. Introduction

The ever-changing environmental conditions seriously affect the development and distribution of higher plants. The spectral characteristics and intensity of light as well as deviation from the respective optimal temperature, are the most important environmental factors that impact in a specific manner every genotype [1]. Light quality and quantity strongly affect photosynthetic processes, carbon metabolism, and plant growth-related parameters at physiological, biochemical, and molecular levels. Receiving less light than required for the proper development suppresses growth and metabolic reactions and causes significant alterations in plant morphology—an increase in plant height, leaf number, and specific leaf area aiming to absorb as much light as possible to ensure the effectivity of photosynthetic reactions [2,3]. At low light intensities plants show lower photosynthetic performance and less dry matter accumulation and are more susceptible to photoinhibition in comparison with those grown at high light intensities [3]. Low-light tolerant species can maintain more efficient photosynthetic rates due to maintenance of high chlorophyll content, enhanced antioxidant enzyme activities, osmotic adjustments, and lower level of lipid peroxidation in comparison with plants that cannot tolerate development at low light availability [3].

Another limiting factor for plant development and productivity is low temperature. Exposure to lower than optimal temperature has a negative effect on water availability; the physicochemical properties of the lipid bilayer of biological membranes are altered, and an imbalance between absorbed excitation energy and metabolic sinks is observed as the rate of enzymatic reactions is restricted at lower than optimal temperature [1,4–6]. To cope with challenging conditions of development at low temperatures, plants increase the accumulation of sugars to prevent dehydration, synthesis of anthocyanins and antioxidant enzymes is activated to mitigate the severity of oxidative stress caused by extreme environmental deviation [6].

The simultaneous treatment by low light and low temperature seriously restricts photosynthetic efficiency, biological carbon, and nitrogen fixation, significantly increases the generation of reactive oxygen species (ROS), and consequently elevates membrane lipid peroxidation [7,8]. Plants exposed to a combined low temperature and weak light stress have thinner and larger leaves compared with the leaves of plants grown under normal growth conditions [8,9]. The response of plants to the simultaneous application of these two stress factors is determined by a complex interaction between light- and temperature-mediated signaling processes expressed in the induction of stress-related genes and synthesis of protective metabolites [10]. Induction of genes involved in antioxidant protection against low light in Arabidopsis occurs at the availability of sufficient light [11]. Furthermore, cold acclimation of plants is more effective when exposed to low but not freezing conditions in the presence of normal or high light than in dark or low light intensities [10]. In addition, induction of some antioxidant enzymes (glutathione reductase and ascorbate peroxidase) is observed in wheat only under normal growth conditions but not at reduced light intensity [12]. Accumulation of stress-protective substances, including polyamines, proline, phenolic substances, etc., is also induced in a light-dependent manner in cold hardened wheat plants [12,13].

In photosynthetic organisms the electron transport system of chloroplasts generates ROS even under optimal conditions that are significantly accelerated when plants are suffering abiotic stress [14,15]. The main types of ROS are superoxide anion ($O2^-$), hydrogen peroxide (H_2O_2), singlet oxygen ($^1O_2$), and hydroxyl radicals (OH^-) [14–16]. Singlet oxygen is generated in the super complex of photosystem II (PSII) after energy exchange between the triplet excited state of chlorophyll in the reaction center and O_2 (^{3}P680*) [17] while O_2^- is generated at the acceptor site of PSI. The generated ROS are extremely harmful to cell components as they can oxidize lipids, proteins, and DNA. Thylakoid membranes, characterized by a high degree of unsaturation of their fatty acid chains, are very vulnerable to oxidative stress [18].

Plants have developed different strategies to alleviate the detrimental effect of oxidative stress, including increased accumulation of antioxidant compounds as ascorbate, tocopherols, and phenolic substances, as well as enhanced activity of antioxidant enzymes such as superoxide dismutase (SOD), ascorbate peroxidase (APX) and catalase (CAT) [15,16].

Phenolic compounds represent a substantial group of secondary metabolites acting as primary antioxidants and free radical scavengers, estimated by FRAP and DPPH assays, respectively, protecting plants from oxidative stress. Due to the aromatic ring, phenols are able to stabilize and relocate the unpaired electrons of their structure, facilitating the donation of hydrogen atoms and electrons from their hydroxyl groups [19].

Anthocyanins are a diverse group of secondary metabolites that belong to the flavonoid family. Accumulation of anthocyanins is related to various types of abiotic stress conditions, especially to high light illumination. Anthocyanins are located in plant cell vacuoles and their main function is to screen the excessively absorbed light. However, some anthocyanins have been shown to possess much higher antioxidant activity in comparison with well-established antioxidants such as ascorbate and tocopherol [20–22].

The antioxidant enzyme SOD is involved in the dismutation of O_2^- to H_2O_2 and O_2, thus providing protection against damaging oxidative action of O_2^- [14,15,23,24].

The dangerous H_2O_2 is relatively stable and can alter the redox state of different cell organelles and molecules but additionally can serve as an abiotic stress signaling molecule in plants [16]. The enzymes APX and CAT provide protection against oxidative stress by scavenging generated H_2O_2 [24–26].

Many horticultural crops, such as cotton, maize, rice, tobacco, tomato, etc., originate from tropical/subtropical regions and require high light supply and high temperatures for development and productivity [27]. Development at suboptimal light and temperature conditions seriously reduces biomass production and yield of tomato [28–30], sweet pepper [31], rice [3], and maize [32]. Tomato (*Solanum lycopersicum* L.) is one of the most important agricultural vegetable crops, used as a source of carotenoids (lycopene) for humans. It is cultivated under controlled conditions during winter and spring to meet the increasing demands of a healthy human diet. Due to its tropical origin development of tomato plants at reduced light supply and suboptimal temperatures seriously reduces plants' growth and production of fruits [29]. However, reported data about the combined effects of low temperature and low light on the development and productivity of tomato plants are rather limited.

Photosynthetic pigments, chlorophylls, and carotenoids are intrinsic components of photosynthetic apparatus responsible for sunlight absorption, transfer, and conversion of solar energy into electrochemical energy. The development of chloroplasts and synthesis of chlorophyll in young plant leaves are negatively influenced by low-light treatment [33]. However, the response of plants to treatment by low light intensity is dependent on the tolerance of the respective genotype to low light, duration of treatment, developmental stage, etc. Low light-tolerant plants increase the content of chlorophyll b and decrease the Chl a/b ratio, indicating increased light-harvesting antenna size that enables the plants to absorb as much light as possible, even at reduced light availability [3].

Carotenoids perform multiple functions, including light harvesting, antioxidant, photoprotective, and structural functions [33]. The preferred conformation of carotenoids in plants is all-*trans*-isomer, but *cis*-isomers are also available, however, in a much lower quantity [34]. The majority of published data concerning carotenoid content discuss the role of carotenoids in the formation of fruits in crops [35–38]. Information about the effect of carotenoid content on the response of plants to low light and low light in combination with low temperature is rather limited. A significant increase of zeaxanthin was detected in pepper (*Capsicum annuum* L.) seedlings after exposure to low light and low temperature, while treatment only by low light led to elevated content of β-carotene [39]. The availability of mutants with altered carotenoid content or ratio between carotenoid species facilitates the unraveling of the contribution of different carotenoid species to the response of higher plants to extreme environmental conditions. In this study we used the *tangerine* mutant of tomato (*Solanum lycopersicum*) that is characterized by a mutation in the CrtISO gene [40]. The *tangerine* mutant contains defective prolycopene isomerase (CRTISO) that is performing the isomerization of tetra-*cis*-lycopene to all-*trans*-lycopene [38,40] and tomato plants accumulate prolycopene instead of all-*trans*-lycopene. The fruits of *tangerine* are orange in color, young leaves are yellowish, and flowers are pale.

The aim of the present report was to evaluate the response of tomato plants, wild type *Ailsa Craig*, and carotenoid mutant *tangerine* (locus t) to short time (5 days) exposure of young plants to simultaneous treatment by low light intensity (125 μmol photons m^{-2} s^{-1}) and suboptimal (15/10 °C day/night) temperature. The ability of plants to recover from the stress in 3 days under control conditions was followed as well. We applied low light treatment as the light and chlorophyll can substitute the role of the defective prolycopene isomerase (CRTISO), as was suggested by Isaacson et al. [40] and Enfissi et al. [37]. The extent of suffered stress by *Ailsa Craig* and *tangerine* plants was monitored by the level of electrolyte leakage, generation of H_2O_2, photosynthetic pigment content, and degree of lipid peroxidation. The protective strategies of tomato plants were evaluated by the antioxidant and antiradical activity of phenolic compounds, accumulation of anthocyanins, and the activities of antioxidant enzymes (SOD, APX and CAT). The investigation was performed on leaves of control, treated, and recovered plants.

2. Results

The effect of treatment by low light (LL) in combination with control (NT) or low (LT) temperature on the pigment content in leaves of *Ailsa Craig* and *tangerine* plants was presented in (Figure 1). At control growth conditions, *Ailsa Craig* plants showed higher pigment content in comparison with *tangerine*. In non-treated mutant plants, the total chlorophyll content (Chl (a+b)) was lower by 15%, being 1.919 ± 0.103 (mg Chl (a+b) g^{-1} FW) in comparison with 2.238 ± 0.038 (mg Chl (a+b) g^{-1} FW) for *Ailsa Craig*. The total chlorophyll content in *Ailsa Craig* leaves decreased after 5 days of treatment by LL NT by 12% and by 23% when LL was combined with LT. However, in *tangerine* leaves, the chlorophyll content was slightly decreased after both treatments but the values were not statistically different in comparison with control plants. Transfer of *Ailsa Craig* and *tangerine* plants to control conditions led to the recovery of chlorophyll content in plants treated by LL NT, while both types of plants that were exposed to LL LT showed chlorophyll content comparable with that of treated ones (Figure 1a). Similar to chlorophyll content the level of carotenoids in control of mutant plants was lower in comparison with *Ailsa Craig*, but by only around 5%. Exposure to both types of treatment, LL NT and LL LT, led to a comparable decline in carotenoid content by 22% for *Ailsa Craig* and, to a lesser extent, for *tangerine* plants—by 10%. After the transfer of plants to control conditions, recovery of carotenoid content was detected after treatment by LL NT. The recovery process was not so effective after exposure to LL LT. *Ailsa Craig* plants showed carotenoid content similar to that of treated ones, while *tangerine* plants demonstrated a further decline in the carotenoid content (Figure 1b). In addition, the ratio Chl a/b in mutant leaves was significantly higher (3.57 ± 0.04) than in *Ailsa Craig* leaves (3.29 ± 0.12), and the difference between both types of plants was maintained during the whole experimental setup (Figure 1c).

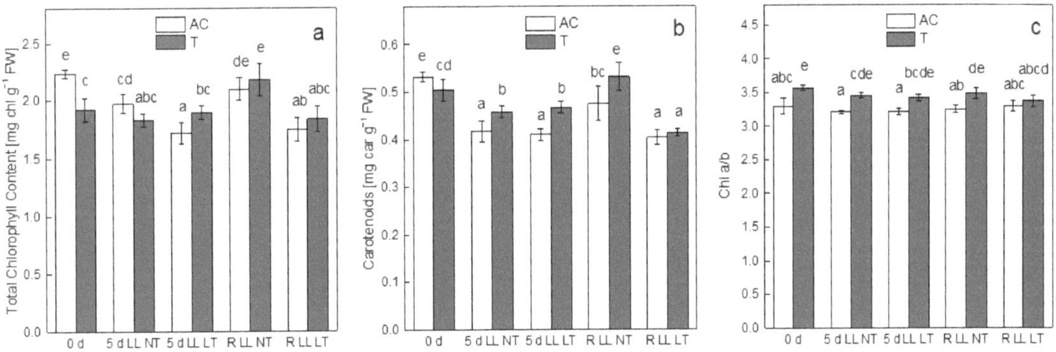

Figure 1. Alterations in photosynthetic pigment content—total chlorophyll content—Chl (a+b) (**a**), carotenoids (**b**) and Chl a/b ratio (**c**) in *Ailsa Craig* (AC) and carotenoid mutant *tangerine* (T) as induced by treatment of whole plants for 5 days by low light illumination (LL) in combination with normal (NT) or low (LT) temperature and after recovery period of 3 days (R). Data are presented as [mg pigment g^{-1} FW]. Mean values ± SE were calculated from two independent experiments and four parallel samples at each time point ($n = 8$). Significant differences between values at $p < 0.05$, as estimated by Fisher's LSD test of ANOVA, were indicated with different letters.

The stability of biological membranes of tomato plants in the course of treatment by LL in combination with NT or LT was followed by the extent of electrolyte leakage from leaves of control, treated, and recovered plants (Figure 2). The values of electrolyte leakage of leaves of both types of control plants were very low, comprising 11% of the total leakage of electrolytes. A statistically significant increase, comparable for *Ailsa Craig* and *tangerine* leaves, was detected only after exposure of plants to combined treatment by LL LT, reaching 15%. The stability of biological membranes after the recovery period was improved and was comparable with that before treatment.

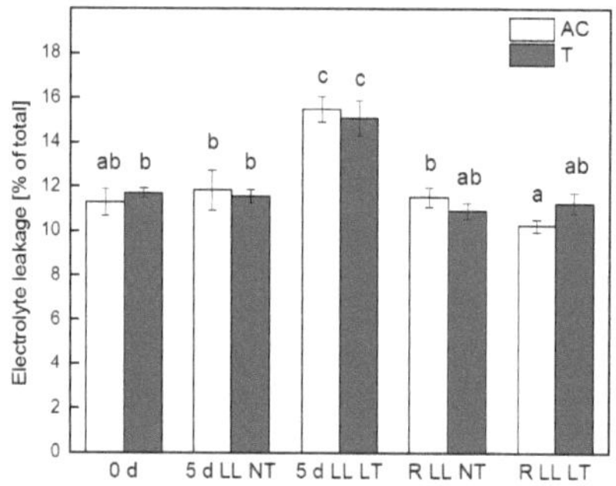

Figure 2. Stability of biological membranes of *Ailsa Craig* and *tangerine* leaves after exposure to low light (LL) in combination with normal (NT) or low (LT) temperature and after recovery period (R) as determined by the extent of electrolyte leakage. Mean values ± SE were calculated from two independent experiments and four parallel samples at each time point (n = 8). Significant differences between values at $p < 0.05$, as estimated by Fisher's LSD test of ANOVA, were indicated with different letters.

The extent of oxidative stress by *Ailsa Craig* plants after LL NT and LL LT treatment was evaluated by the alterations in the level of stress markers MDA and H_2O_2 (Figure 3). The MDA content in tomato plants was increased by exposure to LL reaching 140% at NT and 225% at LT in comparison with control, non-treated *Ailsa Craig* plants. Treatment of *tangerine* plants led to a similar trend of elevation of MDA content, but to a lesser extent, the increase was 128% and 183% after exposure to LL NT or LL LT, respectively. Termination of treatment resulted in a significant reduction of MDA. However, for *Ailsa Craig* and *tangerine* plants the values were higher in comparison with control levels (Figure 3a). Statistically significant elevation of generated H_2O_2 in leaves of *Ailsa Craig* and *tangerine* tomato plants was detected only after treatment by LL NT, by 12%, in comparison with control plants (Figure 3b).

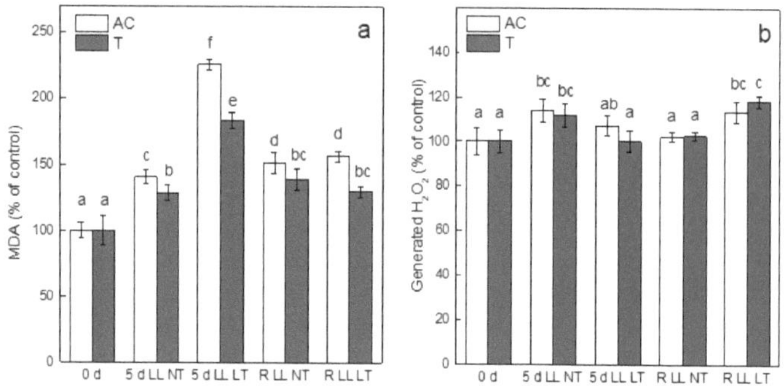

Figure 3. The extent of lipid peroxidation (**a**) and level of H_2O_2 generation (**b**) in leaves of *Ailsa Craig* and *tangerine* plants after exposure for 5 days to low light (LL) and normal (NT) or low (LT) temperature and after recovery period (R). Four parallel samples were processed at every time point of two independent experiments (n = 8). Every bar represents mean values ± SE. Significant differences between values at $p < 0.05$, as estimated by Fisher's LSD test of ANOVA, were indicated with different letters. 100% for MDA—*Ailsa Craig*—27.48 ± 1.64 and for *tangerine* 34.61 ± 3.75 nmol MDA g^{-1} FW; for H_2O_2—*Ailsa Craig*—5.94 ± 0.35 and for *tangerine* 6.02 ± 0.31 μmol H_2O_2 g^{-1} FW.

The antioxidant and antiradical activities of phenolic secondary metabolites in leaves of control, treated and recovered tomato plants, *Ailsa Craig* and *tangerine*, were evaluated by FRAP and DPPH assays, respectively, and expressed as percent from the respective control (Figure 4). Application of LL treatment resulted in around a 10% increase in antioxidant activity for both types of plants (Figure 4a). More significant elevation was detected after combined treatment, LL LT, by 52%, for *Ailsa Craig* and less expressed (by 40%) for *tangerine* plants. After termination of treatment and transfer to control conditions, the antioxidant activity was further increased and values were higher in comparison with that after treatment by LL NT and LL LT. Alterations in the radical scavenging activity in leaves of *Ailsa Craig* and *tangerine* for the entire experimental setup followed the same trend—stronger increase at LL LT in comparison with LL NT treatment and enhanced values after the recovery period, valid for both types of treatment (Figure 4b).

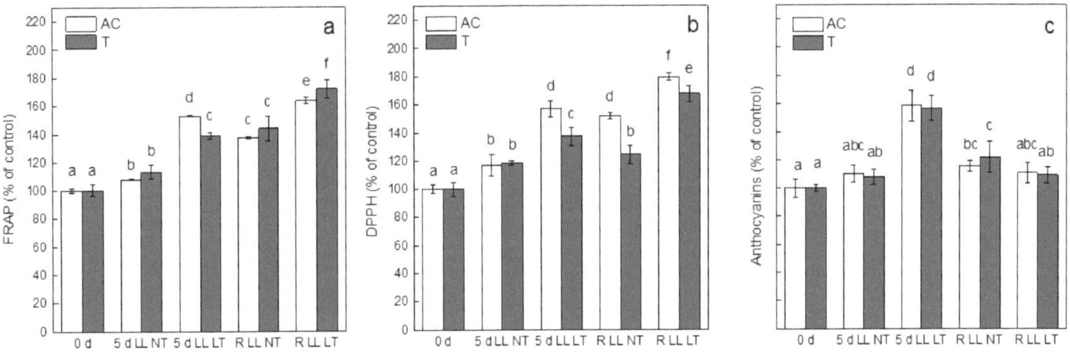

Figure 4. Alterations in total antioxidant (FRAP assay). (**a**) Free radical scavenging activity (DPPH assay) (**b**), and level of anthocyanins (**c**) in leaves of *Ailsa Craig* and *tangerine* plants, control (0 d), exposed for 5 days to low light (LL) and normal (NT) or low (LT) temperature and after recovery at control conditions (R). Mean values ± SE were calculated from two independent experiments with four parallel samples (n = 8). Significant differences between values at $p < 0.05$, as estimated by Fisher's LSD test of ANOVA, are indicated with different letters. one hundred percent for FRAP—*Ailsa Craig*—14.07 ± 0.24 and for *tangerine* 8.94 ± 0.37 µmol F^{2+} g^{-1} FW; 100% for DPPH—*Ailsa Craig*—3.96 ± 0.13 and for *tangerine* 5.48 ± 0.26 µmol Trolox g^{-1} FW. 100% for anthocyanins—*Ailsa Craig*—0.215 ± 0.014 and for *tangerine*—0.293 ± 0.040 nmol g^{-1} FW.

As a response to environmental stress conditions, plants accumulate different antioxidative compounds, including anthocyanins. Tomato plants, *Ailsa Craig* and *tangerine*, responded to treatment by LL in a similar manner with respect to anthocyanin content. An elevation (by 56%) was detected after exposure to LL LT. A smaller increase (by around 10%) was detected after LL NT treatment but was not statistically significant in comparison with controls (Figure 4c). After a recovery period of 3 days, the values of anthocyanins were significantly decreased in plants previously treated by LL LT.

SOD activity was not significantly affected after exposure of *Ailsa Craig* and *tangerine* to LL at both NT and LT. The most obvious increase in SOD activity was observed after 3 days of recovery of *Ailsa Craig* plants that were previously exposed for 5 days to LT LL (Figure 5a).

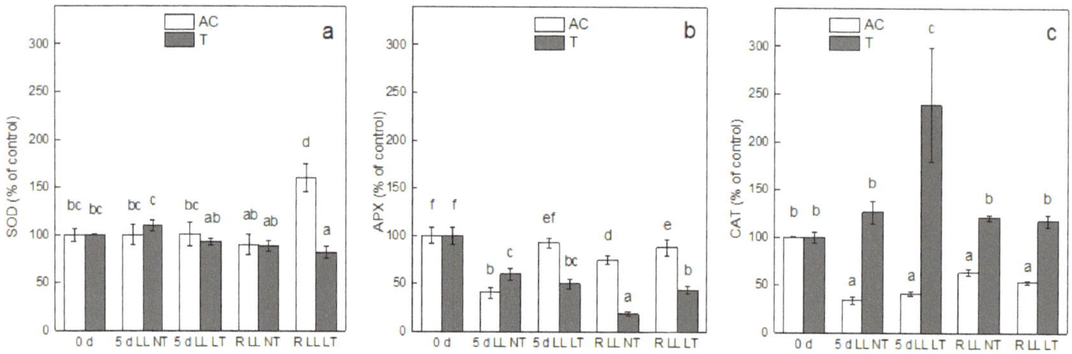

Figure 5. Effect of treatment by low light (LL) combined with normal (NT) or low (LT) temperature of *Ailsa Craig* and *tangerine* plants and after recovery for 3 days (R) on SOD (**a**), APX (**b**), and CAT (**c**) activity. Every bar represents mean values ± SE calculated from two independent experiments and three parallel samples at every time point (n = 6). The same letters within the graph indicate no significant differences assessed by Fisher's LSD test ($p \leq 0.05$) after performing ANOVA. 100% for SOD—*Ailsa Craig*—228.1 ± 15.9 and for *tangerine* 317.9 ± 2.1 U g^{-1} FW; 100% for APX—for *Ailsa Craig* 7474.0 ± 605.7 and for *tangerine* 9955.3 ± 894.1 U mg^{-1} FW min^{-1} and 100% for CAT—*Ailsa Craig*—898.7 ± 7.9 and for *tangerine*—312.9 ± 17.8 nmol H_2O_2 destroyed g^{-1} FW min^{-1}.

In contrast to SOD, APX activity significantly decreased after treatment by LL NT in *Ailsa Craig* and *tangerine* by 60% and 40%, respectively, compared to the controls (Figure 5b). After recovery at control conditions, the APX activity increased in *Ailsa Craig* but remained lower than control, whereas an additional reduction in its activity was observed after recovery of *tangerine*. The APX activity in *Ailsa Craig* was less affected when plants were exposed to LL LT compared to LL NT, whereas similar changes were observed in *tangerine* plants treated at NT and LT under LL. No significant changes in APX were detected after the recovery of plants exposed to LL LT.

CAT activity decreased in *Ailsa Craig* after treatment by LL at NT and LT (about 60%), and regardless of some enhancement, its activity remained lower than control after recovery. In contrast, exposure of *tangerine* to LL resulted in increased CAT activity, especially at LT, and it remained higher than control after recovery.

3. Discussion

The majority of cultivated crops originate from tropical regions and require particular for every respective crop combination of high light intensity and temperature for proper development and high yield. To meet the requirements of the growing world's population for food such types of crops are grown under controlled conditions during fall and winter. Being a thermophilic and photophilic crop, tomato plants suffer the negative impact of receiving less light and lower ambient temperature needed for their development and fruiting during greenhouse cultivation [29]. The published data concerning the development of tomato plants at lower-than-required light illumination and lower-than-optimal temperatures are rather limited. In addition, reports on the response of tomato plants with altered carotenoid content with respect to photosynthetic performance and suffered abiotic stress, as well as antioxidative protective strategies of plants at extreme environmental conditions, are focused mainly on tomato fruits, the source of carotenoids for humans [35–37,40].

Here, we report on the abiotic stress suffered by tomato plants (*Solanum lycopersicum*), wild type *Ailsa Craig*, and carotenoid mutant *tangerine* after treatment for 5 days by low light illumination alone and in combination with low temperature and the applied protective strategy that enables plants to withstand the stress. The ability of tomato plants to recover at control conditions after the applied stress was monitored as well. The aim was to unravel how the altered carotenoid content of *tangerine* mutant that accumulates prolycopene instead of all-*trans*-lycopene due to defective prolycopene isomerase (CRTISO) respond to the applied double stress in comparison with *Ailsa Craig* plants. In a recent investigation, we have shown that the thylakoid membrane structure, primary photosynthetic reactions, and thylakoid membrane fluidity in *tangerine* were affected by low light and suboptimal temperature in a different manner in comparison with *Ailsa Craig*. The treatment by LL LT reduced the photochemical activity of PSII in *Ailsa Craig* to a lesser extent when compared with *tangerine*, but PSII activity in the mutant was affected negatively by LL NT and LL LT to a comparable extent. The fluidity of the *tangerine* lipid phase of the thylakoid membrane was higher in comparison with the *Aisla Craig* membranes. Data indicated that the differences between *Ailsa Craig* and *tangerine* are mainly linked to PSII photochemical activity and its antenna complexes [41].

Photosynthetic piments are indispensable components of photosynthetic organisms that are involved in absorbing and transmitting light energy. The availability of photosynthetic piments and the ratio between different pigment species is an important indicator of chloroplast development and photosynthetic performance [33]. Alterations in the pigment content of higher plants after exposure to environmental stress conditions are dependent on a number of factors, including plant species and their tolerance to the particular treatment [42]. At low light conditions, the development of chloroplasts and the synthesis of pigments in young leaves can be seriously retarded [33]. Results presented indicate that chlorophyll and carotenoid content of young *tangerine* plants under control conditions was lower (by 15% for chlorophyll and by 5% for carotenoids) in comparison with *Ailsa Craig*. However, exposure to LL or to the combined treatment LL LT decreased the piment content of *Ailsa Craig* to a more significant degree in comparison with *tangerine* plants. Both types of plants, exposed only to LL, were able to restore their pigment content during the recovery period; however, the combined treatment was more severe, and plants did not regain the pigment content of control plants. A low light-induced decline in pigment content was also detected in tobacco [43], tomato [30], and purple pakchoi (*Brassica rapa* var Chinensis) [44]. Another difference between *Ailsa Craig* and *tangerine* plants with respect to their pigment content was the higher ratio of Chl a/b in the mutant, suggesting that the light-harvesting antenna size in *tangerine* plants was smaller in comparison with *Ailsa Craig*. It had been reported that exposure to LL of low light tolerant plants resulted in higher Chl b content and lower Chl a/b ratio, thus enabling the tolerant species to absorb as much as possible sunlight [3]. During the entire experimental setup, we did not observe any statistically significant alterations in the Chl a/b ratio for both types of plants, indicating that the antenna size was not altered by treatment by LL or by LL LT, suggesting that *Ailsa Craig* and *tangerine* plants are sensitive to development at low light availability.

ROS are produced in all photosynthetic organisms, even when growing under normal conditions, and are significantly induced when plants face extreme constraints, such as high and low temperatures, high light illumination, salinity, UV, heavy metals, etc., that significantly disturb photosynthetic, biochemical and physiological functions depending on the type and duration of treatment, developmental stage, and respective genotype [1,14,15]. Development at lower than optimal light intensity can also cause oxidative stress [7,8]. The generated at the acceptor side of PSII O_2^- facilitates the formation of hydroxyl radicals (OH^-) that, due to their high redox potential and lack of scavenging enzyme, are able to attack DNA and oxidize proteins and lipids [45,46].

The level of stress-induced accumulation of the end product of peroxidation of unsaturated fatty acid chains in plant membranes, MDA, is a reliable stress indicator for the evaluation of oxidative stress under constrained conditions and damage of biological membranes [47]. The extent of lipid peroxidation in leaves of *Ailsa Craig* and *tangerine* indicated that the oxidative stress suffered by plants was much stronger expressed after exposure to the combined application of LL LT in comparison with treatment for the same period only by LL. It is worth noting that *Ailsa Craig* plants were more sensitive to both types of treatment in comparison with *tangerine* as evidenced by the level of MDA after treatments. Significant reduction of lipid peroxidation after the recovery period was detected only in plants that were subjected to combined stress; however, the values of controls were not reached, suggesting that some stress-related processes were still functioning after termination of stress treatment. It has to be mentioned that the membrane integrity, evaluated by the degree of electrolyte leakage, was disturbed in an identical manner in *Ailsa Craig* and *tangerine* only after exposure to both stress factors (LL LT) and was completely restored after recovery. It can be suggested that the higher level of MDA content is mainly due to the peroxidation of thylakoid membrane lipids that are characterized by a high amount of unsaturated fatty acids [18], while the electrolyte leakage is mainly related to the stability of cell walls.

As a response to abiotic stress, plants have developed different protective mechanisms against oxidative stress, including accumulation of protective metabolites such as proline, tocopherols, phenolic substances, etc., and overexpressing antioxidant enzymes such as SOD, APX, and CAT [14,15]. LL treatment induced increased antioxidant and antiradical activity in leaves of both types of plants that were significantly higher under combined stress (LL LT). It has to be pointed out here that the level of these activities was even higher after the termination of treatments, indicating that some residual stress-related processes remained active during the recovery period, which is in accordance with the high level of MDA after the recovery period.

Another big group of phenolic substances is anthocyanins that, in addition to performing a light screening effect under high light intensity [20–22], have been proven to act as more effective scavengers of generated ROS in comparison with ascorbate and tocopherol [20,21], both in vitro [48] and in vivo [20]. After treatment of *Ailsa Craig* and *tangerine* by LL, a small but not statistically significant increase of anthocyanins was detected in the plants' leaves, which is to be expected as the main role of these metabolites is light screening under high light intensities [20–22]. However, the application of both stress factors, LL LT, induced a significant accumulation of anthocyanins, suggesting that under these conditions, anthocyanins are needed as scavengers of stress-induced ROS [20].

The stress-generated ROS is effectively scavenged by the cascade of superfamilies of antioxidant enzymes SOD, APX, and CAT, SOD being the primary antioxidant defense against O_2^- transferring the radical to H_2O_2, thus providing protection against damaging oxidative action of O_2^- [15,23]. The resulting H_2O_2 is further scavenged by APX and CAT [15,16,24–26]. However, it has been suggested that LL-induced antioxidant enzyme activities are largely dependent on the plant species with respect to its ability to tolerate the respective type of stress. Low light-tolerant species can maintain high antioxidative enzyme activities that provide significant protection against the negative effects of stress-generated ROS on plant physiology and performance, while low light-sensitive varieties fail to provide sufficient antioxidant enzyme protection [3]. The induction of antioxidant enzymes at the application of LL is not unequivocal. An increase in SOD activity and MDA content was detected in the Jinfen 5 variety of tomato (*Solanum lycopersicum*), while a reduction in CAT activity was detected under combined treatment for 10 days by LT and weak light. However, the exposure was longer and at much lower light intensities in comparison with our set up [30]. A decrease in SOD and APX activity was also detected in three tomato genotypes after exposure to LL in combination with LT [49], as well as a decline in APX in purple pakchoi (*Brassica rapa* var. Chinensis) under LL [44]. Under our experimental conditions, we did not observe an increase in antioxidant enzyme activities of SOD, APX, and CAT after exposure to LL in combination with NT or LT, implying the sensitivity of *Ailsa Craig* and *tangerine* plants to LL. An elevation of CAT activity was detected only in *tangerine* at LL application that was stronger expressed when LL was combined with LT. It can be supposed that the lower sensitivity of *tangerine* plants to the treatment, as evidenced by MDA content, can be due to the antioxidant support of the CAT enzyme.

4. Materials and Methods

4.1. Plant Material

In this investigation, we used tomato plants of the *tangerine* mutant (LA3183) and its nearly isogenic wild type "*Ailsa Craig*" (+/+) (LA2838A) that were provided by the Tomato Genetics Resource Center, Davis, CA, USA. Tomato plants were grown as described by Velitchkova et al. [41] and Gerganova et al. [50]. Seeds of *Ailsa Craig* and *tangerine* were initially placed on moist filter paper for 48 h at room temperature and further transferred to pots filled with perlite-containing soil and kept at 4 °C for 4 days. Afterward, the plants were grown in growth chambers (*Fytoscope FS130, Photon Systems Instruments*, Drásov, Czech Republic) for about 22 days. The photocycle was 16 h/8 h (day/night), light intensity was 250 µmol photons m^{-2} s^{-1} (PFD) (NL), temperature 24/22 °C (day/night) (NT), and relative humidity 75%. After the emergence of the third leaf, the plants were exposed for 5 days at low light intensity (125 µmol photons m^{-2} s^{-1}) (LL) in combination with normal (NT) (24/22 °C) or low temperature (LT) (15/10 °C). After the treatment, plants were transferred to control conditions (NL NT) for 3 days of recovery (R). Samples were taken from different leaves at the beginning of each experiment (0 d) after 5 days of treatment (5 d LL NT, 5 d LL LT) and after 3 days of recovery (R LL NT, R LL LT). Two independent experiments were performed, and 4 parallel samples were evaluated at every time point. The conditions that plants were grown before the start of treatment, light illumination, and temperature, were indicated as control conditions throughout the text.

4.2. Determination of Photosynthetic Pigment Content

Pieces of different leaves from control, treated, and recovered plants (40 mg for each sample) of *Ailsa Craig* and *tangerine* mutant were ground with a mortar and pestle with ice-cold 80% (v/v) acetone in dim light [50]. The homogenate was centrifuged in sealed tubes at 4500× g for 15 min at 4 °C. The clear extract was used to spectrophotometrically (*UV-VIS Specord 210 Plus*, Analytic Jena, Jena, Germany) determine the total content of chlorophyll (Chl (a+b)) and carotenoids (Car) using the coefficients and formulas of Lichtenthaler (1987) [51]. Mean values ± SE were calculated from two independent experiments and 4 parallel samples at every time point and expressed as mg pigment g^{-1} FM.

4.3. Electrolyte Leakage

Pieces of leaves from control, treated, and recovered plants (150 mg) were transferred to 15 mL double distilled water. The samples were kept at room temperature and gentle shaking for 24 h. The conductivity of the floating solution was measured with a conductivity meter (*HI5321*, Hanna Instruments, Smithfield, RI, USA). The maximum leakage of the plant tissue was determined after boiling the leaf material for 15 min at 100 °C. The extent of leakage was presented as a percentage of the maximum leakage.

4.4. Lipid Peroxidation and H_2O_2 Content

The extent of lipid peroxidation in leaves of *Ailsa Craig* and *tangerine* plants was determined by the malondialdehyde (MDA) content following the thiobarbituric acid method (TBA) [52]. Pieces of different leaves (100 mg) were homogenized in 3 mL 0.1% (w/v) trichloroacetic acid (TCA) at 4 °C. After centrifugation at 4500× g for 15 min at 4 °C 1 mL of the clear extract was mixed with the same volume 20% TCA containing 0.5% TBA and was boiled in a water bath for 25 min. After cooling, the absorbance at 532 and 600 nm was recorded (*UV–VIS Specord 210 Plus*). The absorbance at 600 nm was read to correct for unspecific turbidity. The amount of formed TBA-reactive metabolites (aldehydes, mainly MDA and endoperoxides) was calculated using the extinction coefficient of 155 mM^{-1} cm^{-1}. The results were expressed on a fresh weight basis [nmol MDA g^{-1} FW].

The supernatant, after centrifugation of homogenized leaves material with 0.1% TCA, was used for the determination of generated H_2O_2 content. To 0.5 mL of the supernatant was added 0.5 mL K phosphate buffer (pH 7.0) and 1 mL 1 M KI. The mixture was kept in the dark for 2 h and was periodically vortex-mixed. The absorbance at 390 nm was spectrophotometrically determined (*UV–VIS Specord 210 Plus*). The amount of H_2O_2 was determined using a standard curve of known H_2O_2 concentrations [53]. Results were expressed on a fresh weight basis [μmol H_2O_2 g^{-1} FW].

4.5. Determination of Anthocyanin Content

Pieces of leaves from different *Ailsa Craig* and *tangerine* plants—control, treated, and recovered plants (50 mg) were homogenized in a 6 mL medium containing ethanol/HCl/H_2O (79/1/20, *v*/*v*/*v*) and centrifuged at 10,000× *g* for 15 min at 4 °C. The clear extract was used to determine anthocyanin content. The absorbance at 535 and 653 nm was recorded. Anthocyanin content was calculated by the formula A535 − 0.24 × A653 [54] and molar extinction coefficient 33,000 M^{-1} cm^{-1} [55]. Results were presented as [μmol g^{-1} FW]. At every time point of each experiment four parallel samples were processed.

4.6. Antioxidant and Free Radical Scavenging Activity of Phenolic Metabolites

The antioxidant activity and free radical scavenging activity of phenolic metabolites in *Ailsa Craig* and *tangerine* plants were accessed as described by Doncheva et al. [56]. Leaf material (500 mg) from control, treated, and recovered plants were ground at 4 °C in the dark with 80% ethanol. The resulting homogenate was centrifuged at 12,000× *g* at 4 °C for 30 min. The supernatant was used for the determination of antioxidant and free radical scavenging activities.

The total antioxidant activity was determined following the ferric-reducing antioxidant power (FRAP) assay [57]. The reduction of ferric ions (Fe^{3+}) to ferrous ions (Fe^{2+}) is performed by the available bioactive substances (antioxidants). To 0.05 mL of the ethanol extract were added 1.5 mL of freshly prepared FRAP reagent and 0.15 mL of distilled water. The reaction was carried out at room temperature. A blue color developed after 15 min, and the absorbance was recorded at 593 nm. FRAP solution served as the blank. The antioxidant activity of the samples was determined using a standard curve with known concentrations of $FeSO_4 \cdot 7H_2O$.

The free radical scavenging activity was evaluated as described by Brand-Williams et al. [58]. DPPH (1,1-Diphenyl-2-picrylhydrazyl) was applied as the free radical source. The discoloration of DPPH solution is dependent on the available in the sample radical species or antioxidants. Freshly prepared DPPH reagent (1.99 mL) was mixed with 0.01 mL of the ethanol extract. The scavenging process was conducted in the dark at ambient temperature for 30 min. In parallel, one test tube with 2 mL of pure DPPH solution (without ethanol extract) was placed for comparison of the degree of discoloration. The absorbance at 515 nm was spectrometrically recorded. Methanol was used as the blank. A standard curve with known Trolox concentrations was used to calculate the free radical scavenging capacity.

4.7. Enzyme Activity Assays

Pieces of leaves (200 mg) from different plants of control, treated, and recovered *Ailsa Craig* and *tangerine* plants were collected and immediately frozen in liquid nitrogen and kept at −80 °C for further evaluation. Leaf material was ground at 4 °C with 1.2 mL 50 mM K phosphate buffer (pH 7.8) supplemented with 0.1 mM EDTA. The homogenate was centrifuged at 15,000× *g* for 20 min at 4 °C. The pellet was resuspended in 0.8 mL of the same buffer and centrifuged again. Both supernatants were combined and used to evaluate the enzyme activities of SOD, APX, and CAT [59].

The activity of superoxide dismutase (SOD; EC 1.15.1.1) was determined according to [60] by monitoring the superoxide radical-induced nitro blue tetrazolium (NBT) in the presence of riboflavin reduction at 560 nm (*UV–VIS Specord 210 Plus*). The reaction mixture contained 50 mM K phosphate buffer (pH 7.8), 0.1% Triton X-100, 0.1 mM EDTA, 0.06 mM NBT, 10 mM methionine, 2 μM riboflavin, and enzyme extract. One unit (U) of SOD was the amount of enzyme that caused 50% inhibition of NBT reduction in light and expressed as [U g^{-1} FW].

The activity of ascorbate peroxidase (L-Ascorbate:H_2O_2 oxidoreductase) (APX; EC 1.11.1.11) was assayed by monitoring the rate of hydrogen peroxide-dependent ascorbate oxidation at 290 nm by *UV–VIS Specord 210 Plus* [61]. The reaction medium contained 25 mM K phosphate buffer (pH 7.0), 0.5 mM ascorbate, 2 mM H_2O_2, 0.1 mM EDTA, and enzyme extract. One U of APX was determined as the decrease of absorbance at 290 nm with 0.001 and was expressed as [U g^{-1} FW min^{-1}].

The catalase (H_2O_2: H_2O_2 oxidoreductase) (CAT; EC 1. 11.1. 6) activity was assayed according to [62] by the time-dependent decline in absorbance at 240 nm (*UV–VIS Specord 210 Plus*) related to the extent of H_2O_2 decomposition at 25 °C in a reaction mixture containing 25 mM K phosphate buffer (pH 7.0), 10 mM H_2O_2 and enzyme extract. The activity of CAT was determined by the amount of degraded H_2O_2 for 1 min and expressed as [nmol H_2O_2 g^{-1} FW min^{-1}].

4.8. Statistics

Results in all figures were presented as mean values ± SE, calculated from two independent experiments with four parallel samples at every time point. Mean values were statistically compared by the Fisher's least significant difference (LSD) test at $p \leq 0.05$ following analysis of variance by ANOVA. All values in every panel were simultaneously statistically compared. Statistically different values were indicated by different letters. A statistical software package (StatGraphics Plus, version 5.1 for Windows, The Plains, VA, USA) was used.

5. Conclusions

Taken together, the results about membrane integrity and lipid peroxidation in the leaves of wild-type *Ailsa Craig* and *tangerine*, a mutant that accumulates prolycopene instead of all-*trans*-lycopene, indicated that the combined action of low light and low temperature was causing much stronger oxidative stress when compared to exposure only to low light. In addition, the high MDA content and high levels of antioxidant and antiradical activities after the recovery period indicated that after 5 days of treatment by low light and low light in the presence of low temperature, there were still active stress-related processes. Protection against generated ROS was provided by increased levels of phenolic compounds, including anthocyanins. The sensitivity of the *tangerine* mutant was less expressed than that of *Ailsa Craig*. Significant protection against oxidative stress in *tangerine* was provided by the high activity of CAT antioxidant enzyme under conditions of low light, combined with control or low temperature.

Author Contributions: Conceptualization, A.V.P., G.M. and K.G.; formal analysis, A.V.P., G.M., M.S. and K.G.; investigation, A.V.P., M.S., P.B., G.M. and K.G.; data curation, A.V.P. and K.G.; writing—original draft preparation, A.V.P.; writing—review and editing, A.V.P., G.M. and K.G.; supervision, A.V.P. and K.G.; project administration, A.V.P.; funding acquisition, A.V.P. All authors have read and agreed to the published version of the manuscript.

Funding: This work was financially supported by the Bulgarian Science Fund under project KP-06-H26/11.

Data Availability Statement: Data are contained within the article.

Acknowledgments: Tomato seeds were obtained from the UC Davis/C.M. Rick Tomato Genetics Resource Center and maintained by the Department of Plant Science, University of California, Davis, CA 95616.

Conflicts of Interest: The authors declare no conflicts of interest.

References

1. Allen, D.J.; Ort, D.R. Impacts of Chilling Temperatures on Photosynthesis in Warm-Climate Plants. *Trends Plant Sci.* **2001**, *6*, 36–42. [CrossRef] [PubMed]
2. Jiang, C.; Johkan, M.; Hohjo, M.; Tsukagoshi, S.; Ebihara, M.; Nakaminami, A.; Maruo, T. Responses of Leaf Photosynthesis, Plant Growth and Fruit Production to Periodic Alteration of Plant Density in Winter Produced Single-Truss Tomatoes. *Hortic. J.* **2017**, *86*, 511–518. [CrossRef]
3. Liu, Q.; Wu, X.; Chen, B.; Ma, J.; Gao, J. Effects of Low Light on Agronomic and Physiological Characteristics of Rice Including Grain Yield and Quality. *Rice Sci.* **2014**, *21*, 243–251. [CrossRef]
4. Popova, A.V.; Dobrev, K.; Velitchkova, M.; Ivanov, A.G. Differential Temperature Effects on Dissipation of Excess Light Energy and Energy Partitioning in *Lut2* Mutant of *Arabidopsis thaliana* under Photoinhibitory Conditions. *Photosynth. Res.* **2019**, *139*, 367–385. [CrossRef] [PubMed]
5. Velitchkova, M.; Borisova, P.; Vasilev, D.; Popova, A.V. Different Impact of High Light on the Response and Recovery of Wild Type and *Lut2* Mutant of *Arabidopsis thaliana* at Low Temperature. *Theor. Exp. Plant Physiol.* **2021**, *33*, 95–111. [CrossRef]
6. Ruelland, E.; Vaultier, M.-N.; Zachowski, A.; Hurry, V. Cold Signalling and Cold Acclimation in Plants. *Adv. Bot. Res.* **2009**, *49*, 35–150. [CrossRef]
7. Höglind, M.; Hanslin, H.M.; Mortensen, L.M. Photosynthesis of *Lolium perenne* L. at Low Temperatures under Low Irradiances. *Environ. Exp. Bot.* **2011**, *70*, 297–304. [CrossRef]
8. Sun, J.; Sui, X.; Huang, H.; Wang, S.; Wei, Y.; Zhang, Z. Low Light Stress Down-Regulated Rubisco Gene Expression and Photosynthetic Capacity During Cucumber (*Cucumis sativus* L.) Leaf Development. *J. Integr. Agric.* **2014**, *13*, 997–1007. [CrossRef]
9. Murchie, E.H.; Hubbart, S.; Peng, S.; Horton, P. Acclimation of Photosynthesis to High Irradiance in Rice: Gene Expression and Interactions with Leaf Development. *J. Exp. Bot.* **2005**, *56*, 449–460. [CrossRef] [PubMed]
10. Janda, T.; Prerostová, S.; Vanková, R.; Darkó, É. Crosstalk between Light- and Temperature-Mediated Processes under Cold and Heat Stress Conditions in Plants. *Int. J. Mol. Sci.* **2021**, *22*, 8602. [CrossRef]
11. Soitamo, A.J.; Piippo, M.; Allahverdiyeva, Y.; Battchikova, N.; Aro, E.-M. Light Has a Specific Role in Modulating Arabidopsis Gene Expression at Low Temperature. *BMC Plant Biol.* **2008**, *8*, 13–20. [CrossRef]
12. Janda, T.; Szalai, G.; Leskó, K.; Yordanova, R.; Apostol, S.; Popova, L.P. Factors Contributing to Enhanced Freezing Tolerance in Wheat during Frost Hardening in the Light. *Phytochemistry* **2007**, *68*, 1674–1682. [CrossRef] [PubMed]
13. Szalai, G.; Pap, M.; Janda, T. Light-Induced Frost Tolerance Differs in Winter and Spring Wheat Plants. *J. Plant Physiol.* **2009**, *166*, 1826–1831. [CrossRef] [PubMed]
14. Asada, K. Production and Scavenging of Reactive Oxygen Species in Chloroplasts and Their Functions. *Plant Physiol.* **2006**, *141*, 391–396. [CrossRef] [PubMed]
15. Sharma, P.; Jha, A.B.; Dubey, R.S.; Pessarakli, M. Reactive Oxygen Species, Oxidative Damage, and Antioxidative Defense Mechanism in Plants under Stressful Conditions. *J. Bot.* **2012**, *2012*, 217037. [CrossRef]
16. Singh, R.; Singh, S.; Parihar, P.; Mishra, R.K.; Tripathi, D.K.; Singh, V.P.; Chauhan, D.K.; Prasad, S.M. Reactive Oxygen Species (ROS): Beneficial Companions of Plants' Developmental Processes. *Front. Plant Sci.* **2016**, *7*, 1299. [CrossRef] [PubMed]
17. Krieger-Liszkay, A. Singlet Oxygen Production in Photosynthesis. *J. Exp. Bot.* **2004**, *56*, 337–346. [CrossRef] [PubMed]
18. Pospíšil, P.; Yamamoto, Y. Damage to Photosystem II by Lipid Peroxidation Products. *Biochim. Biophys. Acta-Gen. Subj.* **2017**, *1861*, 457–466. [CrossRef] [PubMed]
19. Chaves, N.; Santiago, A.; Alías, J.C. Quantification of the Antioxidant Activity of Plant Extracts: Analysis of Sensitivity and Hierarchization Based on the Method Used. *Antioxidants* **2020**, *9*, 76. [CrossRef]
20. Gould, K.S.; McKelvie, J.; Markham, K.R. Do Anthocyanins Function as Antioxidants in Leaves? Imaging of H_2O_2 in Red and Green Leaves after Mechanical Injury. *Plant. Cell Environ.* **2002**, *25*, 1261–1269. [CrossRef]
21. Hernández, I.; Alegre, L.; Van Breusegem, F.; Munné-Bosch, S. How Relevant Are Flavonoids as Antioxidants in Plants? *Trends Plant Sci.* **2009**, *14*, 125–132. [CrossRef] [PubMed]
22. Landi, M.; Tattini, M.; Gould, K.S.; Dabravolski, S.A.; Isayenkov, S.V. Multiple Functional Roles of Anthocyanins in Plant-Environment Interactions. *Environ. Exp. Bot.* **2015**, *119*, 4–17. [CrossRef]
23. Pilon, M.; Ravet, K.; Tapken, W. The Biogenesis and Physiological Function of Chloroplast Superoxide Dismutases. *Biochim. Biophys. Acta-Bioenerg.* **2011**, *1807*, 989–998. [CrossRef] [PubMed]
24. Smirnoff, N.; Arnaud, D. Hydrogen Peroxide Metabolism and Functions in Plants. *New Phytol.* **2019**, *221*, 1197–1214. [CrossRef] [PubMed]
25. Shigeoka, S. Regulation and Function of Ascorbate Peroxidase Isoenzymes. *J. Exp. Bot.* **2002**, *53*, 1305–1319. [CrossRef] [PubMed]
26. Hu, W.H.; Song, X.S.; Shi, K.; Xia, X.J.; Zhou, Y.H.; Yu, J.Q. Changes in Electron Transport, Superoxide Dismutase and Ascorbate Peroxidase Isoenzymes in Chloroplasts and Mitochondria of Cucumber Leaves as Influenced by Chilling. *Photosynthetica* **2008**, *46*, 581–588. [CrossRef]
27. Lyons, J.M. Chilling Injury in Plants. *Annu. Rev. Plant Physiol.* **1973**, *24*, 445–466. [CrossRef]
28. Lu, T.; Yu, H.; Li, Q.; Chai, L.; Jiang, W. Improving Plant Growth and Alleviating Photosynthetic Inhibition and Oxidative Stress From Low-Light Stress With Exogenous GR24 in Tomato (*Solanum lycopersicum* L.) Seedlings. *Front. Plant Sci.* **2019**, *10*, 490. [CrossRef]

29. Shu, S.; Tang, Y.; Yuan, Y.; Sun, J.; Zhong, M.; Guo, S. The Role of 24-Epibrassinolide in the Regulation of Photosynthetic Characteristics and Nitrogen Metabolism of Tomato Seedlings under a Combined Low Temperature and Weak Light Stress. *Plant Physiol. Biochem.* **2016**, *107*, 344–353. [CrossRef]
30. Xiaoa, F.; Yang, Z.; Zhua, L. Low Temperature and Weak Light Affect Greenhouse Tomato Growth and Fruit Quality. *J. Plant Sci.* **2018**, *6*, 16–24. [CrossRef]
31. Sui, X.; Mao, S.; Wang, L.; Zhang, B.; Zhang, Z. Effect of Low Light on the Characteristics of Photosynthesis and Chlorophyll a Fluorescence During Leaf Development of Sweet Pepper. *J. Integr. Agric.* **2012**, *11*, 1633–1643. [CrossRef]
32. Szalai, G.; Majláth, I.; Pál, M.; Gondor, O.K.; Rudnóy, S.; Oláh, C.; Vanková, R.; Kalapos, B.; Janda, T. Janus-Faced Nature of Light in the Cold Acclimation Processes of Maize. *Front. Plant Sci.* **2018**, *9*, 850. [CrossRef] [PubMed]
33. Demmig-Adams, B.; Adams, W.W. Photoprotection in an Ecological Context: The Remarkable Complexity of Thermal Energy Dissipation. *New Phytol.* **2006**, *172*, 11–21. [CrossRef]
34. Khoo, H.-E.; Prasad, K.N.; Kong, K.-W.; Jiang, Y.; Ismail, A. Carotenoids and Their Isomers: Color Pigments in Fruits and Vegetables. *Molecules* **2011**, *16*, 1710–1738. [CrossRef]
35. Guzman, I.; Hamby, S.; Romero, J.; Bosland, P.W.; O'Connell, M.A. Variability of Carotenoid Biosynthesis in Orange Colored *Capsicum* spp. *Plant Sci.* **2010**, *179*, 49–59. [CrossRef]
36. Wahyuni, Y.; Ballester, A.-R.; Sudarmonowati, E.; Bino, R.J.; Bovy, A.G. Metabolite Biodiversity in Pepper (Capsicum) Fruits of Thirty-Two Diverse Accessions: Variation in Health-Related Compounds and Implications for Breeding. *Phytochemistry* **2011**, *72*, 1358–1370. [CrossRef] [PubMed]
37. Enfissi, E.M.A.; Nogueira, M.; Bramley, P.M.; Fraser, P.D. The Regulation of Carotenoid Formation in Tomato Fruit. *Plant J.* **2017**, *89*, 774–788. [CrossRef]
38. Yu, Q.; Ghisla, S.; Hirschberg, J.; Mann, V.; Beyer, P. Plant Carotene Cis-Trans Isomerase CRTISO. *J. Biol. Chem.* **2011**, *286*, 8666–8676. [CrossRef]
39. Zhang, J.F.; Li, J.; Xie, J.M.; Yu, J.; Dawuda, M.M.; Lyv, J.; Tang, Z.Q.; Zhang, J.; Zhang, X.D.; Tang, C.N. Changes in Photosynthesis and Carotenoid Composition of Pepper (*Capsicum annuum* L.) in Response to Low-Light Stress and Low Temperature Combined with Low-Light Stress. *Photosynthetica* **2020**, *58*, 125–136. [CrossRef]
40. Isaacson, T.; Ronen, G.; Zamir, D.; Hirschberg, J. Cloning of Tangerine from Tomato Reveals a Carotenoid Isomerase Essential for the Production of β-Carotene and Xanthophylls in Plants. *Plant Cell* **2002**, *14*, 333–342. [CrossRef]
41. Velitchkova, M.; Stefanov, M.; Popova, A.V. Effect of Low Light on Photosynthetic Performance of Tomato Plants—*Ailsa Craig* and Carotenoid Mutant *Tangerine*. *Plants* **2023**, *12*, 3000. [CrossRef]
42. Yuan, L.; Shu, S.; Sun, J.; Guo, S.; Tezuka, T. Effects of 24-Epibrassinolide on the Photosynthetic Characteristics, Antioxidant System, and Chloroplast Ultrastructure in *Cucumis sativus* L. under $Ca(NO_3)_2$ Stress. *Photosynth. Res.* **2012**, *112*, 205–214. [CrossRef] [PubMed]
43. Wu, X.; Khan, R.; Gao, H.; Liu, H.; Zhang, J.; Ma, X. Low Light Alters the Photosynthesis Process in Cigar Tobacco via Modulation of the Chlorophyll Content, Chlorophyll Fluorescence, and Gene Expression. *Agriculture* **2021**, *11*, 755. [CrossRef]
44. Zhu, H.; Li, X.; Zhai, W.; Liu, Y.; Gao, Q.; Liu, J.; Ren, L.; Chen, H.; Zhu, Y. Effects of Low Light on Photosynthetic Properties, Antioxidant Enzyme Activity, and Anthocyanin Accumulation in Purple Pak-Choi (*Brassica campestris* Ssp. Chinensis Makino). *PLoS ONE* **2017**, *12*, e0179305. [CrossRef]
45. Asada, K.; Takahashi, M. Production and scavenging of active oxygen in photosynthesis. In *Photoinhibition Topics in Photosynthesis*; Kyle, D.J., Osmond, C.B., Arntzen, C.J., Eds.; Elsevier: Amsterdam, The Netherlands, 1987; pp. 227–287.
46. Asada, K. Production and Action of Active Oxygen Species in Photosynthetic Tissues. In *Causes of Photooxidative Stress and Amelioration of Defense Systems in Plants*; Foyer, C.H., Mullineaux, P.M., Eds.; CRC Press: Boca Raton, FL, USA, 1994; pp. 77–104.
47. Sharma, P.; Dubey, R.S. Drought Induces Oxidative Stress and Enhances the Activities of Antioxidant Enzymes in Growing Rice Seedlings. *Plant Growth Regul.* **2005**, *46*, 209–221. [CrossRef]
48. Wang, H.; Cao, G.; Prior, R.L. Oxygen Radical Absorbing Capacity of Anthocyanins. *J. Agric. Food Chem.* **1997**, *45*, 304–309. [CrossRef]
49. Yang, Y.; Dong, L.; Shi, L.; Guo, J.; Jiao, Y.; Xiong, H.; Dickson, R.W.; Shi, A. Effects of Low Temperature and Low Light on Physiology of Tomato Seedlings. *Am. J. Plant Sci.* **2020**, *11*, 162–179. [CrossRef]
50. Gerganova, M.; Popova, A.V.; Stanoeva, D.; Velitchkova, M. Tomato Plants Acclimate Better to Elevated Temperature and High Light than to Treatment with Each Factor Separately. *Plant Physiol. Biochem.* **2016**, *104*, 234–241. [CrossRef]
51. Lichtenthaler, H.K. Chlorophylls and Carotenoids: Pigments of Photosynthetic Membranes. *Methods Enzymol.* **1987**, *148*, 350–382. [CrossRef]
52. Esterbauer, H.; Cheeseman, K.H. Determination of aldehydic lipid peroxidation products: Malonaldehyde and 4-hydroxynonenal. *Methods Enzymol.* **1990**, *186*, 407–421. [CrossRef]
53. Velikova, V.; Yordanov, I.; Edreva, A. Oxidative Stress and Some Antioxidant Systems in Acid Rain-Treated Bean Plants. *Plant Sci.* **2000**, *151*, 59–66. [CrossRef]
54. Murray, J.R.; Hackett, W.P. Dihydroflavonol Reductase Activity in Relation to Differential Anthocyanin Accumulation in Juvenile and Mature Phase *Hedera helix* L. *Plant Physiol.* **1991**, *97*, 343–351. [CrossRef]
55. Hodges, M.D.; Nozzolillo, C. Anthocyanin and Anthocyanoplast Content of Cruciferous Seedlings Subjected to Mineral Nutrient Deficiencies. *J. Plant Physiol.* **1996**, *147*, 749–754. [CrossRef]

56. Doncheva, S.; Moustakas, M.; Ananieva, K.; Chavdarova, M.; Gesheva, E.; Vassilevska, R.; Mateev, P. Plant Response to Lead in the Presence or Absence EDTA in Two Sunflower Genotypes (Cultivated *H. annuus* Cv. 1114 and Interspecific Line *H. annuus* × *H. argophyllus*). *Environ. Sci. Pollut. Res.* **2013**, *20*, 823–833. [CrossRef] [PubMed]
57. Benzie, I.F.F.; Strain, J.J. Ferric reducing/antioxidant power assay: Direct measure of total antioxidant activity of biological fluids and modified version for simultaneous measurement of total antioxidant power and ascorbic acid concentration. *Methods Enzymol.* **1999**, *299*, 15–27. [CrossRef]
58. Brand-Williams, W.; Cuvelier, M.E.; Berset, C. Use of a Free Radical Method to Evaluate Antioxidant Activity. *LWT-Food Sci. Technol.* **1995**, *28*, 25–30. [CrossRef]
59. Elavarthi, S.; Martin, B. Spectrophotometric Assays for Antioxidant Enzymes in Plants. In *Plant Stress Tolerance Methods in Molecular Biology (Methods and Protocols)*; Sunkar, R., Ed.; Humana Press: New York, NY, USA, 2010; Volume 639, pp. 273–280. [CrossRef]
60. Beyer, W.F.; Fridovich, I. Assaying for Superoxide Dismutase Activity: Some Large Consequences of Minor Changes in Conditions. *Anal. Biochem.* **1987**, *161*, 559–566. [CrossRef]
61. Nakano, Y.; Asada, K. Hydrogen Peroxide Is Scavenged by Ascorbate-Specific Peroxidase in Spinach Chloroplasts. *Plant Cell Physiol.* **1981**, *22*, 867–880. [CrossRef]
62. Brennan, T.; Frenkel, C. Involvement of Hydrogen Peroxide in the Regulation of Senescence in Pear. *Plant Physiol.* **1977**, *59*, 411–416. [CrossRef] [PubMed]

plants

MDPI

Article

Exploring Nitric Oxide as a Regulator in Salt Tolerance: Insights into Photosynthetic Efficiency in Maize

Georgi D. Rashkov, Martin A. Stefanov, Ekaterina K. Yotsova, Preslava B. Borisova, Anelia G. Dobrikova and Emilia L. Apostolova *

Institute of Biophysics and Biomedical Engineering, Bulgarian Academy of Sciences, Acad. G. Bonchev Str., Bl. 21, 1113 Sofia, Bulgaria; megajorko@abv.bg (G.D.R.); martin_12.1989@abv.bg (M.A.S.); ekaterina_yotsova@abv.bg (E.K.Y.); pborisova@bio21.bas.bg (P.B.B.); aneli@bio21.bas.bg (A.G.D.)
* Correspondence: emya@bio21.bas.bg or emilia.apostolova@gmail.com

Abstract: The growing issue of salinity is a significant threat to global agriculture, affecting diverse regions worldwide. Nitric oxide (NO) serves as an essential signal molecule in regulating photosynthetic performance under physiological and stress conditions. The present study reveals the protective effects of different concentrations (0–300 μM) of sodium nitroprusside (SNP, a donor of NO) on the functions of the main complexes within the photosynthetic apparatus of maize (*Zea mays* L. Kerala) under salt stress (150 mM NaCl). The data showed that SNP alleviates salt-induced oxidative stress and prevents changes in the fluidity of thylakoid membranes (Laurdan GP) and energy redistribution between the two photosystems (77K chlorophyll fluorescence ratio F_{735}/F_{685}). Chlorophyll fluorescence measurements demonstrated that the foliar spray with SNP under salt stress prevents the decline of photosystem II (PSII) open reaction centers (qP) and improves their efficiency (Φexc), thereby influencing Q_A^- reoxidation. The data also revealed that SNP protects the rate constants for two pathways of Q_A^- reoxidation (k_1 and k_2) from the changes caused by NaCl treatment alone. Additionally, there is a predominance of Q_A^- interaction with plastoquinone in comparison to the recombination of electrons in $Q_A\ Q_B^-$ with the oxygen-evolving complex (OEC). The analysis of flash oxygen evolution showed that SNP treatment prevents a salt-induced 10% increase in PSII centers in the S_0 state, i.e., protects the initial S_0–S_1 state distribution, and the modification of the Mn cluster in the OEC. Moreover, this study demonstrates that SNP-induced defense occurs on both the donor and acceptor sides of the PSII, leading to the protection of overall photosystems performance (PI_{ABS}) and efficient electron transfer from the PSII donor side to the reduction of PSI end electron acceptors (PItotal). This study clearly shows that the optimal protection under salt stress occurs at approximately 50–63 nmoles NO/g FW in leaves, corresponding to foliar spray with 50–150 μM SNP.

Keywords: chlorophyll fluorescence; photosystem II; oxygen-evolving complex; photosynthetic function; membrane fluidity

Citation: Rashkov, G.D.; Stefanov, M.A.; Yotsova, E.K.; Borisova, P.B.; Dobrikova, A.G.; Apostolova, E.L. Exploring Nitric Oxide as a Regulator in Salt Tolerance: Insights into Photosynthetic Efficiency in Maize. *Plants* **2024**, *13*, 1312. https://doi.org/10.3390/plants13101312

Academic Editor: Vladimir V. Kuznetsov

Received: 4 April 2024
Revised: 29 April 2024
Accepted: 8 May 2024
Published: 10 May 2024

1. Introduction

Numerous factors in the natural environment impact the growth and development of plants [1]. Salinity, an abiotic factor, significantly influences crop productivity. Climate changes are causing salinized regions to expand at a rate of 10% annually [2]. Salt stress disrupts plant homeostasis via two mechanisms. Initially, high soil salt concentrations hinder root water absorption [3], leading to stroma closure and reduced photosynthesis efficiency [4]. The second phase of salt stress is triggered by harmful ions that damage the structure and functions of cell membranes, causing growth inhibition and developmental changes in plants [5]. These primary effects directly lead to oxidative stress, characterized by the intense production and accumulation of reactive oxygen species (ROS, such as H_2O_2, $O_2^{\bullet}$, OH, and $^1O_2$), which harm proteins, lipids, and nucleic acids [3].

The chloroplasts serve as the primary sites for generating ROS, a process that relies on the interactions of light and chlorophylls. The major sites in chloroplasts that largely produce ROS are photosystem I (PSI) and photosystem II (PSII) [6]. Increased ROS causes oxidative injury to several photosynthetic enzymes and thylakoid membranes, which in turn reduces CO_2 uptake, slows down plant growth, and ultimately lowers crop yields. The system for scavenging ROS comprises both enzymatic and non-enzymatic antioxidants [7]. The enzymatic antioxidants include superoxide dismutase (SOD), ascorbate peroxidase (APX), guaiacol peroxidase (GPX), catalase (CAT), and glutathione reductase (GR). The activity of these enzymes is essential for mitigating harmful ROS levels within cells [8]. Plants also adapt to elevated salt levels by enhancing their tissue osmotic potential by both inorganic and organic solutes [7,9,10].

Photosynthesis, the primary source of materials and energy for plant growth and development, is significantly affected by salinity [11,12]. This has sparked a growing interest in enhancing photosynthetic tolerance to boost plant yields under stressful conditions [13]. When plants are exposed to salt stress, changes occur in the thylakoid membranes, key components within the chloroplast responsible for photosynthetic light reactions. Salinity alters the number and structural arrangement of chloroplasts, leading to an increase in plastoglobules and a reduction in the thylakoids in granum in the leaf's epidermal chloroplasts [14]. Such observations have been noted in various higher plants, including *Sulla coronaria* [15], *Thellungiella salsuginea* [16], and *Cucumis sativus* [17]. Salt stress also affects photosynthesis by altering different enzymes, leaf pigment content [18], and the structural organization of the pigment–protein complexes in the thylakoid membranes. Changes have been observed in the oxygen-evolving complex (OEC) [19–21], light-harvesting complex of PSII (LHCII) [22–24], and D1 core protein of PSII [25–27]. These alterations impact the electron transfer from Q_A to Q_B, inhibit OEC, and affect the rate of electron transport in the thylakoid membranes [5,28–30].

Nitric oxide (NO) is a molecule with unpaired electrons, making it paramagnetic. It possesses the capability to readily diffuse across membranes [31]. Currently, NO is recognized as a crucial molecule that plays a role in redox signaling. It contributes to the regulation of numerous physiological processes and plays a significant role in regulating plant responses to abiotic stress [32,33]. Sodium nitroprusside (SNP) is a widely used NO donor that is used to study the role of NO in plants under physiological conditions [34] and to simulate the harmful environmental impact [35,36]. It has been shown that NO influences seed germination, maintains water balance, regulates gene expression and osmolyte accumulation, and enhances the activities of antioxidant enzymes in plants under stress. Additionally, NO directly neutralizes ROS under stressful conditions [37,38].

Over the past decade, there has been considerable interest in clarifying the role of NO in the salt tolerance of plants. NO donors can be applied to plants via spraying, incorporation into irrigation water, or injection into leaf apoplasts [32]. The protective role of NO, using an SNP donor under salinity, has been shown in different plants, such as tomato [39], cucumber [40], orange [41], cotton [42], alfalfa [43], apple [44], wheat [45], lentil [46], and sorghum [47]. Previous studies have revealed that the application of SNP under salt stress alleviates salt-induced effects on the stomatal behavior, cell water status, chlorophyll content, membrane damage, and membrane lipid peroxidation, and it also aids in the accumulation of proline, phenolic compounds, and antioxidants in plants [48–50].

Nitric oxide also has a direct effect on the photosynthetic electron transport, with binding sites within PSII. These include the non-heme iron located between the quinone acceptors Q_A and Q_B, the Tyr YD residue, and the Mn cluster of OEC [51,52]. However, data on the influence of NO on the photosynthetic machinery are quite contradictory. It has been observed that the application of SNP (200–1000 μM) led to a decrease in the maximum quantum efficiency of PSII (Fv/Fm) and the photochemical quenching (qP) in leaves of pea and potato plants under non-stress conditions [53,54]. Previous research [55] showed that SNP (100 μM) enhanced the maximum quantum efficiency of PSII (Fv/Fm) and the effective quantum yield of PSII (Φ_{PSII}) during the light-induced greening process in barley seedlings. It has also been shown that NO alleviated the salt-induced changes in the leaf area, plant dry matter production, and pigment content, as well as improved the uptake and transport of numerous macro- and micronutrients [56,57]. It has been demonstrated that NO decreases many of the negative effects of salt stress on plant photosynthetic machinery [58]. Nitric oxide decreased the salt-induced deactivation and breakdown of the PSII reaction center and enhanced their performance in salt-exposed pea, bermudagrass, soybean, and sorghum [47,58–61]. The NO-induced reduction in the harmful effects of high salt concentrations could result from the protection of the photosynthetic pigments, dissipation of surplus energy, increase in the photosystem II (PSII) quantum yield [52], and enhancement of the electron flux to the acceptor side of the photosystem I (PSI) [47]. A recent investigation conducted on *Kandelia obovata* showed that treatment with SNP under salt stress elevated endogenous NO levels, decreased ion toxicity, improved nutrient homeostasis and gas exchange parameters, and stimulated the activities of antioxidant enzymes [62].

Despite the numerous studies on the influence of NO on photosynthesis, knowledge about its defensive effects on the photosynthetic apparatus under salt stress is insufficient. The very harmful effect of salinity on the PSII complex is well known, which corresponds with the inhibition of the electron transport chain. We hypothesize that investigating the influence of SNP, as a donor of NO, at the donor and acceptor side of PSII will provide new information about the protective effect of NO under salt stress in maize plants. In the current research, we examined the impact of different concentrations of SNP on maize plants (*Zea mays* L. Kerala) cultivated in the presence of 150 mM NaCl (severe salinity). This study assessed the primary processes of photosynthesis (with PAM chlorophyll fluorescence and JIP test) and the PSII photochemical activity, revealing the impact of SNP under salt stress on the function of the photosynthetic apparatus. In addition, we measured the pigment composition, membrane stability, and levels of oxidative stress markers. The experimental results clearly show the protective mechanisms of SNP on the photosynthetic apparatus and its functions at applied SNP concentrations of 50 μM and 150 μM. These findings will contribute to offering promising information for developing strategies to increase crop resistance in saline environments.

2. Results

2.1. Pigment Content

The effects of the SNP (25–300 μM) on the pigment composition in maize leaves under salt stress (150 mM NaCl) are shown in Figure 1. The results demonstrate that the treatment with NaCl decreased the amount of chlorophylls (by 40%) and carotenoids (by 43%) (Figure 1). The simultaneous exposure of SNP and NaCl alleviated the salt-induced reduction in the pigment content, but the amounts remained lower than those of the control plants. The protection was better after application of 50 μM and 150 μM SNP in comparison to the lowest (25 μM) and the highest (300 μM) concentrations of SNP (Figure 1). The changes in the pigment composition caused a slight decrease in the Car/Chl ratio after treatment with 150 mM NaCl alone and co-treatment with NaCl and 300 μM SNP (Figure S1).

2.2. Stress Markers

The determination of lipid peroxidation (corresponding to MDA content) and the amount of H_2O_2 was used to evaluate the protective effects of SNP in maize plants under salt stress (150 mM NaCl) (Figure 2). Data showed that the amounts of MDA and H_2O_2 increased by about 64% and 62%, respectively, after treatment with 150 mM NaCl alone. The combined treatment with SNP and NaCl reduced the content of these stress markers (MDA and H_2O_2); however, these amounts were higher than the untreated plants (Figure 2a). The protective effect of SNP is less pronounced in plants treated with a concentration of 300 μM. The accumulation of H_2O_2 in leaves was also visualized histochemically by staining with diaminobenzidine (DAB), which forms a brown precipitate with H_2O_2 in a peroxidase-catalyzed reaction [63]. Data also revealed that H_2O_2 accumulation was in the whole leaf after 150 mM NaCl exposure, while SNP at concentrations of 25 to 150 μM significantly decreased this H_2O_2 accumulation (Figure 2b).

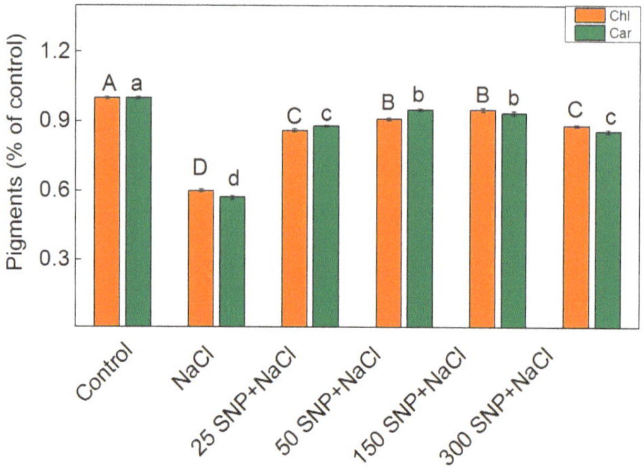

Figure 1. Impact of SNP on the amounts of chlorophylls (Chl) and carotenoids (Car) in maize (*Zea mays* L. Kerala) under salt stress. The control value for the chlorophylls is 49.498 mg/g DW and, for the carotenoids, it is 8.159 mg/g DW. The mean values (±SE) were determined from 8 measurements. Significant differences between variants at $p < 0.05$ are indicated by different letters (uppercase for chlorophyll and lowercase for carotenoid levels).

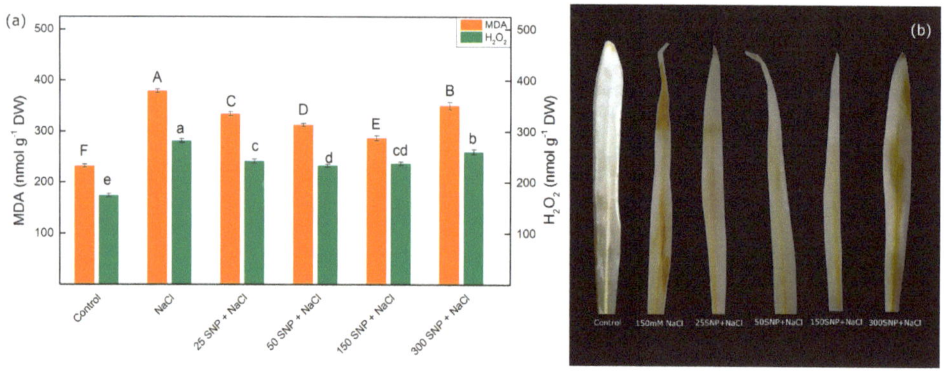

Figure 2. Effects of the SNP c on MDA and H_2O_2 contents under salt stress (150 mM NaCl) (**a**), and visualization of the H_2O_2 accumulation in maize leaves by DAB staining (**b**). Mean values (±SE) were determined from 8 measurements. Different letters indicate significant differences among variants at $p < 0.05$ (uppercase for MDA and lowercase for H_2O_2).

2.3. Membrane Stability Index and Membrane Fluidity

The membrane stability index (MSI) was used as an indicator for the impact of SNP on the membrane stability of maize leaves under salt stress. In comparison to the control, the 150 mM NaCl treatment alone reduced MSI by about 40%. The combined treatment with all studied SNP concentrations and NaCl increased the MSI compared to the NaCl treatment alone (Figure 3). The smallest protective effect was observed after co-treatment with 300 μM SNP and NaCl.

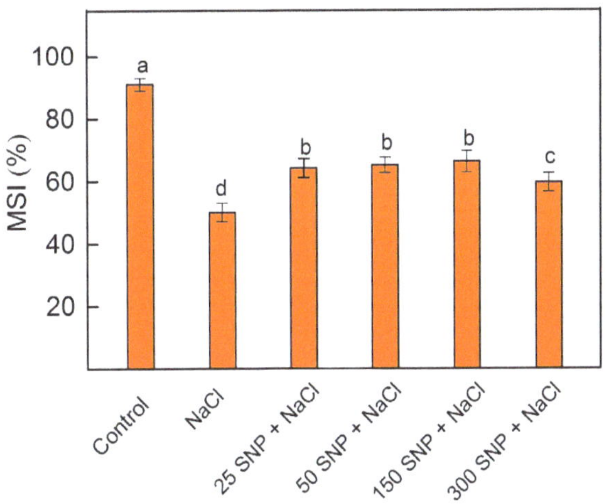

Figure 3. Effects of t SNP on the membrane stability index (MSI) of maize leaves (*Zea mays* L. Kerala). Mean values (±SE) were determined from 8 measurements. Different letters indicate significant differences among variants at $p < 0.05$.

The fluidity of isolated thylakoid membranes from all variants was evaluated by general polarization (GP) of a fluorescent lipophilic membrane dye Laurdan [64]. The experimental results reveal that the salt stress leads to a decrease in the GP value, i.e., an increase in the fluidity of thylakoid membranes. The SNP application fully prevents salt-induced changes in the membrane fluidity, as the GP values of Laurdan were similar to the thylakoid membranes from the control plants (Table 1).

Table 1. Impact of various SNP concentrations on Laurdan GP values and the low-temperature (77 K) fluorescence emission ratio $F_{735}/_{685}$ of isolated thylakoid membranes from leaves of maize (*Zea mays* L. Kerala) grown under salt conditions (150 mM NaCl). The chlorophyll fluorescence was excited at 436 nm. Statistically significant differences at $p < 0.05$ are marked by different letters among the mean values (±SE) in the corresponding column (n = 8).

Variants	GP	$F_{735/685}$
Control	0.478 ± 0.021 [a]	1.25 ± 0.06 [b]
NaCl	0.388 ± 0.018 [b]	1.43 ± 0.06 [a]
25 SNP + NaCl	0.451 ± 0.014 [a]	1.28 ± 0.05 [b]
50 SNP + NaCl	0.485 ± 0.017 [a]	1.28 ± 0.10 [b]
150 SNP + NaCl	0.465 ± 0.016 [a]	1.29 ± 0.06 [b]
300 SNP + NaCl	0.446 ± 0.014 [a]	1.33 ± 0.03 [b]

2.4. Energy Transfer between Pigment–Protein Complexes

Chlorophyll fluorescence spectra at a low temperature of 77 K were employed to assess the transfer of energy between pigment–protein complexes within the thylakoid membranes. The spectra of all studied variants were characterized with bands at 685 nm and 735 nm associated with the PSII complex and PSI complex, respectively [65]. The ratio F_{735}/F_{685} reflects the redistribution of energy between both photosystems. This ratio increased by 14% after treatment with NaCl alone, but its values were similar to the control after co-treatment with all studied SNP concentrations and NaCl (Table 1).

2.5. PAM Chlorophyll Fluorescence

The application of 150 mM NaCl to the maize affected several parameters of PAM chlorophyll fluorescence, including the quantum yield of photochemical to non-photochemical processes (Fv/Fo), the rate of photosynthesis (R_{Fd}), the excitation efficiency of open PSII centers (Φexc), the photochemical quenching (qP), and the excess excitation energy (EXC) (Figure 4). The treatment with NaCl alone led to a twofold reduction in the ratio of the quantum yield of photochemical to non-photochemical processes (Fv/Fo) and the photochemical quenching (qP). Some decrease in the rate of photosynthesis (R_{Fd}, by 24%) and the excitation efficiency of open PSII centers (Φexc, by 35%) was also observed. The SNP application diminished the salt-induced decrease in the Fv/Fo, qP, Φexc, and R_{Fd}. The values of parameters Φexc and R_{Fd} were similar to those of the untreated plants after co-treatment with NaCl and the concentrations of SNP up to 150 μM (Figure 4). Data also revealed that salt treatment increased the excess excitation energy (EXC) by 41% more than the control plants. The treatment with all studied SNP decreased EXC values (Figure S2).

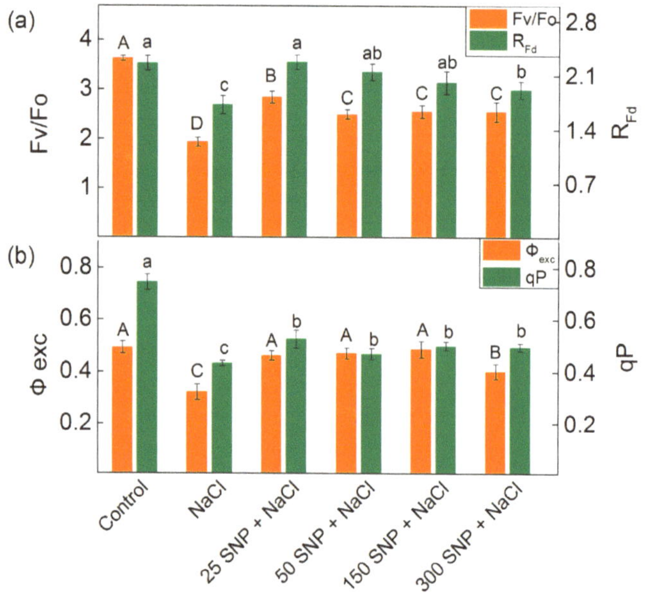

Figure 4. Impact of various SNP concentrations under salinity conditions (150 mM NaCl) on the PAM chlorophyll fluorescence parameters in maize (*Zea mays* L. Kerala). (**a**) Quantum yields of photochemical to non-photochemical processes (Fv/Fo), and the chlorophyll fluorescence decay ratio (R_{Fd}); (**b**) the excitation efficiency of open PSII centers (Φexc) and photochemical quenching (qP). Mean values ± SE were calculated from 8 independent measurements. The different letters indicate significant differences among variants at $p < 0.05$ (uppercase for Fv/Fo and Φexc, lowercase for R_{Fd} and qP).

The dark relaxation of chlorophyll fluorescence after a saturating light pulse in dark adapted leaves in both treated and untreated maize plants can be fitted by two components, with the amplitude A_1 (fast component) and A_2 (slow component) with rate constant k_1 and k_2, respectively. The constant k_1 decreased after treatment with 150 mM NaCl alone, while k_2 was slightly increased. After the co-treatment with SNP and NaCl, the constant k_1 and k_2 were similar to those of the control plants, except k_2 in plants treated with 25 μM SNP. The data revealed also that the ratio of two components (A_1/A_2) increases from 22% to 33% after co-treatment with SNP concentrations up to 150 μM and NaCl compared to the untreated plants (Table 2).

Table 2. Influence of various SNP concentrations under salt stress (150 mM NaCl) on the rate constants (k_1 and k_2) and the ratio of the amplitudes of the fast and slow components (A_1/A_2) of the relaxation of chlorophyll fluorescence after saturating light pulse in dark adapted leaves of maize (*Zea mays* L. Kerala). The different letters among the mean values (±SE) in the corresponding column (n = 8) show the statistical differences at $p < 0.05$.

Variants	k_1 (s^{-1})	k_2 (s^{-1})	A_1/A_2
Control	1.741 ± 0.045 ^{a}	0.085 ± 0.007 ^{b}	6.731 ± 0.230 ^{b}
150 mM NaCl	1.578 ± 0.049 ^{b}	0.107 ± 0.010 ^{a}	6.263 ± 0.221 ^{c}
25 μM SNP + NaCl	1.826 ± 0.099 ^{a}	0.104 ± 0.005 ^{a}	8.929 ± 0.341 ^{a}
50 μM SNP + NaCl	1.895 ± 0.140 ^{a}	0.090 ± 0.008 ^{b}	8.184 ± 0.321^a
150 μM SNP + NaCl	1.903 ± 0.079 ^{a}	0.094 ± 0.008 ^{b}	8.392 ± 0.503 ^{a}
300 μM SNP + NaCl	1.739 ± 0.075 ^{a}	0.085 ± 0.006 ^{b}	6.974 ± 0.409 ^{b}

2.6. Chlorophyll Fluorescence Induction

Selected parameters of the chlorophyll fluorescence induction used to investigate the effects of SNP under salt stress in maize were as follows: ψEo—efficiency of the electron transfer further than Q_A^-; N—maximum turnover of Q_A reduction until Fm reached (corresponding with the size of the plastoquinone pool); Vj—variable fluorescence at the J-step (corresponding with changes in the PSII acceptor side); φPo—maximum quantum yield for primary photochemistry; φRo—quantum yield for reduction of end electron acceptors at the PSI acceptor side; δRo—efficiency with which an electron from the intersystem electron carriers is transferred to reduce end electron acceptors at the PSI acceptor side; PI_{ABS}—performance index for energy conservation from photons absorbed by PSII to the reduction of intersystem electron acceptors; PItotal—performance index for energy conservation from photons absorbed by PSII until the reduction of PSI end electron acceptors; RC/DIo—the reversed parameter of DIo/RC, corresponding with dissipated energy flux per RC; Wk—ratio of the J step to K step, corresponding with the changes in the PSII donor side. Data showed that NaCl treatment leads to a slight increase in the parameters Wk, Vj, and δRo, while the parameters ψEo, N, φPo, φRo, RC/DIo, PI_{ABS}, and PItotal decreased. The effects were more pronounced for parameters Vj, RC/DIo, PI_{ABS}, and PItotal. The foliar application of SNP under salt stress decreased the effects of NaCl on the selected JIP parameters, as the effects were better at 50 μM and 150 μM SNP (Figure 5).

2.7. Photochemical Activity of PSII and Flash Oxygen Evolution

The assessment of the PSII-mediated electron transport, with the electron acceptor BQ ($H_2O \rightarrow BQ$), was conducted to evaluate the photochemical activity of PSII. Data revealed that NaCl treatment, inhibited the PSII-mediated electron transport by 36% (Figure 6). The application of SNP alleviated the impact of NaCl on the PSII activity. This activity was the same as in the control plants after application of concentrations of 50 μM and 150 μM SNP (Figure 6).

The analysis of the flash-induced oxygen yields showed that the active PSII centers in the initial S_0 state (S_0 % = 100 − S_1) and the misses (α) increased significantly after applying salt stress (Table 3). On the other hand, the SNP foliar application mitigated the salt-induced alterations in these kinetic parameters (S_0 and α) (Table 3). The double hits (β) showed no statistically significant differences during all applied treatments.

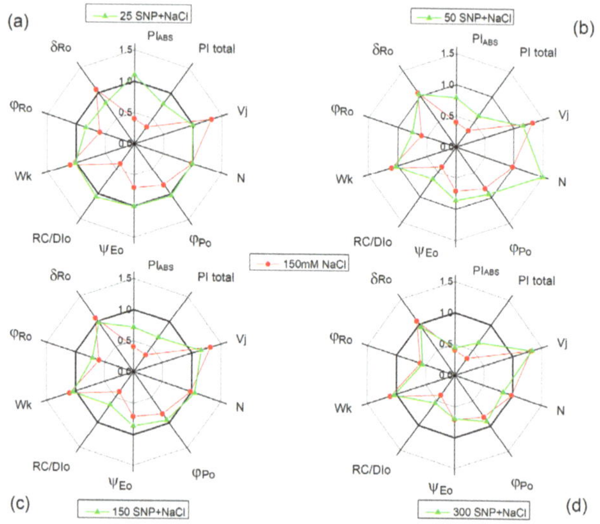

Figure 5. The influence of SNP in the presence of the 150 mM NaCl on the selected JIP parameters in maize (*Zea mays* L. Kerala). Red lines show the influence of 150 mM NaCl. Green lines show the influence of different concentrations of SNP under salt stress: (**a**) 25 μM SNP and 150 mM NaCl; (**b**) 50 μM SNP and 150 mM NaCl; (**c**) 150 μM SNP and 150 mM NaCl; (**d**) 300 μM SNP and 150 mM NaCl. ψEo—efficiency/probability that an electron moves further than Q_A^-; N—maximum turnover of Q_A reduction until Fm reached; Vj—variable fluorescence at the J-step (2 ms); φPo—maximum quantum yield for primary photochemistry; φRo—quantum yield for reduction of end electron acceptors at the PSI acceptor side; δRo—efficiency with which an electron from the intersystem electron carriers is transferred to reduce end electron acceptors at the PSI acceptor side (RE); PI_{ABS}—performance index for energy conservation from photons absorbed by PSII until the reduction of intersystem electron acceptors; PItotal—performance index for energy conservation from photons absorbed by PSII until the reduction of PSI end electron acceptors; RC/DIo—the reversed parameter of DIo/RC—dissipated energy flux per RC (at t = 0); Wk—ratio of the J step to K step. All parameters are normalized to the parameters of the untreated plants.

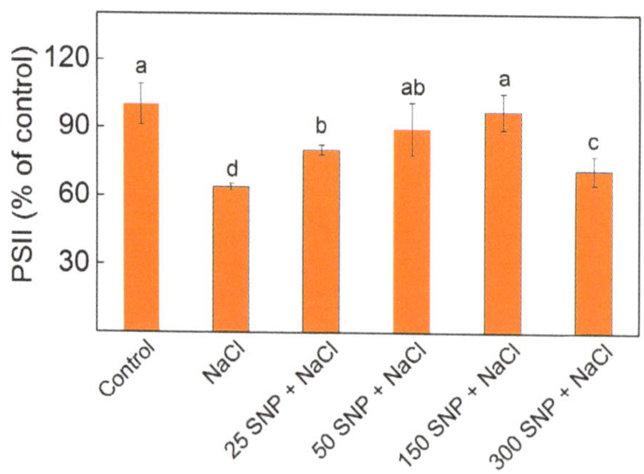

Figure 6. Photochemical activity of PSII ($H_2O \rightarrow BQ$) in isolated thylakoid membranes from leaves of maize (*Zea mays* L. Kerala) after exposure to various SNP concentrations and 150 mM NaCl. The values are expressed as a percentage of the respective control. Significant differences between variants at $p < 0.05$ are indicated by different letters.

Table 3. Effects of 150 mM NaCl treatment and different SNP concentrations on the kinetic parameters of the flash-induced oxygen yields: S_0—the PSII centers in the initial reduced state (S_0 % = 100 − S_1) in the darkness, misses (α), and double hits (β). Significant differences between variants at $p < 0.05$ are indicated by different letters.

Variants	S_0 (%)	α (%)	β (%)
Control	24.5 ± 1.3 [b]	20.3 ± 1.8 [c]	6.2 ± 1.3 [a]
150 mM NaCl	34.1 ± 2.7 [a]	28.8 ± 1.5 [a]	7.5 ± 1.7 [a]
25 SNP + NaCl	25.3 ± 1.2 [b]	24.3 ± 2.1 [b]	6.8 ± 1.4 [a]
50 SNP + NaCl	23.3 ± 1.7 [b]	25.7 ± 1.2 [b]	5.9 ± 1.8 [a]
150 SNP + NaCl	23,1 ± 1.9 [b]	24.3 ± 2.5 [b]	6.4 ± 1.7 [a]
300 SNP + NaCl	24.1 ± 1.4 [b]	25.3 ± 1.3 [b]	6.9 ± 1.4 [a]

2.8. NO Content

The data indicated that the application of NaCl led to an increase in the NO amount by 17% when compared to the untreated plants (Figure 7). The co-treatment with SNP under salt stress caused an additional rise in NO content depending on the applied SNP concentration (Figure 7).

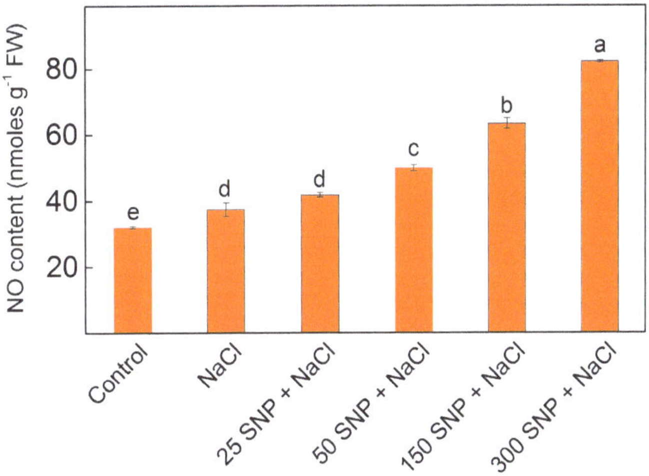

Figure 7. NO levels in leaves of maize (*Zea mays* L. Kerala) after treatment with different SNP concentrations in the presence of 150 mM NaCl. Significant differences between treatments at $p < 0.05$ are denoted by different letters.

2.9. Principal Component Analysis

Principal component analysis (PCA) revealed that the first two components explain 99.43% of the data variability (Figure S3). The control maize, positioned in the first quadrant, shows a negative correlation with the EXC parameter describing the dissipation of excess energy, which is located in the third quadrant. A strong positive correlation was determined for the parameters EXC and $F_{735/685}$ and the maize treated with 150 mM NaCl, located in the lowest point of the second quadrant. A more pronounced positive correlation of the photochemical to non-photochemical processes (Fv/Fo) (first quadrant, far from PC1 and PC2 axes) and a weaker one for the photosynthetic rate (R_{Fd}) (second quadrant near the PC1 axis) was found with respect to the control maize in comparison to the other plant variants (150 mM NaCl and 150 μM SNP + 150 mM NaCl), located in the second quadrant. The variables located in the fourth quadrant are related to pigment content (Car, Chl, Car/Chl), membrane fluidity (Laurdan GP), and photochemical activity (qP, Φexc), and have an insignificant contribution to the changes occurring in all the maize variants, with a weakly pronounced positive correlation with the control maize plants and a negative correlation with the maize treated with 150 mM NaCl alone.

3. Discussion

Despite extensive research, the precise role of NO in plant survival under adverse environmental conditions has not been fully understood, particularly regarding its defensive function related to the operation of various pigment–protein complexes in photosynthetic membranes. In this study, we reveal new evidence regarding the impact of SNP, acting as a NO donor, on different components of the photosynthetic apparatus under salt stress.

Data in this study demonstrated a reduction in photosynthetic pigments (Chl and Car) similar to previous studies on wheat, sorghum, pea, barley, and other plant species [66–71]. This negative impact on the pigments is due to an inhibition of the pigment biosynthesis and/or an enhancement of their degradation [72]. The protective effect of 50 μM and 150 μM SNP on the photosynthetic pigments (Figure 1) may result from stimulating Chl and Car biosynthesis observed previously [34]. Additionally, SNP could counteract the salt-induced negative effect on their biosynthesis, which is accompanied by enhancement of the LHCII accumulation [55,73]. The data also demonstrated that foliar spray with SNP prevents, under salt stress, the decrease in the Car/Chl ratio at concentrations up to 150 μM SNP (Figure S1). Considering, the important role of carotenoids as antioxidants involved in the protection of thylakoid membranes from oxidative stress at a high salinity [74], it could be suggested that their enhancement is also one of the protective mechanisms of SNP under salt stress.

The enhancement of the H_2O_2 content (Figure 2) and other ROS species caused lipid peroxidation and disruption of the membrane structures [75,76]. The high accumulation of the MDA under salt stress revealed an enhancement of the lipid peroxidation (Figure 2). In addition to the lipid peroxidation, the salt stress also leads to changes in the lipid composition [77]. The salt-induced changes in the lipids and the proteins cause structural changes in the thylakoid membranes, corresponding with the reduction in thylakoids in grana regions [14,16,78], which is accompanied by a decrease in the MSI (Figure 3).

Furthermore, the observed decrease in Laurdan GP values for the thylakoid membranes isolated from NaCl-treated plants compared to controls (Table 1) clearly indicates an increase in membrane fluidity. It has been suggested that the more fluid thylakoid membranes may facilitate the diffusion of the LHCII complex from PSII to PSI complexes [79,80], which is confirmed by an increase in the fluorescence emission ratio (F_{735}/F_{685}) after NaCl treatment (Table 1). Our results reveal that the application of the SNP not only decreased the salt-induced oxidative stress (i.e., changes in the amounts of H_2O_2 and MDA) but also prevented the membrane structural alterations, such as changes in the membrane fluidity and energy transfer between both photosystems (Figure 3 and Table 1). In addition, current data demonstrated that SNP foliar spray (at concentrations of 25 to 150 μM) alleviated salt-induced oxidative stress to a certain extent. This reduction was evident through the decreased amounts of H_2O_2 and MDA in salt-stressed leaves, thus preventing salt-induced damage of the membranes (Figure 3).

Previous studies [7,12,17,30,81–84], as well as the data in the present study, revealed a strong influence of salinity on the function of the photosynthetic apparatus. The analysis of the chlorophyll fluorescence curves at room temperature demonstrated that the salt treatment led to a decrease in the ratio of the photochemical to non-photochemical processes (Fv/Fo), the photochemical quenching (qP), the excitation efficiency of open PSII centers (Φexc), and the rate of photosynthesis (R_{Fd}) (Figure 4). This impact on the PSII function is due to changes in the acceptor and donor side of the complex [30,85,86]. At the same time, the parameter EXC showed an increase in the energy losses (Figure S2), which was accompanied with a decrease in the efficiency of PSII under NaCl treatment alone (Figure 4). The salt treatment also led to a decrease in the parameter RC/DIo (the reversed parameter of the dissipated energy flux per RC, DIo/RC), showing an increase in the dissipated energy

(Figure 5). Dissipation of the excess light corresponds with a decrease in ROS formation, acting as a photoprotective mechanism [87].

To assess the salt-induced changes and the protective mechanisms of SNP on the PSII acceptor side, we analyzed the chlorophyll fluorescence signals following a saturating light pulse [88,89]. Data revealed that the alterations in the PSII functions were a result of salt-induced impacts on both pathways of Q_A^- reoxidation: one involving plastoquinone and the other via recombination of electrons in Q_A Q_B^- with oxidized S_2 (or S_3) states of the OEC (Table 2). However, the influence on the Q_A functions decreased the efficiency of the electron movement further than Q_A^- (ψEo parameter) and the probability that an electron from the intersystem electron carriers is transferred to reduce end electron acceptors at the PSI acceptor side (Figure 5). The performance index on the absorption base (PI_{ABS}), commonly used to assess overall PSII performance [83,90], was strongly influenced by NaCl (Figure 5), corresponding to the inhibition of PSII-mediated electron transport (Figure 6). Moreover, PItotal was also reduced under salt stress, which suggests delayed performance from the PSII electron donor side to the reduction of the PSI end electron acceptors.

In SNP-treated plants under salt stress, Q_A interaction with plastoquinone prevailed, resulting in an increased A_1/A_2 ratio (Table 2). These changes were associated with an increase in open PSII centers (qP) and their efficiency (Φexc), improving the photosynthetic performance (PI_{ABS}) and performance of the electron transport reduction of the PSI end electron acceptors along with stimulation of the photosynthesis rate (R_{Fd}) (Figures 4 and 5). Simultaneously, there was an elevation in excess excitation energy (EXC) (Figure S2). All these observations corresponded with the inhibition of PSII-mediated electron transport in the presence of the exogenous acceptor BQ, after NaCl treatment alone and with full protection, at applied concentrations of 50 μM and 150 μM SNP (Figure 6).

More detailed information regarding the influence of NaCl on the PSII donor side was obtained by analyzing kinetic parameters of the flash-induced oxygen evolution without exogenous acceptors, i.e., electrons are accepted from the plastoquinone (PQ) (Table 3). The salt treatment resulted in changes in the initial S_0–S_1 state distribution due to an increase in the number of active PSII centers in the most reduced S_0 states (Mn^{2+}, Mn^{3+}, Mn^{4+}, Mn^{4+}). This increase indicates a modification in the Mn_4Ca cluster within the OEC [91,92]. A similar influence of salt stress on PSII centers in the initial S_0 state in darkness has also been shown in barley plants [30,93]. The observed increase in the S_0 state corresponds with an increase in the misses (α). The application of SNP under salt stress fully prevented the salt-induced alterations in the initial S_0–S_1 state distribution of the PSII complexes. Therefore, the SNP protection on the PSII donor side is most probably due to the prevention of the salt-induced modification of the Mn_4Ca cluster of OEC.

These effects of NO on the membrane integrity and function of the photosynthetic apparatus could also result from direct neutralization of the ROS, enhancement of antioxidant enzyme activities, and increased accumulation of osmolyte compounds [38]. The application of SNP protected against membrane damage, as well as the salt-induced changes in membrane fluidity, Q_A reoxidation, and modification of the Mn clusters of the OEC. By preventing the salt-induced changes in both donor and acceptor sides of PSII, the PSII performance and overall function of the photosynthetic apparatus were improved. The data also revealed that better protection of the thylakoid membranes is achieved at a concentration of NO up to 63 nmoles/g FW (at 150 μM SNP).

4. Materials and Methods

4.1. Plant Growth Conditions and Treatment

The seeds of maize (*Zea mays* L. Kerala) were obtained from Euralis Ltd. in Lescar, France. The plants were cultivated hydroponically in a climate chamber under controlled conditions: 25 °C (daily)/22 °C (night) temperature, 150 μmol photons/m^2 s light intensity, a 12 h light/dark photoperiod, and 70% air humidity. Two-week-old maize plants were foliar-sprayed with different concentrations of SNP (25 μM, 50 μM, 150 μM, and 300 μM) 24 h before the addition of 150 mM NaCl in the Hoagland solution. The solution of SNP can be released as free nitric oxide (NO) or NO^+ and free CN^- or CN radicals [53,94]. The measurements were performed 5 days after addition of NaCl in nutrient solution. The concentrations of NaCl and SNP are based on a preliminary study [34,47]. The following variants were studied: control (without SNP and NaCl), NaCl (150 mM NaCl), 25 SNP + NaCl (25 μM SNP and 150 mM NaCl), 50 SNP + NaCl (50 μM SNP and 150 mM NaCl), 150 SNP + NaCl (150 μM SNP and 150 mM NaCl), 300 SNP + NaCl (300 μM SNP and 150 mM NaCl). Two independent experiments were conducted in four replications for each variant (four boxes with three plants in each). The measurements were made on the fully expanded leaves.

4.2. *Leaf Pigment Content*

The method of Lichtenthaler (1987) was used to determine the amount of chlorophylls and carotenoids. Leaf tissue (0.03 g) was cut into small pieces and grinded with 8 mL of 80% acetone in dark and cold. After centrifugation at 4500× *g* for 10 min at 0–4 °C, the supernatant was measured spectrophotometrically (Specord 210 Plus, Ed. 2010, Analytik Jena AG, Jena, Germany) at 663.2, 646.8, and 470 nm and the pigment content totals Chl and Car were calculated using Lichtenthaler's equations [95].

4.3. *Stress Markers, Membrane Stability Index*

The fully expanded leaves were taken from different variants to estimate the content of hydrogen peroxide (H_2O_2) and malondialdehyde (MDA) following the procedure described in [34]. The H_2O_2 and MDA amounts were calculated by recording the absorbance at 390 nm and 532 nm, respectively, using Specord 210 Plus (Edition 2010; Analytik Jena AG, Jena, Germany), and the values are expressed as nmol per g DW.

The visualization of the H_2O_2 accumulation in maize leaves was performed with the dye diaminobenzidine (DAB) because peroxidase catalyzed the reaction of DAB with H_2O_2 to form a brown polymer as previously described in [63]. Several fresh leaves were soaked in DAB solution (1 mg/mL) and incubated at room temperature overnight in the dark. The leaves were then placed in boiling ethanol (95%) to remove the background.

The membrane stability index (MSI) for maize leaves was evaluated based on the electrolyte conductivity (EC) as described previously in [34].The MSI values were calculated as MSI (%) = [1 − (EC1/EC2)] × 100, where EC1 and EC2 are the measured electrolyte conductivities of the leaf sample solutions after incubation for 24 h at 20 °C and after boiling for 30 min, respectively.

4.4. *Thylakoid Membrane Fluidity*

The fluidity of thylakoid membranes was followed by a fluorescence polarization study with a fluorescent lipophilic dye Laurdan (6-Dodecanoyl-2-dimethylaminonaphthalene, Sigma-Aldrich, St. Louis, MO, USA) as described previously in [64,96]. The isolated thylakoid membranes with a concentration of 15 μg Chl/mL were incubated with 30 μM Laurdan, using 1 mM stock solution dissolved in dimethyl sulfoxide, (DMSO, Sigma-Aldrich, St. Louis, MO, USA) for 30–40 min at room temperature in the dark. The steady-state fluorescence polarization was determined using a spectrofluorometer JASCO FP8300 (Jasco, Tokyo, Japan). Fluorescence was excited at 390 nm and registered at 460 and 515 nm with a 10 nm emission slit width using a quartz cuvette of 1 cm path length according to [64]. The general polarization (GP) of Laurdan fluorescence was determined as $GP = (I_{460} - I_{515})/(I_{460} + I_{515})$, where I_{460} is the fluorescence intensity at 460 nm (characteristic for tightly packed membrane lipids) and I_{515} is the fluorescence intensity at 515 nm (characteristic for less tightly packed lipids). The lower GP values point to an increased fluidity, i.e., membranes with a less ordered fluid phase [64].

4.5. *Low-Temperature (77 K) Fluorescence Measurements*

Low-temperature (77K) chlorophyll fluorescence emission spectra were obtained using a spectrofluorometer (Jobin Yvon JY3, Division d'Instruments S.A., Longjumeau, France) equipped with a nitrogen device. The samples were quickly frozen in liquid nitrogen. The measurements were made as in [30]. Data analysis and graphing software (Origin version 9.0, OriginLab Corporation, Northampton, MA, USA) was employed to analyze emission spectra registered after the excitation of Chl *a* (at 436 nm). The spectra of all studied variants were characterized with two fluorescence bands at 685 nm (for PSII) and 735 nm (for PSI). The fluorescence emission ratio $F_{735/685}$ was calculated. This ratio characterized the energy redistribution between both photosystems [30,65].

4.6. Room-Temperature Chlorophyll Fluorescence

Pulse-modulated amplitude (PAM) chlorophyll fluorescence was measured on dark-adapted (for 15 min) leaves using a fluorimeter (H.Walz, Effeltrich, Germany, model PAM 101–103). The measurements were made as in [97]. The minimal fluorescence level (Fo) was recorded at a frequency of 1.6 kHz and a measuring light of 0.110 µmol photons/m^2s PFD. The maximal fluorescence levels for the dark-adapted state (Fm) and light-adapted state (Fm′) were determined using saturated pulse light of 3000 photons µmol/m^2s for 0.8 s. The photosynthetic process was triggered by exposing the plants to actinic light with an intensity of 150 µmol photons/m^2s. The PAM chlorophyll fluorescence parameters calculated to assess the impact of SNP under salt stress on the function of the photosynthetic apparatus functions are as follows [98]: Fv/Fo—the ratio of photochemical to non-photochemical processes; qP—the coefficient of photochemical quenching [98]; Φexc—the excitation efficiency of open PSII centers [99]; and EXC—the excitation excess energy [100]. The parameter R_{Fd}—the chlorophyll fluorescence decay ratio after saturating light pulse (3000 µmol photons/m^2s) in dark-adapted leaves—was determined as described in [84].

The additional information about the effects of SNP under salt stress on the PSII complex gives the dark relaxation of chlorophyll fluorescence by a signal after saturating light pulse in dark-adapted leaves. Analysis of the curves provides details about the electron transfer from Q_A to plastoquinone [29]. Fluorescence signals can be fitted by two components. The ratio of the fast (A_1) and slow (A_2) components (A_1/A_2) and rate constants k_1 and k_2 were determined as in [101]. The constants k_1 and k_2 are related to Q_A—reoxidation pathways [101].

Chlorophyll fluorescence induction curves were obtained and measured with a Handy PEA+ device (Hansatech, Norfolk, UK), as described in [47]. The leaves were adapted in dark for 30 min. The light pulse intensity was 3000 µmol photons/m^2s. The following JIP parameters were determined [29]: PI_{ABS} and PItotal—the performance indexes, Vj—relative variable fluorescence at the J step, N—maximum turnovers of Q_A reduction until Fm was reached, ϕ_{Po}—the maximum quantum yield of primary photochemistry, ψ_{Eo}—moves an electron into the electron transport chain beyond Q_A^-, RC/DIo—the reversed parameter of DIo/RC—dissipated energy flux per RC (at t = 0), Wk—the ratio of K phase to J phase, ϕ_{Ro}—quantum yield of reduction of end electron acceptors at the PSI acceptor side, δ_{Ro}—efficiency with which/probability that an electron from the intersystem electron carriers moves to reduce end electron acceptors at the PSI acceptor side. Fully developed leaves (the middle area of the third and fourth leaves) were used for all fluorescence analyses.

4.7. Isolation of Thylakoid Membranes

Thylakoid membranes were isolated from leaves of maize plants as described in [102]. For measurements, the thylakoid membranes were suspended in 20 mM HEPES (pH 7.6), 0.4 M sucrose, 5 mM $MgCl_2$, 10 mM NaCl. The Chl content in thylakoid membranes was extracted with 80% (*v*/*v*) acetone and was assessed using Lichtenthaler's equations [95].

4.8. Photochemical Activity of PSII

The photochemical activity of PSII was assessed for the PSII-mediated electron transport as in [30]. The measurements were made on the isolated thylakoid membranes (25 µg Chl/mL) using a Clark-type electrode (Model DW1, Hansatech, Instruments Ltd., Norfolk, UK). The reaction medium for PSII-mediated electron transport ($H_2O \rightarrow BQ$)

contained 20 mM MES (pH 6.5), 0.4 M sucrose, 5 mM $MgCl_2$, 10 mM NaCl, 0.4 mM BQ (1,4-benzoquinone).

4.9. Oxygen Evolution Measurements

The flash-induced oxygen yields under short flash illumination by saturating (4 J) and short (10 s) periodic flash sequences applied on the thylakoid membranes' suspension (150 μg Chl/mL) were determined using a polarographic oxygen rate electrode (Joliot-type) without the addition of artificial electron acceptors as described previously [91]. The analysis of the flash-induced oxygen yields was performed using the model of Kok [103]. According to this model, the cooperation of five oxidation states of OEC (S_0–S_4) in the same PSII centers is required for the production of one oxygen molecule. In the darkness, only the S_0 and S_1 states are stable. The calculations of the used parameters were made as described in [91]: S_0—percentage of active oxygen-evolving PSII centers in the most reduced (S_0) state in darkness, i.e., the initial S_0–S_1 state distribution ($S_0 + S_1 = 100\%$), misses (α), and double hits (β).

4.10. NO Content

The NO content was determined following the method of [104]. The leaves were homogenized in an acetic acid buffer with low pH supplemented with zinc acetate. After centrifugation, the supernatant was neutralized with charcoal, followed by the addition of Griess reagent. The NO content (nmol/g FW) was determined using a calibration curve generated with sodium nitrite as a standard.

4.11. Principal Component Analysis

Principal component analysis (PCA), a multivariate statistical method, was employed to reduce a vast array of measured parameters into the most informative ones [105]. PCA was utilized to explore the impact of SNP under salt stress (150 mM NaCl) on the fluorescence parameters and their correlations with bio-chemical parameters. For categorizing/classifying the variations in response to salt stress and SNP application, a clustering algorithm was implemented [106]. PCA multivariate statistical analysis was conducted, and graphical representations of PCA were generated using Originlab 9 software for data analysis and graphing (OriginLab Corporation, Northampton, MA, USA).

4.12. Statistical Analysis

Differences among the various t were assessed by Student's *t*-test or one-way ANOVA with post hoc Tukey's test. The mean values (n = 8) were considered statistically significant at least for $p < 0.05$.

5. Conclusions

In summary, the data in the present study revealed the protection of thylakoid membranes in maize after foliar spray with SNP under salt stress. The protection is due to a decrease in the stress markers (H_2O_2 and MDA) preventing the salt-induced changes in membrane fluidity and energy transfer between the pigment–protein complexes within the photosynthetic apparatus. The effects of SNP were accompanied by an increase in the PSII open reaction center and tier efficiency, as well as the prevention of salt-induced alterations on both the donor and acceptor sides of PSII. The data demonstrated that co-treatment with SNP and NaCl leads to a decrease in salt-induced changes in the rate constants of two pathways of Q_A reoxidation: one involving plastoquinone and the other involving recombination on Q_AQ_B- with oxidized S_2 (or S_3) states of the OEC. Moreover, the application of SNP under salt stress predominantly favors reoxidation through interaction with plastoquinone. The application of the SNP also prevented the modification of the Mn clusters of the OEC at high salt concentrations and improved the oxygen-evolving activity. The data also revealed that SNP provided better protection under salt stress at concentrations between 50 μM and 150 μM or an amount of NO equivalent to 50–63 nmoles/g FW in leaves.

Supplementary Materials: The following supporting information can be downloaded at: https://www.mdpi.com/article/10.3390/plants13101312/s1, Figure S1. Impact of SNP on the total chlorophyll-to-carotenoid ratio of maize (*Zea mays* L. Kerala) under salt stress. Figure S2. Effects of different SNP levels on excess excitation energy (EXC) in the leaves of one maize variety (*Zea mays* L. Kerala) during salt stress. Figure S3. Principal component analysis (PCA) shows variation in the selected parameters after treatment with NaCl alone and co-treatment SNP and NaCl. Table S1. Variable contributions (loadings) for the principal component analysis model in Figure S1.

Author Contributions: Conceptualization, E.L.A.; methodology, M.A.S., G.D.R., E.K.Y., P.B.B., A.G.D. and E.L.A.; software, M.A.S.; validation, E.L.A.; formal analysis, E.L.A.; investigation, M.A.S., G.D.R., E.K.Y., P.B.B. and A.G.D.; writing—original draft preparation, E.L.A.; writing—review and editing, G.D.R., M.A.S., A.G.D. and E.L.A.; visualization, G.D.R. and M.A.S.; supervision and project administration, E.L.A. All authors have read and agreed to the published version of the manuscript.

Funding: This research was funded by the Bulgarian Science Research Fund, КП-06-Н36/9, 2019.

Data Availability Statement: Data are contained within the article and Supplementary Materials.

Conflicts of Interest: The authors declare no conflicts of interest.

References

1. Munns, R.; Gilliham, M. Salinity tolerance of crops—What is the cost? *New Phytol.* **2015**, *208*, 668–673. [CrossRef] [PubMed]
2. Shrivastava, P.; Kumar, R. Soil salinity: A serious environmental issue and plant growth promoting bacteria as one of the tools for its alleviation. *Saudi J. Biol. Sci.* **2015**, *22*, 123–131. [CrossRef] [PubMed]
3. Trifunović-Momčilov, M.; Stamenković, N.; Đurić, M.; Milošević, S.; Marković, M.; Giba, Z.; Subotić, A. Role of sodium nitroprusside on potential mitigation of salt stress in centaury (*Centaurium erythraea* Rafn) shoots grown in vitro. *Life* **2023**, *13*, 154. [CrossRef] [PubMed]
4. Bose, J.; Munns, R.; Shabala, S.; Gilliham, M.; Pogson, B.; Tyerman, S.D. Chloroplast function and ion regulation in plants growing on saline soils: Lessons from halophytes. *J. Exp. Bot.* **2017**, *68*, 3129–3143. [CrossRef] [PubMed]
5. Hao, S.; Wang, Y.; Yan, Y.; Liu, Y.; Wang, J.; Chen, S. A review on plant responses to salt stress and their mechanisms of salt resistance. *Horticulturae* **2021**, *7*, 132. [CrossRef]
6. Hasanuzzaman, M.; Raihan, M.R.H.; Masud, A.A.C.; Rahman, K.; Nowroz, F.; Rahman, M.; Nahar, K.; Fujita, M. Regulation of reactive oxygen species and antioxidant defense in plants under salinity. *Int. J. Mol. Sci.* **2021**, *22*, 9326. [CrossRef] [PubMed]
7. Hameed, A.; Ahmed, M.Z.; Hussain, T.; Aziz, I.; Ahmad, N.; Gul, B.; Nielsen, B.L. Effects of salinity stress on chloroplast structure and function. *Cells* **2021**, *10*, 2023. [CrossRef] [PubMed]
8. Ozden, M.; Demirel, U.; Kahraman, A. Effects of proline on antioxidant system in leaves of grapevine (*Vitis vinifera* L.) exposed to oxidative stress by H_2O_2. *Sci. Hortic.* **2009**, *119*, 163–168. [CrossRef]
9. Banu, M.N.A.; Hoque, M.A.; Watanabe-Sugimoto, M.; Matsuoka, K.; Nakamura, Y.; Shimoishi, Y.; Murata, Y. Proline and glycinebetaine induce antioxidant defense gene expression and suppress cell death in cultured tobacco cells under salt stress. *J. Plant Physiol.* **2009**, *166*, 146–156. [CrossRef]
10. ÇELİK, Ö.; ATAK, Ç. The effect of salt stress on antioxidative enzymes and proline content of two Turkish tobacco varieties. *Turk. J. Biol.* **2012**, *36*, 339–356. [CrossRef]
11. Papadakis, I.E.; Sotiras, M.I.; Landi, M.; Ladikou, E.; Oikonomou, A.; Psychoyou, M.; Fasseas, C. Salinity alters plant's allometry and sugar metabolism, and impairs the photosynthetic process and photosystem II efficiency in Eriobotrya japonica plants. *Agrochimica* **2019**, *63*, 27–42. [CrossRef]
12. Stefanov, M.; Biswal, A.K.; Misra, M.; Misra, A.N.; Apostolova, E.L. Responses of Photosynthetic Apparatus to Salt Stress: Structure, Function, and Protection. In *Handbook of Plant and Crop Stress*, 4th ed.; Pessarakli, M., Ed.; Taylor & Francis CRC Press: New York, NY, USA, 2019; pp. 233–250, ISBN 9781351104609.
13. Morales, F.; Ancín, M.; Fakhet, D.; González-Torralba, J.; Gámez, A.L.; Seminario, A.; Soba, D.; Ben Mariem, S.; Garriga, M.; Aranjuelo, I. Photosynthetic metabolism under stressful growth conditions as a bases for crop Breeding and yield improvement. *Plants* **2020**, *9*, 88. [CrossRef] [PubMed]
14. Barhoumi, Z.; Djebali, W.; Chaïbi, W.; Abdelly, C.; Smaoui, A. Salt impact on photosynthesis and leaf ultrastructure of Aeluropus littoralis. *J. Plant Res.* **2007**, *120*, 529–537. [CrossRef] [PubMed]
15. Bejaoui, F.; Salas, J.J.; Nouairi, I.; Smaoui, A.; Abdelly, C.; Martínez-Force, E.; Youssef, N. Ben Changes in chloroplast lipid contents and chloroplast ultrastructure in Sulla carnosa and Sulla coronaria leaves under salt stress. *J. Plant Physiol.* **2016**, *198*, 32–38. [CrossRef] [PubMed]

16. Goussi, R.; Manaa, A.; Derbali, W.; Cantamessa, S.; Abdelly, C.; Barbato, R. Comparative analysis of salt stress, duration and intensity, on the chloroplast ultrastructure and photosynthetic apparatus in Thellungiella salsuginea. *J. Photochem. Photobiol. B Biol.* **2018**, *183*, 275–287. [CrossRef] [PubMed]
17. Shu, S.; Guo, S.R.; Sun, J.; Yuan, L.Y. Effects of salt stress on the structure and function of the photosynthetic apparatus in *Cucumis sativus* and its protection by exogenous putrescine. *Physiol. Plant.* **2012**, *146*, 285–296. [CrossRef] [PubMed]
18. Muhammad, I.; Shalmani, A.; Ali, M.; Yang, Q.-H.; Ahmad, H.; Li, F.B. Mechanisms regulating the dynamics of photosynthesis under abiotic stresses. *Front. Plant Sci.* **2021**, *11*, 615942. [CrossRef] [PubMed]
19. Ji, X.; Cheng, J.; Gong, D.; Zhao, X.; Qi, Y.; Su, Y.; Ma, W. The effect of NaCl stress on photosynthetic efficiency and lipid production in freshwater microalga—*Scenedesmus obliquus* XJ002. *Sci. Total Environ.* **2018**, *633*, 593–599. [CrossRef] [PubMed]
20. Dąbrowski, P.; Baczewska, A.H.; Pawluśkiewicz, B.; Paunov, M.; Alexantrov, V.; Goltsev, V.; Kalaji, M.H. Prompt chlorophyll a fluorescence as a rapid tool for diagnostic changes in PSII structure inhibited by salt stress in Perennial ryegrass. *J. Photochem. Photobiol. B* **2016**, *157*, 22–31. [CrossRef]
21. Caruso, G.; Cavaliere, C.; Guarino, C.; Gubbiotti, R.; Foglia, P.; Laganà, A. Identification of changes in Triticum durum L. leaf proteome in response to salt stress by two-dimensional electrophoresis and MALDI-TOF mass spectrometry. *Anal. Bioanal. Chem.* **2008**, *391*, 381–390. [CrossRef]
22. Huang, L.; Li, Z.; Liu, Q.; Pu, G.; Zhang, Y.; Li, J. Research on the adaptive mechanism of photosynthetic apparatus under salt stress: New directions to increase crop yield in saline soils. *Ann. Appl. Biol.* **2019**, *175*, 1–17. [CrossRef]
23. Pang, Q.; Chen, S.; Dai, S.; Chen, Y.; Wang, Y.; Yan, X. Comparative proteomics of salt tolerance in *Arabidopsis thaliana* and *Thellungiella halophila*. *J. Proteome Res.* **2010**, *9*, 2584–2599. [CrossRef]
24. Liu, Z.; Zou, L.; Chen, C.; Zhao, H.; Yan, Y.; Wang, C.; Liu, X. ITRAQ-based quantitative proteomic analysis of salt stress in *Spica Prunellae*. *Sci. Rep.* **2019**, *9*, 9590. [CrossRef]
25. Ashraf, M.; Harris, P.J.C. Photosynthesis under stressful environments: An overview. *Photosynthetica* **2013**, *51*, 163–190. [CrossRef]
26. Pandey, D.M.; Choi, I.; Yeo, U.-D. Photosystem 2-activity and thylakoid membrane polypeptides of in vitro cultured *Chrysanthemum* as affected by NaCl. *Biol. Plant.* **2009**, *53*, 329–333. [CrossRef]
27. Mishra, S.K.; Subrahmanyam, D.; Singhal, G.S. Interrelationship between salt and light stress on primary processes of photosynthesis. *J. Plant Physiol.* **1991**, *138*, 92–96. [CrossRef]
28. Ioannidis, N.E.; Sfichi, L.; Kotzabasis, K. Putrescine stimulates chemiosmotic ATP synthesis. *Biochim. Biophys. Acta—Bioenerg.* **2006**, *1757*, 821–828. [CrossRef]
29. Stefanov, M.A.; Rashkov, G.D.; Apostolova, E.L. Assessment of the photosynthetic apparatus functions by chlorophyll fluorescence and P700 absorbance in C3 and C4 plants under physiological conditions and under salt stress. *Int. J. Mol. Sci.* **2022**, *23*, 3768. [CrossRef]
30. Stefanov, M.A.; Rashkov, G.D.; Yotsova, E.K.; Dobrikova, A.G.; Apostolova, E.L. Impact of salinity on the energy transfer between pigment–protein complexes in photosynthetic apparatus, functions of the oxygen-evolving complex and photochemical activities of photosystem II and photosystem I in two *Paulownia* lines. *Int. J. Mol. Sci.* **2023**, *24*, 3108. [CrossRef]
31. Simontacchi, M.; Galatro, A.; Ramos-Artuso, F.; Santa-María, G.E. Plant survival in achanging environment: The role of nitric oxide in plant responses to abiotic stress. *Front. Plant Sci.* **2015**, *6*, 155160. [CrossRef]
32. Nabi, R.B.S.; Tayade, R.; Hussain, A.; Kulkarni, K.P.; Imran, Q.M.; Mun, B.-G.; Yun, B.-W. Nitric oxide regulates plant responses to drought, salinity, and heavy metal stress. *Environ. Exp. Bot.* **2019**, *161*, 120–133. [CrossRef]
33. Wani, K.I.; Naeem, M.; Castroverde, C.D.M.; Kalaji, H.M.; Albaqami, M.; Aftab, T. Molecular mechanisms of nitric oxide (NO) signaling and reactive oxygen species (ROS) homeostasis during abiotic stresses in plants. *Int. J. Mol. Sci.* **2021**, *22*, 9656. [CrossRef]
34. Rashkov, G.D.; Stefanov, M.A.; Yotsova, E.K.; Borisova, P.B.; Dobrikova, A.G.; Apostolova, E.L. Impact of sodium nitroprusside on the photosynthetic performance of maize and sorghum. *Plants* **2024**, *13*, 118. [CrossRef]
35. Ullah, F.; Saqib, S.; Khan, W.; Ayaz, A.; Batool, A.; Wang, W.-Y.; Xiong, Y.-C. The multifaceted role of sodium nitroprusside in plants: Crosstalk with phytohormones under normal and stressful conditions. *Plant Growth Regul.* **2024**. [CrossRef]
36. Jabeen, Z.; Fayyaz, H.A.; Irshad, F.; Hussain, N.; Hassan, M.N.; Li, J.; Rehman, S.; Haider, W.; Yasmin, H.; Mumtaz, S.; et al. Sodium nitroprusside application improves morphological and physiological attributes of soybean (*Glycine max* L.) under salinity stress. *PLoS ONE* **2021**, *16*, e0248207. [CrossRef]
37. Sohag, A.A.M.; Tahjib-Ul-Arif, M.; Afrin, S.; Khan, M.K.; Hannan, M.A.; Skalicky, M.; Mortuza, M.G.; Brestic, M.; Hossain, M.A.; Murata, Y. Insights into nitric oxide-mediated water balance, antioxidant defence and mineral homeostasis in rice (*Oryza sativa* L.) under chilling stress. *Nitric Oxide* **2020**, *100–101*, 7–16. [CrossRef]
38. Allagulova, C.R.; Lubyanova, A.R.; Avalbaev, A.M. Multiple ways of nitric oxide production in plants and its functional activity under abiotic stress conditions. *Int. J. Mol. Sci.* **2023**, *24*, 11637. [CrossRef]
39. Hayat, S.; Yadav, S.; Wani, A.S.; Irfan, M.; Alyemini, M.N.; Ahmad, A. Impact of sodium nitroprusside on nitrate reductase, proline content, and antioxidant system in tomato under salinity stress. *Hortic. Environ. Biotechnol.* **2012**, *53*, 362–367. [CrossRef]
40. Lin, Y.; Liu, Z.; Shi, Q.; Wang, X.; Wei, M.; Yang, F. Exogenous nitric oxide (NO) increased antioxidant capacity of cucumber hypocotyl and radicle under salt stress. *Sci. Hortic.* **2012**, *142*, 118–127. [CrossRef]

41. Tanou, G.; Filippou, P.; Belghazi, M.; Job, D.; Diamantidis, G.; Fotopoulos, V.; Molassiotis, A. Oxidative and nitrosative-based signaling and associated post-translational modifications orchestrate the acclimation of citrus plants to salinity stress. *Plant J.* **2012**, *72*, 585–599. [CrossRef]
42. Dong, Y.; Jinc, S.; Liu, S.; Xu, L.; Kong, J. Effects of exogenous nitric oxide on growth of cotton seedlings under NaCl stress. *J. Soil Sci. Plant Nutr.* **2014**, *14*, 1–13. [CrossRef]
43. Jian, W.; Zhang, D.; Zhu, F.; Wang, S.; Pu, X.; Deng, X.-G.; Luo, S.-S.; Lin, H. Alternative oxidase pathway is involved in the exogenous SNP-elevated tolerance of Medicago truncatula to salt stress. *J. Plant Physiol.* **2016**, *193*, 79–87. [CrossRef] [PubMed]
44. Aras, S.; Keles, H.; Eşitken, A. SNP mitigates malignant salt effects on apple plants. *Erwerbs-Obstbau* **2020**, *62*, 107–115. [CrossRef]
45. Sehar, Z.; Masood, A.; Khan, N.A. Nitric oxide reverses glucose-mediated photosynthetic repression in wheat (*Triticum aestivum* L.) under salt stress. *Environ. Exp. Bot.* **2019**, *161*, 277–289. [CrossRef]
46. Yasir, T.A.; Khan, A.; Skalicky, M.; Wasaya, A.; Rehmani, M.I.A.; Sarwar, N.; Mubeen, K.; Aziz, M.; Hassan, M.M.; Hassan, F.A.S.; et al. Exogenous sodium nitroprusside mitigates salt stress in lentil (*Lens culinaris* Medik.) by affecting the growth, yield, and biochemical properties. *Molecules* **2021**, *26*, 2576. [CrossRef]
47. Stefanov, M.A.; Rashkov, G.D.; Yotsova, E.K.; Borisova, P.B.; Dobrikova, A.G.; Apostolova, E.L. Protective effects of sodium nitroprusside on photosynthetic performance of Sorghum bicolor L. under salt stress. *Plants* **2023**, *12*, 832. [CrossRef]
48. Kolbert, Z.; Szőllősi, R.; Feigl, G.; Kónya, Z.; Rónavári, A. Nitric oxide signalling in plant nanobiology: Current status and perspectives. *J. Exp. Bot.* **2021**, *72*, 928–940. [CrossRef] [PubMed]
49. Hajihashemi, S.; Skalicky, M.; Brestic, M.; Pavla, V. Effect of sodium nitroprusside on physiological and anatomical features of salt-stressed *Raphanus sativus*. *Plant Physiol. Biochem.* **2021**, *169*, 160–170. [CrossRef] [PubMed]
50. Hajihashemi, S.; Skalicky, M.; Brestic, M.; Pavla, V. Cross-talk between nitric oxide, hydrogen peroxide and calcium in salt-stressed *Chenopodium quinoa* Willd. At seed germination stage. *Plant Physiol. Biochem.* **2020**, *154*, 657–664. [CrossRef]
51. Vladkova, R.; Dobrikova, A.G.; Singh, R.; Misra, A.N.; Apostolova, E. Photoelectron transport ability of chloroplast thylakoid membranes treated with NO donor SNP: Changes in flash oxygen evolution and chlorophyll fluorescence. *Nitric Oxide* **2011**, *24*, 84–90. [CrossRef]
52. Misra, A.N.; Vladkova, R.; Singh, R.; Misra, M.; Dobrikova, A.G.; Apostolova, E.L. Action and target sites of nitric oxide in chloroplasts. *Nitric Oxide—Biol. Chem.* **2014**, *39*, 35–45. [CrossRef] [PubMed]
53. Wodala, B.; Deák, Z.; Vass, I.; Erdei, L.; Altorjay, I.; Horváth, F. In vivo target sites of Nitric oxide in photosynthetic electron transport as studied by chlorophyll fluorescence in pea leaves. *Plant Physiol.* **2008**, *146*, 1920–1927. [CrossRef] [PubMed]
54. Yang, J.D.; Zhao, H.L.; Zhang, T.H.; Yun, J.F. Effects of exogenous nitric oxide on photochemical activity of photosystem II in potato leaf tissue under non-stress condition. *Acta Bot. Sin.* **2004**, *46*, 1009–1014.
55. Zhang, L.; Wang, Y.; Zhao, L.; Shi, S.; Zhang, L. Involvement of nitric oxide in light-mediated greening of barley seedlings. *J. Plant Physiol.* **2006**, *163*, 818–826. [CrossRef] [PubMed]
56. Fatma, M.; Khan, N.A. Nitric Oxide Protects Photosynthetic Capacity Inhibition by Salinity in Indian Mustard. *J. Funct. Environ. Bot.* **2014**, *4*, 106–116. [CrossRef]
57. Dong, F.; Simon, J.; Rienks, M.; Lindermayr, C.; Rennenberg, H.; Näsholm, T. Effects of rhizopheric nitric oxide (NO) on N uptake in Fagus sylvatica seedlings depend on soil CO2 concentration, soil N availability and N source. *Tree Physiol.* **2015**, *35*, 910–920. [CrossRef] [PubMed]
58. Shang, J.X.; Li, X.; Li, C.; Zhao, L. The role of nitric oxide in plant responses to salt stress. *Int. J. Mol. Sci.* **2022**, *23*, 6167. [CrossRef] [PubMed]
59. Dadasoglu, E.; Ekinci, M.; Kul, R.; Shams, M.; Turan, M.; Yildirim, E. Nitric oxide enhances salt tolerance through regulating antioxidant enzyme activity and nutrient uptake in pea. *Legum. Res.* **2021**, *44*, 41–45. [CrossRef]
60. Liu, A.; Fan, J.; Gitau, M.M.; Chen, L.; Fu, J. Nitric oxide involvement in bermudagrass response to salt stress. *J. Am. Soc. Hortic. Sci.* **2016**, *141*, 425–433. [CrossRef]
61. Yüzbaşioğlu, E.A.; Öz, G.C. Relationship between salt stress and nitric oxide in *Glycine max* L. (soybean). *Fresenius Environ. Bull.* **2010**, *19*, 425–431.
62. Hasanuzzaman, M.; Inafuku, M.; Nahar, K.; Fujita, M.; Oku, H. Nitric Oxide Regulates Plant Growth, Physiology, Antioxidant Defense, and Ion Homeostasis to Confer Salt Tolerance in the Mangrove Species, Kandelia obovata. *Antioxidants* **2021**, *10*, 611. [CrossRef]
63. Daudi, A.; O'Brien, J. Detection of Hydrogen Peroxide by DAB Staining in Arabidopsis Leaves. *Bio-Protoc.* **2012**, *2*, e263. [CrossRef]
64. Szilágyi, A.; Selstam, E.; Åkerlund, H.-E. Laurdan fluorescence spectroscopy in the thylakoid bilayer: The effect of violaxanthin to zeaxanthin conversion on the galactolipid dominated lipid environment. *Biochim. Biophys. Acta—Biomembr.* **2008**, *1778*, 348–355. [CrossRef] [PubMed]
65. Strasser, R.J.; Tsimilli-Michael, M.; Srivastava, A. Analysis of the Chlorophyll a Fluorescence Transient. In *Chlorophyll a Fluorescence. Advances in Photosynthesis and Respiration*; Papageorgiou, G.C., Govindjee, Eds.; Springer: Dordrecht, The Netherlands, 2004; pp. 321–362.
66. Yin, X.; Hu, Y.; Zhao, Y.; Meng, L.; Zhang, X.; Liu, H.; Wang, L.; Cui, G. Effects of exogenous nitric oxide on wild barley (*Hordeum brevisubulatum*) under salt stress. *Biotechnol. Biotechnol. Equip.* **2021**, *35*, 2005–2016. [CrossRef]

67. Seleiman, M.F.; Aslam, M.T.; Alhammad, B.A.; Hassan, M.U.; Maqbool, R.; Chattha, M.U.; Khan, I.; Gitari, H.I.; Uslu, O.S.; Roy, R.; et al. Salinity stress in wheat: Effects, mechanisms and management strategies. *Phyton* **2022**, *91*, 667–694. [CrossRef]
68. Jusovic, M.; Velitchkova, M.Y.; Misheva, S.P.; Börner, A.; Apostolova, E.L.; Dobrikova, A.G. Photosynthetic responses of a wheat mutant (Rht-B1c) with altered DELLA proteins to salt stress. *J. Plant Growth Regul.* **2018**, *37*, 645–656. [CrossRef]
69. Rajabi Dehnavi, A.; Zahedi, M.; Piernik, A. Understanding salinity stress responses in sorghum: Exploring genotype variability and salt tolerance mechanisms. *Front. Plant Sci.* **2024**, *14*, 1296286. [CrossRef] [PubMed]
70. Popova, A.V.; Borisova, P.; Vasilev, D. Response of pea plants (*Pisum sativum* cv. Ran 1) to NaCl treatment in regard to Membrane Stability and photosynthetic Activity. *Plants* **2023**, *12*, 324. [CrossRef]
71. Youssef, M.H.M.; Raafat, A.; El-Yazied, A.A.; Selim, S.; Azab, E.; Khojah, E.; El Nahhas, N.; Ibrahim, M.F.M. Exogenous application of alpha-lipoic acid mitigates salt-induced oxidative damage in sorghum plants through regulation growth, leaf pigments, ionic homeostasis, antioxidant enzymes, and expression of salt stress responsive genes. *Plants* **2021**, *10*, 2519. [CrossRef]
72. Sudhir, P.; Murthy, S.D.S. Effects of salt stress on basic processes of photosynthesis. *Photosynthetica* **2004**, *42*, 481–486. [CrossRef]
73. Galatro, A.; Puntarulo, S.; Guiamet, J.J.; Simontacchi, M. Chloroplast functionality has a positive effect on nitric oxide level in soybean cotyledons. *Plant Physiol. Biochem.* **2013**, *66*, 26–33. [CrossRef]
74. Edge, R.; McGarvey, D.J.; Truscott, T.G. The carotenoids as anti-oxidants—A review. *J. Photochem. Photobiol. B Biol.* **1997**, *41*, 189–200. [CrossRef]
75. Gill, S.S.; Tuteja, N. Reactive oxygen species and antioxidant machinery in abiotic stress tolerance in crop plants. *Plant Physiol. Biochem.* **2010**, *48*, 909–930. [CrossRef] [PubMed]
76. Khorobrykh, S.; Havurinne, V.; Mattila, H.; Tyystjärvi, E. Oxygen and ROS in photosynthesis. *Plants* **2020**, *9*, 91. [CrossRef]
77. Chalbi, N.; Hessini, K.; Gandour, M.; Mohamed, S.N.; Smaoui, A.; Abdelly, C.; Youssef, N. Ben Are changes in membrane lipids and fatty acid composition related to salt-stress resistance in wild and cultivated barley? *J. Plant Nutr. Soil Sci.* **2013**, *176*, 138–147. [CrossRef]
78. Parida, A.K.; Das, A.B. Salt tolerance and salinity effects on plants: A review. *Ecotoxicol. Environ. Saf.* **2005**, *60*, 324–349. [CrossRef]
79. Kirchhoff, H. Structure-function relationships in photosynthetic membranes: Challenges and emerging fields. *Plant Sci.* **2018**, *266*, 76–82. [CrossRef] [PubMed]
80. Tikkanen, M.; Aro, E.-M. Thylakoid protein phosphorylation in dynamic regulation of photosystem II in higher plants. *Biochim. Biophys. Acta—Bioenerg.* **2012**, *1817*, 232–238. [CrossRef]
81. Zhao, C.; Zhang, H.; Song, C.; Zhu, J.-K.; Shabala, S. Mechanisms of plant responses and adaptation to soil salinity. *Innovation* **2020**, *1*, 100017. [CrossRef]
82. Jajoo, A. Changes in Photosystem II in response to salt stress. In *Ecophysiology and Responses of Plants under Salt Stress*; Springer: New York, NY, USA, 2013; pp. 149–168, ISBN 9781461447. [CrossRef]
83. Kalaji, H.M.; Rastogi, A.; Živčák, M.; Brestic, M.; Daszkowska-Golec, A.; Sitko, K.; Alsharafa, K.Y.; Lotfi, R.; Stypiński, P.; Samborska, I.A.; et al. Prompt chlorophyll fluorescence as a tool for crop phenotyping: An example of barley landraces exposed to various abiotic stress factors. *Photosynthetica* **2018**, *56*, 953–961. [CrossRef]
84. Stefanov, M.A.; Rashkov, G.D.; Yotsova, E.K.; Borisova, P.B.; Dobrikova, A.G.; Apostolova, E.L. Different sensitivity levels of the photosynthetic apparatus in *Zea mays* L. and *Sorghum bicolor* L. under salt stress. *Plants* **2021**, *10*, 1469. [CrossRef] [PubMed]
85. Salim Akhter, M.; Noreen, S.; Mahmood, S.; Ashraf, M.; Abdullah Alsahli, A.; Ahmad, P. Influence of salinity stress on PSII in barley (*Hordeum vulgare* L.) genotypes, probed by chlorophyll-a fluorescence. *J. King Saud Univ.—Sci.* **2021**, *33*, 101239. [CrossRef]
86. Ran, X.; Wang, X.; Gao, X.; Liang, H.; Liu, B.; Huang, X. Effects of salt stress on the photosynthetic physiology and mineral ion absorption and distribution in white willow (*Salix alba* L.). *PLoS ONE* **2021**, *16*, e0260086. [CrossRef]
87. Moustakas, M. Plant photochemistry, reactive oxygen species, and photoprotection. *Photochem* **2021**, *2*, 5–8. [CrossRef]
88. Deák, Z.; Sass, L.; Kiss, É.; Vass, I. Characterization of wave phenomena in the relaxation of flash-induced chlorophyll fluorescence yield in cyanobacteria. *Biochim. Biophys. Acta—Bioenerg.* **2014**, *1837*, 1522–1532. [CrossRef] [PubMed]
89. Shirao, M.; Kuroki, S.; Kaneko, K.; Kinjo, Y.; Tsuyama, M.; Förster, B.; Takahashi, S.; Badger, M.R. Gymnosperms have increased capacity for electron leakage to oxygen (Mehler and PTOX reactions) in photosynthesis compared with angiosperms. *Plant Cell Physiol.* **2013**, *54*, 1152–1163. [CrossRef] [PubMed]
90. Lima-Moro, A.; Bertoli, S.C.; Braga-Reis, I.; Moro, E.; Ziliani, R.R.; Spolaor, B.O.; de Freitas, Í.R.; dos Santos, B.L. Photosynthetic activity and OJIP fluorescence with the application of a nutritional solution. *Acta Physiol. Plant.* **2022**, *44*, 67. [CrossRef]
91. Ivanova, P.I.; Dobrikova, A.G.; Taneva, S.G.; Apostolova, E.L. Sensitivity of the photosynthetic apparatus to UV-A radiation: Role of light-harvesting complex II–photosystem II supercomplex organization. *Radiat. Environ. Biophys.* **2008**, *47*, 169–177. [CrossRef] [PubMed]
92. Amin, M. Predicting the oxidation states of Mn ions in the oxygen-evolving complex of photosystem II using supervised and unsupervised machine learning. *Photosynth. Res.* **2023**, *156*, 89–100. [CrossRef]
93. Maslenkova, L.; Gambarova, N.; Zeinalov, Y. NaCl-induced changes in oxygen evolving activity and thylakoid membrane patterns of barley plants. Adaptation to salinity. *Bulg. J. Plant Physiol.* **1995**, *21*, 29–35.
94. Shishido, S.M.; de Oliveira, M.G. Photosensitivity of aqueous sodium nitroprusside solutions: Nitric oxide release versus cyanide toxicity. *Prog. React. Kinet. Mech.* **2001**, *26*, 239–261. [CrossRef]
95. Lichtenthaler, H.K. Chlorophylls and carotenoids: Pigments of photosynthetic biomembranes. *Methods Enzymol.* **1987**, *148*, 350–382. [CrossRef]

96. Scheinpflug, K.; Krylova, O.; Strahl, H. Measurement of Cell Membrane Fluidity by Laurdan GP: Fluorescence Spectroscopy and Microscopy. *Antibiotics* **2017**, *1520*, 159–174.
97. Stefanov, M.; Rashkov, G.; Borisova, P.; Apostolova, E. Sensitivity of the photosynthetic apparatus in maize and sorghum under different drought levels. *Plants* **2023**, *12*, 1863. [CrossRef] [PubMed]
98. Roháček, K. Chlorophyll fluorescence parameters: The definitions, photosynthetic meaning, and mutual relationships. *Photosynthetica* **2002**, *40*, 13–29. [CrossRef]
99. Nasraoui-Hajaji, A.; Gharbi, F.; Ghorbel, M.H.; Gouia, H. Cadmium stress effects on photosynthesis and PSII efficiency in tomato grown on NO_3^- or NH_4^+ as nitrogen source. *Acta Bot. Gall.* **2010**, *157*, 101–115. [CrossRef]
100. Moustakas, M.; Dobrikova, A.; Sperdouli, I.; Hanć, A.; Adamakis, I.-D.S.; Moustaka, J.; Apostolova, E. A hormetic spatiotemporal photosystem II response mechanism of Salvia to excess zinc exposure. *Int. J. Mol. Sci.* **2022**, *23*, 11232. [CrossRef] [PubMed]
101. Stefanov, M.; Yotsova, E.; Gesheva, E.; Dimitrova, V.; Markovska, Y.; Doncheva, S.; Apostolova, E.L. Role of flavonoids and proline in the protection of photosynthetic apparatus in *Paulownia* under salt stress. *S. Afr. J. Bot.* **2021**, *139*, 246–253. [CrossRef]
102. Nie, G.Y.; Baker, N.R. Modifications to thylakoid composition during development of maize leaves at low growth temperatures. *Plant Physiol.* **1991**, *95*, 184–191. [CrossRef] [PubMed]
103. Kok, B.; Forbush, B.; McGloin, M. Cooperation of charges in photosynthetic O2 evolution-I. A linear four step mechanism. *Photochem. Photobiol.* **1970**, *11*, 457–475. [CrossRef]
104. Fatma, M.; Masood, A.; Per, T.S.; Khan, N.A. Nitric Oxide Alleviates Salt Stress Inhibited Photosynthetic Performance by Interacting with Sulfur Assimilation in Mustard. *Front. Plant Sci.* **2016**, *7*, 181343. [CrossRef] [PubMed]
105. Kalaji, H.M.; Schansker, G.; Brestic, M.; Bussotti, F.; Calatayud, A.; Ferroni, L.; Goltsev, V.; Guidi, L.; Jajoo, A.; Li, P.; et al. *Frequently Asked Questions about Chlorophyll Fluorescence, the Sequel*; Springer: Berlin/Heidelberg, Germany, 2017; Volume 132, ISBN 1112001603.
106. Bussotti, F.; Gerosa, G.; Digrado, A.; Pollastrini, M. Selection of chlorophyll fluorescence parameters as indicators of photosynthetic efficiency in large scale plant ecological studies. *Ecol. Indic.* **2020**, *108*, 105686. [CrossRef]

plants

Article

Genome-Wide Identification and Analysis of *ZF-HD* Gene Family in Moso Bamboo (*Phyllostachys edulis*)

Feiyi Huang [1], Jiaxin Wang [1] and Chao Tang [2,*]

[1] Co-Innovation Center for Sustainable Forestry in Southern China, Bamboo Research Institute, College of Life Sciences, Nanjing Forestry University, Nanjing 210037, China
[2] State Key Laboratory of Crop Genetics & Germplasm Enhancement and Utilization, Centre of Pear Engineering Technology Research, Nanjing Agricultural University, Nanjing 210095, China
* Correspondence: 2019201@njau.edu.cn

Abstract: Zinc finger-homeodomain (ZF-HD) proteins play essential roles in plant growth, development and stress responses. However, knowledge of the expression and evolutionary history of *ZF-HD* genes in moso bamboo remains limited. In this study, a total of 24 *ZF-HD* genes were found unevenly distributed on 12 chromosomes in moso bamboo (*Phyllostachys edulis*). Phylogenetic analysis indicated that *PeZF-HD*s were divided into two subfamilies: ZHD and MIF. The ZHD subfamily genes were further classified into seven groups according to their orthologous relationships among the rice and *Arabidopsis ZF-HD* gene family. The gene structures and conserved motifs of *PeZF-HD*s were analyzed. Whole-genome duplication (WGD) or segmental duplication promoted the evolution and expansion of the moso bamboo *ZF-HD* gene family. Ka/Ks ratios suggested that the twenty-four duplication pairs had undergone purifying selection. Promoter analysis showed that most *PeZF-HD*s contained *cis*-elements associated with stress responses and hormones. Expression analysis demonstrated that many *PeZF-HD*s were responsive to abiotic stress treatment. Overall, this work investigated *PeZF-HD* genes in moso bamboo using bioinformatic approaches. The evolutionary research on gene structure, motif distribution and *cis*-regulatory elements indicated that *PeZF-HD*s play distinct roles in biological processes, which provides a theoretical basis for exploring the physiological functions of *ZF-HD*s and selecting candidate stress-related genes in moso bamboo.

Keywords: moso bamboo; genome-wide; ZF-HD; abiotic stress; expression pattern

Citation: Huang, F.; Wang, J.; Tang, C. Genome-Wide Identification and Analysis of *ZF-HD* Gene Family in Moso Bamboo (*Phyllostachys edulis*). *Plants* **2023**, *12*, 4064. https://doi.org/10.3390/plants12234064

Academic Editors: Vladimir V. Kuznetsov and Emilia Apostolova

Received: 19 July 2023
Revised: 5 October 2023
Accepted: 23 October 2023
Published: 3 December 2023

1. Introduction

Plants often face various abiotic stresses during their life cycle, including extreme temperatures, drought and salinity, which seriously limit plant growth, development, quality and yield [1]. The transcription factor is a key kind of regulatory protein which can bind to the promoters of downstream genes and regulate their expression to adapt to multiple stresses [2]. As plant-specific transcription factors, zinc finger-homeodomain (ZF-HD) proteins have aroused much interest as they play vital regulatory roles in plant growth and development, as well as various stress responses [3].

The ZF-HD protein contains the conserved ZF-HD_dimer domain (PF04770) which consists of two highly conserved domains: the N-terminal C2H2-type zinc finger (ZF) and the C-terminal homeodomain (HD) [4]. The ZF domain can not only bind DNA but enhance the protein–DNA interactions controlled by the HD domain to either enhance or repress transcription [5]. As a DNA-binding domain, the HD has approximately 60 conserved amino acids, which fold into a three-helix structure and specifically connect to DNA [6]. HD-containing proteins are also involved in protein–protein interactions and other regulatory functions. Based on the size, location and structure of the HD domain and its association with other domains, HD-containing proteins can be divided into six distinct families: finger-like domain-associated HD (PHD finger), WUSCHEL-related homeobox

(WOX), knotted-related homeobox (KNOX), zinc finger motif-associated HD (ZF-HD), leucine zipper-associated HD (HD-Zip) and bell-type HD [7]. The zinc finger structure contains a single zinc ion surrounded by two pairs of conserved cysteine and/or histidine residues [8]. According to the number, spacing pattern and nature of zinc-binding residues, zinc finger structures are classified into different types, such as C2H2 and C3H [9]. ZF-HD proteins can be divided into ZF-HD and mini zinc finger (MIF) groups based on phylogeny. MIF groups only contain the ZF domain without the HD domain, and are involved in integrating signals from GA, auxin, abscisic acid (ABA) and brassinosteroid [10]. However, the origins of and evolutionary relationship between these two groups remains unclear.

ZF-HD proteins were first discovered in the C4 plant *Flaveria* [5]. Subsequently, *ZF-HD* genes were reported in many plant species, such as *Arabidopsis*, rice, maize and tomato [11–14]. There has been increasing evidence suggesting that ZF-HD genes play important roles in responding to adverse stresses, except for plant growth and development [12]. Multiple *ZF-HDs* in *Arabidopsis* are involved in stress responses to adversity, and similar findings have also been reported in wheat and tomato [15–17]. *AtZF-HD1* can bind to the promoter of early response to dehydration stress 1 (ERD1) and its expression is inducible by drought, salinity and ABA. An overexpression of *ZF-HD1* and *NAC* enhances drought resistance in *Arabidopsis* [15]. A silencing of *SlZF-HD13* decreases the salt and drought tolerance of tomato [17]. In rice, four *ZF-HDs* can bind to a *DREB1B* promoter and activate its expression under drought and cold stress [12]. Most *BraZF-HDs* are induced by abiotic stress, indicating their important roles in controlling stress response [18]. Except for in the model plant, the identification and functions of *ZF-HDs* have also been broadly elucidated. There have been, respectively, 49, 50, 22, and 32 *ZF-HD* genes identified in four cotton genomes, in which gene duplication drove the expansion of the *ZF-HD* gene family during the divergence of the *Gossypium* species [19]. Twenty *FtZF-HD* genes have been identified in Tartary buckwheat (*Fagopyrum tataricum*), which have been further divided into five subfamilies [20]. In quinoa, 23 *CqZF-HDs* were detected in its genome, and *CqZF-HD14* was able to promote its resistance to drought by regulating the expressions of *CqNAC79* and *CqHIPP34* [21]. Although *ZF-HD* genes have been widely studied in many plant species, reports on *ZF-HD* genes in moso bamboo (*Phyllostachys edulis*) are limited.

Moso bamboo belongs to the Bambusoideae subfamily of the Poaceae family and is an important non-timber forest product. It is the most widespread bamboo species in China and has the highest economic, ecological and ornamental value [22]. Due to its propensity for fast growth, it may contribute to water and soil conservation and climate regulation [23]. Disadvantageous environmental and climatic conditions restrict the growth and development of moso bamboo, leading to severe yield losses. Recently, the genome of moso bamboo has been published, providing the foundation for a genome-wide analysis of *ZF-HD* genes in moso bamboo [24]. Herein, 24 putative *PeZF-HD* genes were identified and a comprehensive analysis was performed, including analyses of physicochemical properties, phylogenetic relationships, gene structure, conserved motifs, chromosomal localization, cis-elements, expression profiling and evolutionary analysis. These results will improve the understanding of the *PeZF-HD* gene family and provide the key candidate genes related to abiotic stress in moso bamboo.

2. Results

2.1. Identification of ZF-HD Genes in Moso Bamboo

P. edulis ZF-HD genes were screened by BLASTP and HMM. After excluding the redundant sequences, SMART was performed to confirm the presence of ZF-HD_dimer domains (Table S3). A total of 24 ZF-HD proteins were identified as *PeZF-HD* genes, and were named PeZF-HD1–PeZF-HD24. The characteristics of these *PeZF-HDs* are shown in Table 1, including the length of CDS and protein, molecular weight, theoretical isoelectric point, exon number, chromosome location and subcellular localization. The length of CDS of *PeZF-HD* genes varied from 252 bp to 1212 bp. The protein sequence lengths ranged from 83 to 403 amino acids, the molecular weights spanned 8.59 to 43.36 kDa, and the isoelectric points had a range of 5.90 to 10.30. The number of exons in *PeZF-HD* genes ranged from one to three. The predicted subcellular localization results indicated that all PeZF-HD proteins were located in the nucleus.

Table 1. Detailed information of ZF-HDs identified in moso bamboo genome.

Gene Name	CDS (bp)	Size (aa)	Mass (KDa)	pI	Domain	Exon Number	Location	Predicted Localization
PeZF-HD1	816	271	29.36	9.41	ZF-HD	1	chromosomes 12:13827712:13828527:+	Nucleus
PeZF-HD2	801	266	28.61	9.05	ZF-HD	1	chromosomes 3:49715743:49719049:+	Nucleus
PeZF-HD3	762	253	27.04	8.50	ZF-HD	2	chromosomes 11:13377647:13378939:+	Nucleus
PeZF-HD4	672	223	23.46	8.78	ZF-HD	2	chromosomes 18:35699501:35700292:+	Nucleus
PeZF-HD5	717	238	24.80	6.36	ZF-HD	2	chromosomes 13:57959543:57960412:+	Nucleus
PeZF-HD6	1164	387	40.65	8.59	ZF-HD	1	chromosomes 13:15365059:15366222:-	Nucleus
PeZF-HD7	867	288	29.90	6.75	ZF-HD	1	chromosomes 22:13492044:13492910:+	Nucleus
PeZF-HD8	633	210	22.79	8.41	ZF-HD	2	chromosomes 3:15339955:15340674:+	Nucleus
PeZF-HD9	804	267	27.82	8.49	ZF-HD	1	chromosomes 24:36670935:36671738:-	Nucleus
PeZF-HD10	684	227	24.51	8.99	ZF-HD	2	chromosomes 9:63441848:63443360:+	Nucleus
PeZF-HD11	1206	401	43.36	6.58	ZF-HD	1	chromosomes 3:87050975:87052180:-	Nucleus
PeZF-HD12	1212	403	43.21	6.35	ZF-HD	1	chromosomes 17:17852519:17853730:+	Nucleus
PeZF-HD13	300	99	10.57	8.60	ZF-HD	2	chromosomes 13:2989657:2990222:-	Nucleus
PeZF-HD14	279	92	10.17	6.79	ZF-HD	1	chromosomes 18:45423575:45424757:-	Nucleus
PeZF-HD15	417	138	15.17	10.30	ZF-HD	2	chromosomes 11:2971936:2974744:+	Nucleus
PeZF-HD16	462	153	16.81	9.06	ZF-HD	2	chromosomes 12:2567371:2568681:+	Nucleus
PeZF-HD17	519	172	18.32	8.95	ZF-HD	2	chromosomes 2:29976915:29981334:+	Nucleus
PeZF-HD18	252	83	8.59	6.86	ZF-HD	2	chromosomes 22:21025823:21026158:-	Nucleus
PeZF-HD19	348	115	11.41	5.90	ZF-HD	2	chromosomes 16:73535364:73537228:+	Nucleus
PeZF-HD20	1065	354	36.91	7.26	ZF-HD	1	chromosomes 3:24470226:24471290:+	Nucleus
PeZF-HD21	1035	344	35.64	7.71	ZF-HD	1	chromosomes 22:20913172:20914206:+	Nucleus
PeZF-HD22	1041	346	35.73	6.97	ZF-HD	1	chromosomes 13:64992849:64993889:+	Nucleus
PeZF-HD23	957	318	33.41	6.73	ZF-HD	2	chromosomes 18:45322937:45323998:+	Nucleus
PeZF-HD24	801	266	27.63	8.14	ZF-HD	1	chromosomes 23:37483279:37484079:+	Nucleus

2.2. Phylogenetic Analysis of the Moso Bamboo ZF-HD Gene Family

To study the evolutionary relationships of PeZF-HDs, a phylogenetic tree consisting of the ZF-HD protein sequences from moso bamboo, rice and *Arabidopsis* was constructed using the Maximum Likelihood (ML) method (Figure 1). The phylogenetic distribution revealed that the PeZF-HD family can be divided into two major groups: MIF and ZHD. ZHD included 43 genes and could be subdivided into seven subgroups: ZHD I, ZHD II, ZHD III, ZHD IV, ZHD V, ZHD VI and ZHD VII. The quantitative distribution of each group of ZF-HD proteins in moso bamboo, rice and *Arabidopsis* were counted (Table S4). Compared to MIF, ZHD had more members in moso bamboo, rice and *Arabidopsis*. The number of ZHD V genes was the highest, with four *PeZF-HD*, two *OsZF-HD* and five *AtZF-HD* genes. Phylogenetic analysis of ZF-HDs in three species showed that PeZF-HDs shared more sequence similarity with OsZF-HDs than with AtZF-HDs (Figure 1).

2.3. Conserved Motifs and Gene Structure Analysis of the Moso Bamboo ZF-HD Gene Family

To investigate the structural diversity of the *PeZF-HD* genes, a phylogenetic tree using only the full-length PeZF-HD protein sequences was constructed. Phylogenetic analysis showed that the PeZF-HDs could be categorized into six classes, which was consistent with the phylogenetic tree data among moso bamboo, rice and *Arabidopsis* (Figure 2A). MEME was used to determine differences in the PeZF-HD proteins. Eight conserved motifs were identified and named motifs 1–8 (Figure 2B). Motif 1 was the largest motif and was composed of 60 amino acids. Motif 2 and Motif 3, which were the components of the ZF-HD_dimer domain, were the most common motifs. Motif 2 was present in all PeZF-HDs, whereas Motif 3 was absent in PeZF-HD23. Motif 5 was present in ZHD IV and ZHD V. Motifs 7 and 8 were found exclusively in ZHD V, whereas Motif 4 was detected specifically in MIF. PeZF-HDs within the same class showed similar motif distributions, suggesting that they have functional similarities. Furthermore, the motif distribution in MIF differed greatly from that in the other classes, as it had only two to three motifs (motifs 2, 3 and 4) and the shortest sequence length. These motif distribution differences were probably results of the functional diversity of *PeZF-HDs*.

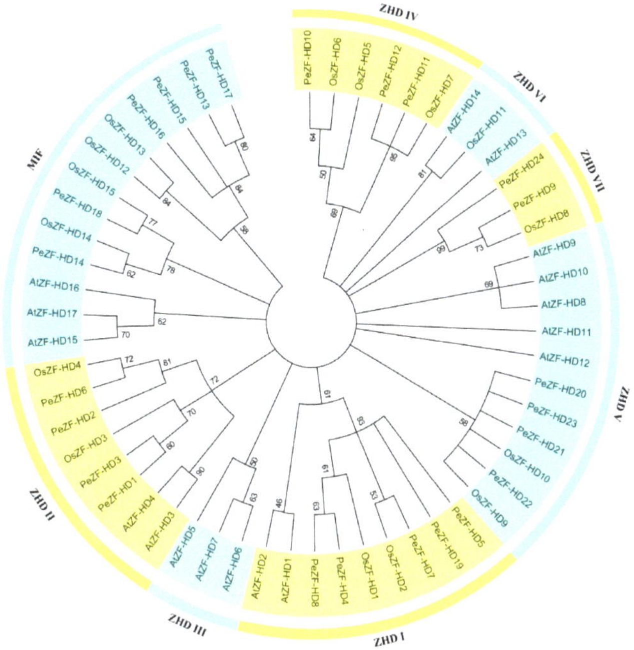

Figure 1. Phylogenetic analysis of ZF-HD proteins from rice, *Arabidopsis* and moso bamboo. The tree was built using the Maximum Likelihood (ML) method with 1000 bootstrap replicates. The branched lines of the subtrees are colored to indicate different subgroups.

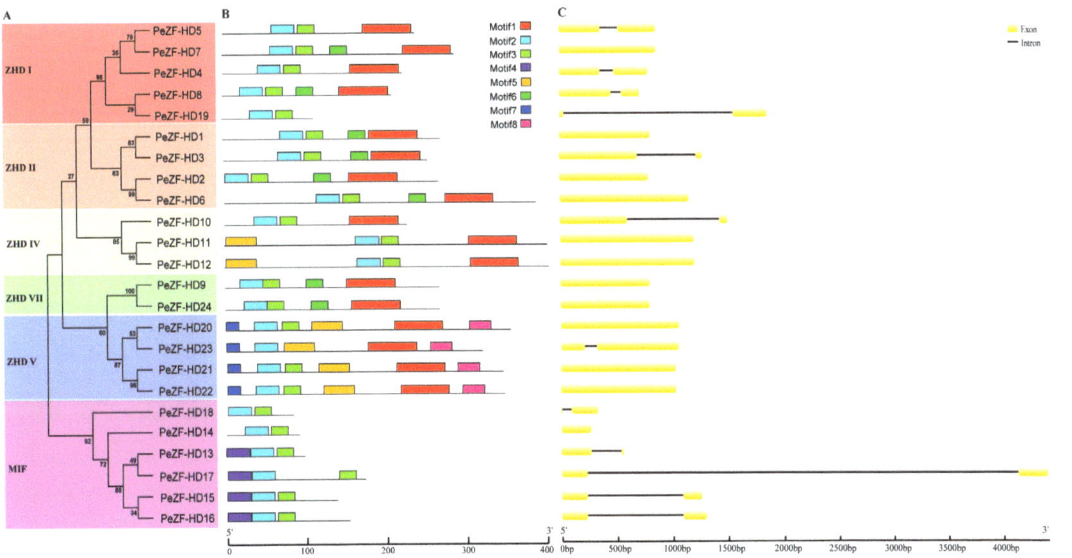

Figure 2. Phylogenetic relationships, gene structures and conserved domains of *PeZF-HDs*. (**A**) ML phylogenetic tree of PeZF-HDs. (**B**) Motif distributions in PeZF-HDs. The colored boxes represent motifs 1–8. Scale bar: 100 amino acids. (**C**) Exon–intron distribution of *PeZF-HDs*. The black lines and yellow boxes represent introns and exons, respectively. Scale bar: 500 bp.

The exon–intron organizations were analyzed with an online GSDS tool. As shown in Figure 2C, half of the *PeZF-HDs* were intronless and the remaining half had one intron. Most *PeZF-HDs* in the same class had similar intron–exon structures. For example, the *PeZF-HDs* in ZHD V (except for *PeZF-HD23*) all contained no introns. However, the intron–exon structure was not always conserved in most sister gene pairs. For instance, *PeZF-HD5/-7* and *PeZF-HD1/-3* had different numbers of introns and exons. Moreover, all *PeZF-HDs* were less than 2 kb in length, except for *PeZF-HD17* (4.4 kb).

2.4. Chromosome Distribution and Gene Duplication of the PeZF-HD Genes

The 24 *PeZF-HDs* were unevenly distributed across the 12 chromosome scaffolds based on their location in the moso bamboo genome (Figure 3). Chromosomes 3 and 13 contained the largest number of *PeZF-HD* members (four and four genes, respectively), followed by chromosomes 18 and 22 (three and three genes, respectively). Both chromosomes 11 and 12 contained two genes, while chromosomes 2, 9, 16, 17, 23 and 24 possessed only one gene, and no *ZF-HDs* were present on the twelve remaining moso bamboo chromosomes. Eighteen *PeZF-HDs* were located on the positive-strand chromosomes, while the remaining six genes were positioned on the negative-strand chromosomes. To further investigate the mechanism of the *PeZF-HD* gene family expansion, the potential duplication events of *PeZF-HDs* were identified with the MCScanX program. All *PeZF-HDs* appeared to have arisen from WGD or segmental duplications, except for *PeZF-HD19*, which underwent dispersed duplication (Table S5). No tandem duplication event was found in *PeZF-HDs*.

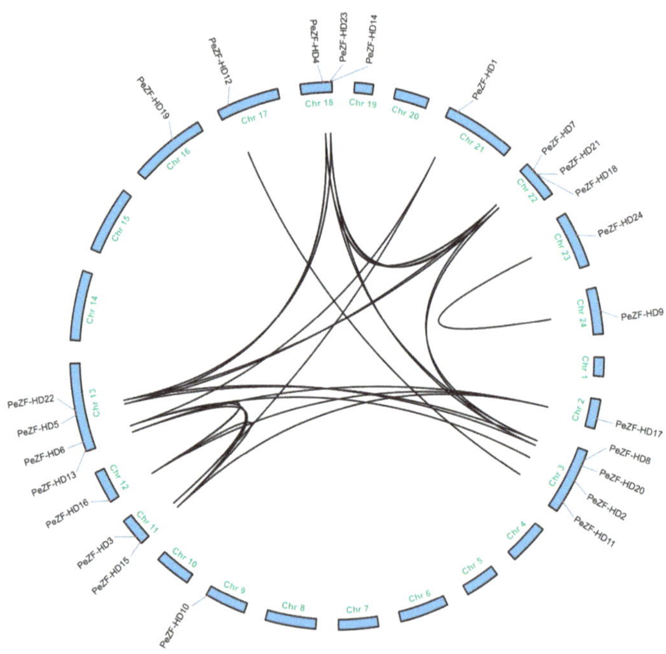

Figure 3. Chromosome distribution and syntenic relationships of *PeZF-HDs*. Chromosomes are shown in a circular form. The duplicate *PeZF-HD* gene pairs are connected by black lines.

To estimate the selection constraints of the duplicated *PeZF-HD* genes, the Ka, Ks replacement rate and Ka/Ks ratios of the twenty-four duplication pairs were calculated and listed in Table S6. The Ka/Ks ratios of these paralogous pairs ranged from 0.106 to 0.679, with an average value of 0.375, suggesting that these genes have undergone strong purifying selection during evolution. The Ks values of the *PeZF-HD* gene pairs varied from 0.099 to 0.639, indicating that a large-scale *PeZF-HD* gene duplication event occurred as early as 7.65–49.14 million years ago (MYA). The divergence for most *PeZF-HD* gene pairs (16 of 24) was about 16.51 to 49.14 MYA, which differs from the moso bamboo whole-genome duplication 7–12 MYA, suggesting that the *PeZF-HD* gene underwent other ancient whole-genome duplication events earlier.

2.5. Analysis of Cis-Regulatory Elements in PeZF-HD Promoters

To explore the potential functions of *PeZF-HDs*, the cis-regulatory elements at the promoter regions were investigated. Besides basic elements, cis-elements responsive to light, development, environmental stress and hormones were found in *PeZF-HD* promoters (Figure 4 and Table S1). A light-responsive element, with the largest proportion in *PeZF-HD* promoters, was represented by as many as 22 types. Each *PeZF-HD* promoter contained between 3 and 10 types. The elements related to development were also analyzed, such as the CAT-box, CCGTCC-box, RY-element, O2-site and GCN4 motif. With regard to environmental stress-related elements, we mainly investigated the elements related to low temperature stress (LTR), drought induction (MBS), salicylic acid response (TCA-element and SARE), and defense and stress response (TC-rich repeats). Interestingly, all *PeZF-HD* promoters contained the MYB element. Numerous elements in the *PeZF-HD* promoters were related to the response to hormones, like ABA (ABRE), MeJA (TGACG-motif and CGTCA-motif), auxin (AuxRR-core, TGA-box and TGA-element) and gibberellic acid (P-box, GAREmotif and TATC-box). Among them, the ABRE was the most abundant element in the *PeZF-HD* promoters, as it was found in 22 promoters. These results indicate that *PeZF-HDs* might be associated with moso bamboo development and have different roles in response to different stresses and hormone signals.

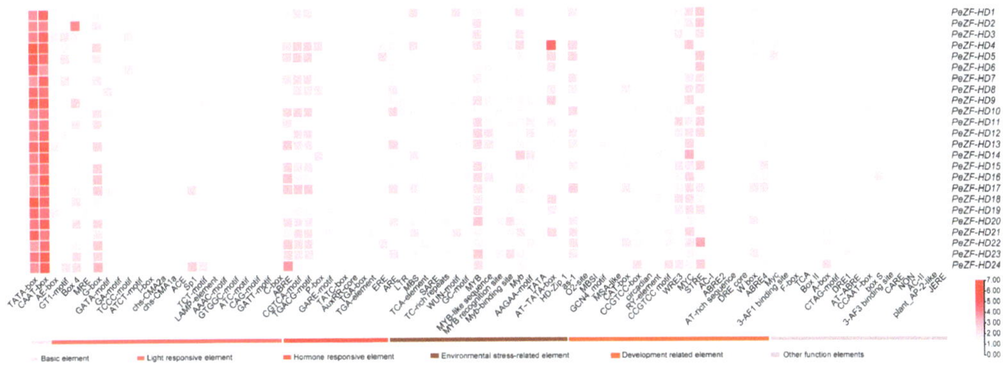

Figure 4. The cis-acting elements of *PeZF-HDs* promoters. The closed boxes of different colors represent different kinds of cis-acting elements.

2.6. Expression Profiles of PeZF-HD Genes against Different Abiotic Stresses

The analysis of potential cis-acting elements indicated that *PeZF-HDs* may play roles in plant responses to multiple environmental stresses. Based on the published transcriptome data, twelve *PeZF-HDs* with significantly different expressions were screened. Their expression profiles under cold, salt and drought stresses were determined by qPCR analysis (Figure 5). For moso bamboo seedlings treated with cold stress, the expression of *PeZF-HD2*, *4* and *20* displayed significant up-regulation, while rapid down-regulation was observed for *PeZF-HD5* and *22*. Under salt stress, *PeZF-HD14* was repressed, whereas *PeZF-HD4*, *10* and *13* were induced. In the drought stress, the expression of *PeZF-HD5* and *14* was down-regulated and *PeZF-HD4*, *10*, *20* and 22 were up-regulated. *PeZF-HD4* was significantly up-regulated under cold, salt and drought stresses. The expression of *PeZF-HD14* was reduced and *PeZF-HD10* was induced in both drought and salt stresses, while that of *PeZF-HD5* was down-regulated and *PeZF-HD20* was up-regulated in both cold and drought stresses. Interestingly, some *PeZF-HDs* exhibited inverse gene expression profiles in response to cold, salt and drought treatments. For instance, *PeZF-HD22* was reduced under cold treatment, but was induced under salt and drought treatment. Moreover, the expression of *PeZF-HD22* first decreased and then increased under cold, salt or drought treatments. The plant hormone ABA acts in plant stress signal response and plant growth. Thus, the expression profiles of the twelve *PeZF-HD*s under ABA treatment were also analyzed. Most of the *PeZF-HD*s signals were dramatically up-regulated in response to ABA. Eight of twelve *PeZF-HD*s (*PeZF-HD1*, *2*, *4*, *10*, *13*, *20*, *21* and *23*) were up-regulated to varying degrees with this treatment. We concluded the potentially functional *ZF-HD* genes of different species in response to various environmental stresses (Table S7), and the expressions of most *PeZF-HD*s showed similar responses to various environmental stresses in different species. However, the diverse *PeZF-HDs* may differ in their modes of regulation in moso bamboo under abiotic stress.

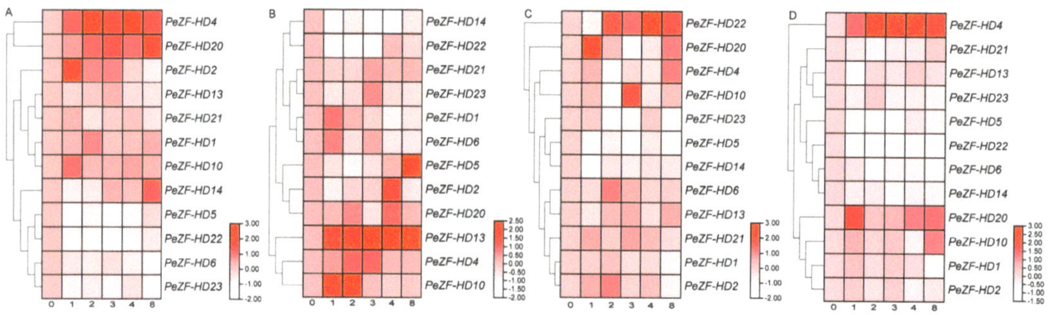

Figure 5. Heatmap and cluster of *PeZF-HD*s expressions under cold (**A**), Nacl (**B**), PEG (**C**) and ABA (**D**) treatment. The relative expression levels of *PeZF-HD*s under these treatments are shown by fold-change values and converted to log2 format compared to the control. The x-axis represents the days of treatment. The color scale at the right of each heatmap represents relative expression levels. The qPCR results were obtained based on three biological and three technical replicates.

3. Discussion

Plants often face adverse environmental conditions during their lifetime, such as drought, high salinity and cold, which affect physiological and biochemical processes and thus inhibit plant growth and development. To this end, plants have evolved a sophisticated stress resistance system regulated by a series of genes. To date, ZF-HDs have attracted increasing attention in applied and plant basic sciences. ZF-HD family genes are only existent in plants and play crucial roles in plant development and abiotic stress response [3]. The structural characteristics and functions of the ZF-HD gene family have been studied in various plants, such as *Arabidopsis*, rice, maize, tomato and Chinese cabbage [11–14,18]. However, ZF-HD genes have not been studied in moso bamboo.

Based on the moso bamboo genome, 24 *ZF-HD* genes (18 *ZHD* genes and 6 *MIF* genes) were identified and their physical and chemical properties, conserved motifs, gene evolution and expression were analyzed in this study. The amino acid length of ZF-HDs in moso bamboo ranged from 83 to 403, which was similar to that in rice (105–417). The number of *PeZF-HD*s was more than that of rice and *Arabidopsis*, similar to that of maize and tomato, and less than that of Chinese cabbage [11–14,18]. The genome size of moso bamboo (2021 Mb) is larger than that of *Arabidopsis* (164 Mb), rice (441 Mb), tomato (950 Mb) and Chinese cabbage (283.8 Mb), and comparable to that of maize (2300 Mb) [25–29]. These results indicate that the number of *ZF-HD*s has no relationship to the plant's genome size, but may be related to a gene duplication event (Table S5).

*PeZF-HD*s were classified into eight groups based on sequence homology and classification related to rice and *Arabidopsis*, different from the seven subfamilies in alfalfa and cucumber [30,31]. Phylogenetic analysis showed that all PeZF-HD subgroups had at least one homolog of rice ZF-HDs (Figure 1), indicating that the PeZF-HDs shared more sequence similarity with OsZF-HDs, and the subgroups of ZF-HDs were usually conserved in monocotyledonous plants. The groups of *PeZF-HDs* lacked the ZHD III and ZHD VI groups according to classification in relation to rice and *Arabidopsis*. Rice and *Arabidopsis* ZF-HDs were found in ZHD VI, but only *Arabidopsis* ZF-HDs were found in ZHD III, suggesting the protein divergence between monocots and dicots. The number of each group in moso bamboo was different compared to rice and *Arabidopsis*. *ZF-HDs* in plants with different gene numbers in different groups may be under different evolutionary constraints. Furthermore, MIF proteins of moso bamboo, rice and *Arabidopsis* form a phylogenetically distinct clade with ZHD proteins, indicating that the structural divergence between MIF and ZHD genes may be derived from ZHDs after losing the HD domain or may originate from MIFs after gaining the HD domain [4].

Conserved motifs and structural analysis showed that *PeZF-HDs* in the same class were similar in structure (Figure 2), suggesting that PeZF-HD members may be functionally conserved during their evolution. Motifs 2 and 3 were the essential components of the ZF-HD_dimer domain and existed in all 24 *PeZF-HDs*, implying that these two motifs were crucial for *PeZF-HDs* functions. The intron–exon analysis exhibited that half of *PeZF-HDs* lacked introns, and the remaining twelve *PeZF-HDs* had one intron, which reflected intron gain and was almost consistent with the *ZF-HDs* in tomato [13]; yet, most plant *ZF-HDs* are intronless [11,12,14,18]. Precise and complete intron gains and losses stimulate the production of new genes, based on previous reports [32]. The variable intron–exon structures of *ZF-HDs* in moso bamboo compared with that in other plants suggested that there existed structural divergence among *ZF-HDs* in moso bamboo. Moreover, the similar exon–intron organization in different subfamilies suggests that these genes were highly conserved during evolution. Thus, different motifs and exon–intron organizations may provide the structural basis for the diverse functions of PeZF-HDs. These results, combined with phylogenetic analysis, confirmed the reliability of group classification and the similar biological functions of *PeZF-HDs* within the same group.

Gene duplication patterns may reveal the generational types of genes and evolutionary ways of gene functions [33]. Almost all *PeZF-HDs* (except for *PeZF-HD19*) were duplicated through WGD or segmental duplication events, suggesting that WGD or segmental duplication played a key role in the expansion of *PeZF-HDs*, which was similar to the *ZF-HDs* in other plants, like tomato and Chinese cabbage [13,18]. The Ka/Ks values of paralogous genes were high at 0.106-0.679 (the average value was 0.375), implying purifying selection and a powerful selection constraint in *PeZF-HDs* (Table S6). Intra-genomic collinearity analysis of *PeZF-HDs* was performed and 24 pairs of duplicated genes were found in the moso bamboo genome (Figure 3). Most duplication pairs had conserved motifs and gene structures, and only a few had a certain degree of differentiation (Figure 2). For example, *PeZF-HD1* was intronless, while its corresponding *PeZF-HD3* had one intron, which could be due to the gaining of a single intron in the gene structure during evolution. Most divergence times of the *PeZF-HD* gene pairs occurred from 16.51 to 49.14 MYA, which differs from the recent whole-genome duplication time of moso bamboo (7-12 MYA) [20], suggesting that the effects of other earlier ancient whole-genome duplication events also influenced the expansion of *PeZF-HD* genes.

The different *cis*-elements on promoters likely led to functional differentiation in *PeZF-HDs*. Elements related to stress responses were found in *PeZF-HD* promoters, including MYB, MBS, LTR, TCA-element, TC-rich repeats and ABRE (Table S1), which are directly related to ABA, drought, cold and salt stress responses. *ZF-HD* genes were reported to be involved in various abiotic stresses and ABA treatment, such as salt and drought stress [12,17,19,34]. Moreover, *ZF-HD* genes in alfalfa also showed similarly decreased expression levels under alkaline stress [31]. Previous studies investigated the expression profiles of *ZF-HD* genes under multiple stress conditions with transcriptome analysis and qPCR [11,19,31]. In this study, the qPCR results showed that *PeZF-HDs* 5, 4 and 6 were differentially expressed under cold, salt and PEG stresses, respectively, which also verified the results of the *cis*-elements analysis. Most differentially expressed *PeZF-HDs* were also significantly changed under ABA treatment (Figure 5). Particularly, overlapping responses of *PeZF-HDs* to multiple stresses were found, like *PeZF-HD10*, which was induced under drought and salt stresses, helping the potential hub genes in moso bamboo to acclimate to stressful conditions. The overexpression of *AtZHD1* improved drought tolerance in *Arabidopsis* [15] and the overexpression of *CqZF-HD14* enhanced the defense system of quinoa under drought [21]. *PeZF-HD4*, the homologous gene to *AtZF-HD1*, was up-regulated under cold, salt, PEG and ABA treatments. *PeZF-HD5*, another homologous gene to *AtZF-HD1*, was up-regulated under salt treatment. *AtZHD4* was induced under cold, salt and drought treatments [35]. The expression of *AtZHD4* homolog *PeZF-HD2* also increased under cold, salt and drought treatments. These results showed that *PeZF-HDs* may play important roles in response to various abiotic stresses, most of which may be mediated by ABA signaling. The above results provide an information base for selecting candidate genes and promote further functional investigations of biotic stress resistance in moso bamboo.

4. Materials and Methods

4.1. Plant Material and Treatments

The moso bamboo (*Phyllostachys edulis*) was selected for this experiment as it has completed whole-genome sequencing. The germinated seeds were planted in pots with medium (soil and vermiculite, 1:1) in a growth room under a controlled environment (16 h light at 23 °C/8 h dark at 18 °C). Two-month-old seedlings were treated with 20% PEG6000, 100 μM ABA and 250 mM NaCl, respectively. For the low temperature treatment, the seedlings were transferred to 4 °C conditions. The leaf samples were collected along a continuous time course (0, 1, 2, 3, 4, 8 h) with three biological replicates and then stored at −80 °C. Total RNA was extracted by the RNAsimple Total RNA Kit (Tiangen, Beijing, China) and residual genomic DNA was removed with DNase I (Qiagen, Valencia, CA, USA) following the manufacturer's instructions.

4.2. Identification of ZF-HD Genes in Moso Bamboo

The *P. edulis* genome sequence was downloaded from the Bamboo Genome Database (http://202.127.18.221/bamboo/index.php, accessed on 8 April 2023). The ZF-HD gene sequences of *Oryza* and *Arabidopsis* were obtained from the rice genome annotation project and the *Arabidopsis* information resource based on published studies, respectively [4,11]. Members of the *PeZF-HD* genes were identified by BLASTP searches and Hidden Markov Model (HMM) searches. Protein sequences of *Oryza* and *Arabidopsis ZF-HDs* were used as queries to make BLASTP searches (E value $< 1 \times 10^{-10}$). The annotation file of ZF-HD_dimer domain (PF04770) was applied to build the HMM profile against the *P. edulis* protein dataset (E value $< 1 \times 10^{-10}$). The ZF-HD proteins were further identified by InterProScan (https://www.ebi.ac.uk/interpro/, accessed on 8 April 2023). A total of 24 *PeZF-HD* genes were identified in the moso bamboo genome and redundant sequences were removed.

4.3. Sequence Analysis

The molecular weight, length and isoelectric point of PeZF-HD proteins were analyzed by the ExPASy website (https://web.expasy.org, accessed on 20 April 2023) (Table 1). The CDS and genomic sequences of *PeZF-HD* were aligned to analyze the exon–intron structures using the GSDS online program [36]. The protein sequences were analyzed by the MEME program (http:/meme.nbcr.net/meme/intro.html, accessed on 20 April 2023) with the following parameters: the maximum number of motifs was 8, and the width of the optimum motif was 6–300. The subcellular location of the putative PeZF-HD proteins was predicted by Plant-mPLoc (http://www.csbio.sjtu.edu.cn/bioinf/plant-multi/, accessed on 20 April 2023).

4.4. Phylogenetic Analysis and Classification

The ZF-HD protein sequences of rice and *Arabidopsis* were obtained from the TIGR-Rice Genome Annotation Project and *Arabidopsis* Information resource, respectively [37,38]. The full-length amino acid sequences were aligned using Clustal X according to our previous report [39]. The phylogenetic tree was built using MEGA 7.0 software with the Maximum Likelihood (ML) algorithm and 1000 bootstrap replicates.

4.5. Analysis of Cis-Element in the Promoter Regions

To predict *cis*-regulatory elements in promoters of *PeZF-HD* genes, 2000 bp upstream sequences of the translational start codon were detected by PlantCARE (http://bioinformatics.psb.ugent.be/webtools/plantcare/html/, accessed on 10 May 2023). The heat map was made using TBtools by counting the *cis*-regulatory elements on each promoter (Table S1) [40].

4.6. Chromosome Localization and Gene Duplication

The structural and positional information of *PeZF-HD*s on the chromosomes of moso bamboo was downloaded from the moso bamboo genome database [24]. The positions of *PeZF-HD*s on the moso bamboo chromosomes were drawn using the MapGene2Chromosome2 web tool (http://mg2c.iask.in/, accessed on 19 May 2023) [41]. The syntenic relationships of the *ZF-HD*s in moso bamboo, rice and *Arabidopsis* were analyzed using MCScanX with default parameters [42]. The results of the chromosomal location and synteny relationships were visualized by Circos software (http://mkweb.bcgsc.ca/circos) [43]. The duplication events of the *PeZF-HD*s were analyzed using MCScanX following the default parameters.

4.7. Calculation of Ka/Ks Ratios

The non-synonymous (Ka) and synonymous (Ks) replacement rates and Ka/Ks values of the duplicated *ZF-HD* gene pairs were calculated using KaKs_Calculator software v2.0 [44]. Evolutionary divergence times of the moso bamboo *ZF-HD* gene family were calculated by the bamboo-specific divergence time formula ($T = Ks/2\lambda$, $\lambda = 6.5 \times 10^{-9}$). The selection pressure of the duplicate gene pairs was determined by the Ka/Ks ratio [45].

4.8. Quantitative Real-Time PCR Expression Analysis

Total RNA extraction, cDNA synthesis and qPCR reaction were performed according to our protocol [39]. To confirm the specificity of the qPCR reactions, a melting curve was implemented. The *PeNTB* gene was used as an internal standard [46]. Data were calculated by the $2^{-\Delta\Delta CT}$ method [47]. The heat map was made by TBtools with a normalized row scale [40]. The expression results were analyzed by three independent biological replicates. Primers are shown in Table S2. The published transcriptome data of moso bamboo under PEG, NaCl and ABA treatments were obtained from the Gene Expression Omnibus database with the accession number GSE169067 (http://www.ncbi.nlm.nih.gov/geo, accessed on 10 April 2023).

5. Conclusions

In this study, genome-wide detection and analyses of the *PeZF-HD* gene family were conducted in moso bamboo. The twenty-four PeZF-HDs were classified into two main groups and eight subgroups based on the grouping principles of the rice and *Arabidopsis* ZF-HD family, which was further supported by their similar motif compositions and exon–intron structures. WGD or segmental duplications were indicated to have been the major driving force for the expansion of *PeZF-HD*s. Multiple *PeZF-HD* expressions were stimulated by various types of abiotic stress. Our results provide a valuable basis for the functional study of ZF-HD genes and facilitate the identification of candidate genes in moso bamboo stress response networks, which may shed a light on moso bamboo cultivation.

Supplementary Materials: The following supporting information can be downloaded at: https://www.mdpi.com/article/10.3390/plants12234064/s1, Table S1: Detailed information of *ZF-HDs* identified in moso bamboo genome; Table S2: Catergory information of cis-elements in *PeZF-HDs* promoters; Table S3: Primers used in the paper; Table S4: Identification of conserved domains in PeZF-HD proteins using the SMART tool; Table S5: Quantitative distribution of each group of *ZF-HDs* in moso bamboo, rice and *Arabidopsis*; Table S6: The mode of gene duplication in *PeZF-HDs*; Table S7: The KaKs ratios and estimated divergence times for duplicate gene pairs of *PeZF-HDs*. References [48,49] are cited in the supplementary materials.

Author Contributions: Conceptualization, methodology and software, F.H. and C.T.; Data curation, J.W.; writing—original draft preparation, F.H.; writing—review and editing, F.H., J.W. and C.T. All authors have read and agreed to the published version of the manuscript.

Funding: This work was supported by grants from the National Natural Science Foundation of China (32102410) and the Natural Science Foundation of Jiangsu province (BK20210394).

Data Availability Statement: The data is contained within the manuscript and Supplementary Materials.

Conflicts of Interest: The authors declare no conflict of interest.

References

1. Pereira, A. Plant abiotic stress challenges from the changing environment. *Front. Plant Sci.* **2016**, *7*, 1123. [CrossRef] [PubMed]
2. Amorim, L.L.B.; da Fonseca Dos Santos, R.; Neto, J.P.B.; Guida-Santos, M.; Crovella, S.; Benko-Iseppon, A.M. Transcription factors involved in plant resistance to pathogens. *Curr. Protein Pept. Sci.* **2017**, *18*, 335–351. [CrossRef] [PubMed]
3. Tan, Q.K.; Irish, V.F. The Arabidopsis zinc finger-homeodomain genes encode proteins with unique biochemical properties that are coordinately expressed during floral development. *Plant Physiol.* **2006**, *140*, 1095–1108. [CrossRef] [PubMed]
4. Hu, W.; dePamphilis, C.W.; Ma, H. Phylogenetic analysis of the plant-specific zinc finger-homeobox and mini zinc finger gene families. *J. Integr. Plant Biol.* **2008**, *50*, 1031–1045. [CrossRef] [PubMed]
5. Windhövel, A.; Hein, I.; Dabrowa, R.; Stockhaus, J. Characterization of a novel class of plant homeodomain proteins that bind to the C4 phosphoenolpyruvate carboxylase gene of Flaveria trinervia. *Plant Mol. Biol.* **2001**, *45*, 201–214. [CrossRef]
6. Mukherjee, K.; Brocchieri, L.; Bürglin, T.R. A comprehensive classification and evolutionary analysis of plant homeobox genes. *Mol. Biol. Evol.* **2009**, *26*, 2775–2794. [CrossRef]
7. Ariel, F.D.; Manavella, P.A.; Dezar, C.A.; Chan, R.L. The true story of the HD-Zip family. *Trends Plant Sci.* **2007**, *12*, 419–426. [CrossRef]
8. Klug, A.; Schwabe, J.W. Protein motifs 5. Zinc fingers. *FASEB J.* **1995**, *9*, 597–604. [CrossRef]
9. Krishna, S.S.; Majumdar, I.; Grishin, N.V. Structural classification of zinc fingers: Survey and summary. *Nucleic Acids Res.* **2003**, *31*, 532–550. [CrossRef]
10. Hu, W.; Ma, H. Characterization of a novel putative zinc finger gene MIF1: Involvement in multiple hormonal regulation of Arabidopsis development. *Plant J.* **2006**, *45*, 399–422. [CrossRef]
11. Jain, M.; Tyagi, A.K.; Khurana, J.P. Genome-wide identification, classification, evolutionary expansion and expression analyses of homeobox genes in rice. *FEBS J.* **2008**, *275*, 2845–2861. [CrossRef] [PubMed]
12. Figueiredo, D.D.; Barros, P.M.; Cordeiro, A.M.; Serra, T.S.; Lourenço, T.; Chander, S.; Oliveira, M.M.; Saibo, N.J. Seven zinc-finger transcription factors are novel regulators of the stress responsive gene *OsDREB1B*. *J. Exp. Bot.* **2012**, *63*, 3643–3656. [CrossRef] [PubMed]
13. Khatun, K.; Nath, U.K.; Robin, A.H.K.; Park, J.I.; Lee, D.J.; Kim, M.B.; Kim, C.K.; Lim, K.B.; Nou, I.S.; Chung, M.Y. Genome-wide analysis and expression profiling of zinc finger homeodomain (ZHD) family genes reveal likely roles in organ development and stress responses in tomato. *BMC Genom.* **2017**, *18*, 695. [CrossRef] [PubMed]
14. Islam, M.A.U.; Nupur, J.A.; Khalid, M.H.B.; Din, A.M.U.; Shafiq, M.; Alshegaihi, R.M.; Ali, Q.; Kamran, Z.; Manzoor, M.; Haider, M.S.; et al. Genome-wide identification and in silico analysis of ZF-HD transcription factor genes in *Zea mays* L. *Genes* **2022**, *13*, 2112. [CrossRef] [PubMed]
15. Tran, L.S.; Nakashima, K.; Sakuma, Y.; Osakabe, Y.; Qin, F.; Simpson, S.D.; Maruyama, K.; Fujita, Y.; Shinozaki, K.; Yamaguchi-Shinozaki, K. Co-expression of the stress-inducible zinc finger homeodomain ZFHD1 and NAC transcription factors enhances expression of the *ERD1* gene in *Arabidopsis*. *Plant J.* **2007**, *49*, 46–63. [CrossRef] [PubMed]
16. Niu, H.; Xia, P.; Hu, Y.; Zhan, C.; Li, Y.; Gong, S.; Li, Y.; Ma, D. Genome-wide identification of ZF-HD gene family in Triticum aestivum: Molecular evolution mechanism and function analysis. *PLoS ONE* **2021**, *16*, e0256579. [CrossRef]
17. Zhao, T.T.; Wang, Z.Y.; Bao, Y.F.; Zhang, X.C.; Yang, H.H.; Dong, Z.Y.; Jiang, J.B.; Zhang, H.; Li, J.F.; Chen, Q.S.; et al. Downregulation of SL-ZH13 transcription factor gene expression decreases drought tolerance of tomato. *J. Integr. Agric.* **2019**, *18*, 1579–1586. [CrossRef]

18. Wang, W.; Wu, P.; Li, Y.; Hou, X. Genome-wide analysis and expression patterns of ZF-HD transcription factors under different developmental tissues and abiotic stresses in Chinese cabbage. *Mol. Genet. Genom.* **2016**, *291*, 1451–1464. [CrossRef]
19. Xing, L.; Peng, K.; Xue, S.; Yuan, W.; Zhu, B.; Zhao, P.; Wu, H.; Cheng, Y.; Fang, M.; Liu, Z. Genome-wide analysis of zinc finger-homeodomain (ZF-HD) transcription factors in diploid and tetraploid cotton. *Funct. Integr. Genom.* **2022**, *22*, 1269–1281. [CrossRef]
20. Liu, M.; Wang, X.; Sun, W.; Ma, Z.; Zheng, T.; Huang, L.; Wu, Q.; Tang, Z.; Bu, T.; Li, C.; et al. Genome-wide investigation of the *ZF-HD* gene family in Tartary buckwheat (*Fagopyrum tataricum*). *BMC Plant Biol.* **2019**, *19*, 248. [CrossRef]
21. Sun, W.; Wei, J.; Wu, G.; Xu, H.; Chen, Y.; Yao, M.; Zhan, J.; Yan, J.; Wu, N.; Chen, H.; et al. CqZF-HD14 enhances drought tolerance in quinoa seedlings through interaction with CqHIPP34 and CqNAC79. *Plant Sci.* **2022**, *323*, 111406. [CrossRef] [PubMed]
22. Peng, Z.; Lu, Y.; Li, L.; Zhao, Q.; Feng, Q.; Gao, Z.; Lu, H.; Hu, T.; Yao, N.; Liu, K.; et al. The draft genome of the fast-growing non-timber forest species moso bamboo (*Phyllostachys heterocycla*). *Nat. Genet.* **2013**, *45*, 456–461. [CrossRef] [PubMed]
23. Wei, Q.; Guo, L.; Jiao, C.; Fei, Z.; Chen, M.; Cao, J.; Ding, Y.; Yuan, Q. Characterization of the developmental dynamics of the elongation of a bamboo internode during the fast growth stage. *Tree Physiol.* **2019**, *39*, 1201–1214. [CrossRef]
24. Zhao, H.; Gao, Z.; Wang, L.; Wang, J.; Wang, S.; Fei, B.; Chen, C.; Shi, C.; Liu, X.; Zhang, H.; et al. Chromosome-level reference genome and alternative splicing atlas of moso bamboo (*Phyllostachys edulis*). *GigaScience* **2018**, *7*, giy115. [CrossRef] [PubMed]
25. Duvick, J.; Fu, A.; Muppirala, U.; Sabharwal, M.; Wilkerson, M.D.; Lawrence, C.J.; Lushbough, C.; Brendel, V. PlantGDB: A resource for comparative plant genomics. *Nucleic Acids Res.* **2008**, *36* (Suppl. 1), D959–D965. [CrossRef] [PubMed]
26. Burr, B. Mapping and sequencing the rice genome. *Plant Cell* **2002**, *14*, 521–523. [CrossRef]
27. Filichkin, S.A.; Priest, H.D.; Givan, S.A.; Shen, R.; Bryant, D.W.; Fox, S.E.; Wong, W.K.; Mockler, T.C. Genome-wide mapping of alternative splicing in Arabidopsis thaliana. *Genome Res.* **2010**, *20*, 45–58. [CrossRef]
28. Van der Hoeven, R.; Ronning, C.; Giovannoni, J.; Martin, G.; Tanksley, S. Deductions about the number, organization, and evolution of genes in the tomato genome based on analysis of a large expressed sequence tag collection and selective genomic sequencing. *Plant Cell* **2002**, *14*, 1441–1456. [CrossRef]
29. Wang, X.; Wang, H.; Wang, J.; Sun, R.; Wu, J.; Liu, S.; Bai, Y.; Mun, J.H.; Bancroft, I.; Cheng, F.; et al. The genome of the mesopolyploid crop species *Brassica rapa*. *Nat. Genet.* **2011**, *43*, 1035–1039. [CrossRef]
30. Lai, W.; Zhu, C.; Hu, Z.; Liu, S.; Wu, H.; Zhou, Y. Identification and transcriptional analysis of zinc finger-homeodomain (ZF-HD) family genes in cucumber. *Biochem. Genet.* **2021**, *59*, 884–901. [CrossRef]
31. He, K.; Li, C.; Zhang, Z.; Zhan, L.; Cong, C.; Zhang, D.; Cai, H. Genome-wide investigation of the *ZF-HD* gene family in two varieties of alfalfa (*Medicago sativa* L.) and its expression pattern under alkaline stress. *BMC Genom.* **2022**, *23*, 150. [CrossRef] [PubMed]
32. Roy, S.W.; Penny, D. Patterns of intron loss and gain in plants: Intron loss-dominated evolution and genome-wide comparison of *O. sativa* and *A. thaliana*. *Mol. Biol. Evol.* **2007**, *24*, 171–181. [CrossRef] [PubMed]
33. Wang, T.; Hu, J.; Ma, X.; Li, C.; Yang, Q.; Feng, S.; Li, M.; Li, N.; Song, X. Identification, evolution and expression analyses of whole genome-wide TLP gene family in *Brassica napus*. *BMC Genom.* **2020**, *21*, 264. [CrossRef]
34. Shi, B.; Haq, I.; Fiaz, S.; Alharthi, B.; Xu, M.; Wang, J.; Hou, W.; Feng, X. Genome-wide identification and expression analysis of the ZF-HD gene family in pea (*Pisum sativum* L.). *Front. Genet.* **2022**, *13*, 1089375. [CrossRef] [PubMed]
35. Barth, O.; Vogt, S.; Uhlemann, R.; Zschiesche, W.; Humbeck, K. Stress induced and nuclear localized HIPP26 from Arabidopsis thaliana interacts via its heavy metal associated domain with the drought stress related zinc finger transcription factor ATHB29. *Plant Mol. Biol.* **2009**, *69*, 213–226. [CrossRef] [PubMed]
36. Hu, B.; Jin, J.; Guo, A.Y.; Zhang, H.; Luo, J.; Gao, G. GSDS 2.0: An upgraded gene feature visualization server. *Bioinformatics* **2015**, *31*, 1296–1297. [CrossRef]
37. Yuan, Q.; Ouyang, S.; Liu, J.; Suh, B.; Cheung, F.; Sultana, R.; Lee, D.; Quackenbush, J.; Buell, C.R. The TIGR rice genome annotation resource: Annotating the rice genome and creating resources for plant biologists. *Nucleic Acids Res.* **2003**, *31*, 229–233. [CrossRef]
38. Swarbreck, D.; Wilks, C.; Lamesch, P.; Berardini, T.Z.; Garcia-Hernandez, M.; Foerster, H.; Li, D.; Meyer, T.; Muller, R.; Ploetz, L.; et al. The Arabidopsis Information Resource (TAIR): Gene structure and function annotation. *Nucleic Acids Res.* **2008**, *36*, D1009–D1014. [CrossRef]
39. Huang, F.; Liu, T.; Hou, X. Isolation and functional characterization of a floral repressor, BcMAF1, from Pak-choi (*Brassica rapa* ssp. chinensis). *Front. Plant Sci.* **2018**, *9*, 290. [CrossRef]
40. Chen, C.; Chen, H.; Zhang, Y.; Thomas, H.R.; Frank, M.H.; He, Y.; Xia, R. TBtools: An integrative toolkit developed for interactive analyses of big biological data. *Mol. Plant* **2020**, *13*, 1194–1202. [CrossRef]
41. Jiangtao, C.; Yingzhen, K.; Qian, W.; Yuhe, S.; Daping, G.; Jing, L.; Guanshan, L. MapGene2Chrom, a tool to draw gene physical map based on Perl and SVG languages. *Yi chuan* **2015**, *37*, 91–97.
42. Wang, Y.; Tang, H.; Debarry, J.D.; Tan, X.; Li, J.; Wang, X.; Lee, T.H.; Jin, H.; Marler, B.; Guo, H.; et al. MCScanX: A toolkit for detection and evolutionary analysis of gene synteny and collinearity. *Nucleic Acids Res.* **2012**, *40*, e49. [CrossRef] [PubMed]
43. Krzywinski, M.; Schein, J.; Birol, I.; Connors, J.; Gascoyne, R.; Horsman, D.; Jones, S.J.; Marra, M.A. Circos: An information aesthetic for comparative genomics. *Genome Res.* **2009**, *19*, 1639–1645. [CrossRef] [PubMed]
44. Zhang, Z.; Li, J.; Zhao, X.Q.; Wang, J.; Wong, G.K.; Yu, J. KaKs_Calculator: Calculating Ka and Ks through model selection and model averaging. *Proteom. Bioinf.* **2006**, *4*, 259–263. [CrossRef]

45. Lynch, M.; Conery, J.S. The evolutionary fate and consequences of duplicate genes. *Science* **2000**, *290*, 1151–1155. [CrossRef] [PubMed]
46. Fan, C.; Ma, J.; Guo, Q.; Li, X.; Wang, H.; Lu, M. Selection of reference genes for quantitative real-time PCR in bamboo (*Phyllostachys edulis*). *PLoS ONE* **2013**, *8*, e56573. [CrossRef]
47. Livak, K.J.; Schmittgen, T.D. Analysis of relative gene expression data using real-time quantitative PCR and the $2^{-\Delta\Delta CT}$ Method. *Methods* **2001**, *25*, 402–408. [CrossRef]
48. Bai, Y.; Kissoudis, C.; Yan, Z.; Visser, R.G.; van der Linden, G. Plant behaviour under combined stress: Tomato responses to combined salinity and pathogen stress. *Plant J.* **2018**, *93*, 781–793. [CrossRef]
49. Luo, H.; Song, F.; Zheng, Z. Overexpression in transgenic tobacco reveals different roles for the rice homeodomain gene OsBIHD1 in biotic and abiotic stress responses. *J. Exp. Bot.* **2005**, *56*, 2673–2682. [CrossRef]

plants

Article

Genome-Wide Identification and Characterization of the NAC Gene Family and Its Involvement in Cold Response in *Dendrobium officinale*

Qianyu Yang [1], Zhihui Li [1], Xiao Wang [1,2], Chunqian Jiang [3], Feihong Liu [1], Yuxin Nian [1], Xiaoyun Fu [1], Guangzhu Zhou [1], Lei Liu [3] and Hui Wang [3,*]

[1] College of Forestry, Shenyang Agricultural University, Shenhe District, Shenyang 110866, China; yangdayu0055@163.com (Q.Y.); wx1970218@163.com (X.W.); lfh-303@163.com (F.L.); 2004500017@syau.edu.cn (Y.N.)
[2] Institute of Applied Ecology, Chinese Academy of Sciences, Shenyang 110016, China
[3] Key Laboratory of Tree Breeding and Cultivation of National Forestry and Grassland Administration, Research Institute of Forestry, Chinese Academy of Forestry, Beijing 100091, China; liulei519@caf.ac.cn (L.L.)
* Correspondence: drwanghui@caf.ac.cn

Abstract: The NAC (NAM, ATAF1/2 and CUC2) gene family is one of the largest plant-specific transcription factor families, functioning as crucial regulators in diverse biological processes such as plant growth and development as well as biotic and abiotic stress responses. Although it has been widely characterized in many plants, the significance of the NAC family in *Dendrobium officinale* remained elusive up to now. In this study, a genome-wide search method was conducted to identify NAC genes in *Dendrobium officinale* (DoNACs) and a total of 110 putative DoNACs were obtained. Phylogenetic analysis classified them into 15 subfamilies according to the nomenclature in Arabidopsis and rice. The members in the subfamilies shared more similar gene structures and conversed protein domain compositions. Furthermore, the expression profiles of these DoNACs were investigated in diverse tissues and under cold stress by RNA-seq data. Then, a total of five up-regulated and five down-regulated, cold-responsive DoNACs were validated through QRT-PCR analysis, demonstrating they were involved in regulating cold stress response. Additionally, the subcellular localization of two down-regulated candidates (DoNAC39 and DoNAC58) was demonstrated to be localized in the nuclei. This study reported the genomic organization, protein domain compositions and expression patterns of the NAC family in *Dendrobium officinale*, which provided targets for further functional studies of DoNACs and also contributed to the dissection of the role of NAC in regulating cold tolerance in *Dendrobium officinale*.

Keywords: cold stress; *Dendrobium officinale*; expression profiles; NAC gene family

Citation: Yang, Q.; Li, Z.; Wang, X.; Jiang, C.; Liu, F.; Nian, Y.; Fu, X.; Zhou, G.; Liu, L.; Wang, H. Genome-Wide Identification and Characterization of the NAC Gene Family and Its Involvement in Cold Response in *Dendrobium officinale*. *Plants* **2023**, *12*, 3626. https://doi.org/10.3390/plants12203626

Academic Editor: Emilia Apostolova

Received: 28 July 2023
Revised: 21 September 2023
Accepted: 9 October 2023
Published: 20 October 2023

1. Introduction

The NAC gene family is one of the largest plant-specific transcription factor families, playing a crucial role in regulating growth and development as well as responses to diverse biotic and abiotic stresses in plants [1,2]. The NAC family is named according to three conserved domains, including petunia no apical meristem (NAM), ATAF1/2 and cup-shaped cotyledon (CUC) [3,4]. The NAM gene was firstly cloned and functionally characterized in petunias, which was found to determine the position of the shoot apical meristem and primordia or in plants [5]. Subsequently, the ATAF1/2 and CUC genes were successively found and identified [6]. Generally, NAC proteins have a typical and conserved N-terminal region (NAC domain), together with a variable C terminal [7]. The NAC domain is about 150 AA in length, comprising five subdomains (A–E), of which subdomains A, C and D are rather conserved, while subdomains B and E are highly viable with special functions [1,8–10]. At the same time, the C-terminal region is mainly involved in transcriptional activation or repression to achieve its regulation function (TAR or TRR) [11,12].

Extensive studies have demonstrated that NAC transcription factors are widely involved in various biological processes, such as cell division, shoot apical meristem, flower development and fruit ripening, as well as leaf senescence and so on [12–14]. *NAC29* and *NAC21* were reported to play an indispensable role in cellulose synthesis in rice [15]. Over-expression of *OsNAC2* promoted shoot branching and also increased the growth of rice tiller buds [16]. In Arabidopsis, *ANAC046* and *ANAC087* regulated the programmed cell death of lateral roots [17] and *NAP* functioned as an important regulator to control leaf senescence and floral morphogenesis [18,19]. Meanwhile, the NAC family has also been found to participate in the response to diverse stresses [13,20]. It is reported that *SNAC2* functioned to regulate cold and salt tolerance in rice [21]. Over-expression of *ONAC022* significantly improved drought and salt tolerance in rice [22]. In tomatoes, the NAC protein JUNGBRUNNEN1 can enhance drought tolerance by activating the expression of stress-induced genes [23]. In Arabidopsis, *ANAC013* and *ANAC017* regulated the oxidative stress response by mediating mitochondrial feedback signals [24,25]. *MaNAC1* interacted with *MaCBF1* to modulate cold tolerance in bananas [26]. In light of its importance, the NAC gene family has been systematically investigated in many plant species at the whole-genome level, including Arabidopsis [1], rice [1], maize [27], soybeans [28] and wild emmer wheat [29], some of which have been well characterized functionally [30].

Dendrobium officinale is a well-known medicinal and ornamental herb belonging to the orchid family Dendrobium genus, which originated from East Asia and Southeast Asia, mainly in China, Japan, India and other regions [31]. *D. officinale* contains rich bioactive components such as polysaccharides, dendrobium alkaloids, flavonoids and so on, and is used as a precious Chinese herbal medicine with a medicinal-use history of more than 2000 years [32]. Due to its rareness and significant medicinal value, the production of *D. officinale* not only provides the necessary plant material for traditional Chinese medicine, but also brings huge economic benefits to accelerate rural revitalization. However, abiotic stresses, especially cold stress, can severely impair the production of *Dendrobium officinale*, which reduces the yield of *D. officinale* and also damages the medicinal value [33]. Thus, it is urgent to breed for cold-tolerant *D. officinale* by making use of the elite genes associated with cold tolerance. Although previous studies have revealed that NAC genes played crucial roles in controlling cold tolerance in different plants [12,34], the significance of NAC in *D. officinale* remained elusive up to now. In this study, we performed a genome-wide search of the NAC family in *D. officinale* at the genome scale and the phylogenetic relationship and a conserved motif of the putative DoNACs were investigated. Then, their expression patterns were detected in diverse tissues and under cold stress by RNA-seq data, and 10 cold-responsive candidates were further validated by QPCR analysis. It is the first study to identify the NAC gene family in *D. officinale*, which provides potential targets for further functional study and contributes to improving the cold tolerance in *D. officinale* and beyond.

2. Results

2.1. Identification of the NAC Gene Family in D. officinale

To globally obtain the NAC genes in the *D. officinale* genome, a genome-wide search method was conducted using both HMMSearch and BLASTP methods. After removing redundant sequences and confirming the conserved NAC or NAM domains, a total of 110 putative NAC genes were identified and designated as *DoNAC1* to *DoNAC110* (Table 1). All of them contained a conserved NAC domain (PF01849) or NAM domain (PF02365), indicating the accuracy of the prediction. The protein length of these *DoNACs* ranged from 97 (*DoNAC47*) to 1174 AA (*DoNAC13*) with an average length of 292 AA. Their molecular weight (MW) ranged from 11.14 (*DoNAC47*) to 129.50 (*DoNAC13*) kDa with an average of 33.20 kDa and the pI value ranged from 4.19 (*DoNAC93*) to 10.28 (*DoNAC3*) with an average of 7.29. These diverge protein characteristics suggested their divergent functions. Finally, their subcellular localization was predicted and results showed that most of the DoNACs were located in the nucleus, although some of them were predicted into cytoplasm, indicating that DoNAC acted as transcription factors to play regulatory roles in biological processes in *D. officinale* (Table 1).

Table 1. The NAC genes identified in *D. officinale*.

Gene ID	Gene Name	Protein Size (AA)	pI	Molecular Weight (kDa)	Subcellular
Dendrobium_GLEAN_10145674	DoNAC1	403	5.83	45.14922	Nuclear
Dendrobium_GLEAN_10132016	DoNAC2	214	8.96	24.05615	Nuclear
Dendrobium_GLEAN_10127752	DoNAC3	201	10.28	23.17154	Mitochondrial
Dendrobium_GLEAN_10125186	DoNAC4	341	6.88	39.36333	Nuclear
Dendrobium_GLEAN_10125242	DoNAC5	105	5	11.45904	Cytoplasmic
Dendrobium_GLEAN_10121576	DoNAC6	338	7.11	38.55624	Nuclear
Dendrobium_GLEAN_10119429	DoNAC7	313	6.55	36.06974	Nuclear
Dendrobium_GLEAN_10119030	DoNAC8	125	9.54	14.32841	Mitochondrial
Dendrobium_GLEAN_10116337	DoNAC9	280	5.81	32.51057	Cytoplasmic
Dendrobium_GLEAN_10115867	DoNAC10	347	4.33	39.12014	Nuclear
Dendrobium_GLEAN_10114825	DoNAC11	195	5.28	22.19483	Cytoplasmic
Dendrobium_GLEAN_10113712	DoNAC12	709	4.63	78.28001	Nuclear
Dendrobium_GLEAN_10113713	DoNAC13	1174	4.78	129.50276	Nuclear
Dendrobium_GLEAN_10110893	DoNAC14	297	9.1	34.35212	Nuclear
Dendrobium_GLEAN_10110543	DoNAC15	398	6.68	45.2573	Nuclear
Dendrobium_GLEAN_10110544	DoNAC16	119	6.29	14.24564	Cytoplasmic
Dendrobium_GLEAN_10110681	DoNAC17	341	6.32	38.85693	Nuclear
Dendrobium_GLEAN_10109120	DoNAC18	251	9.36	28.12291	Nuclear
Dendrobium_GLEAN_10106022	DoNAC19	117	9.68	13.13935	Extracellular
Dendrobium_GLEAN_10104194	DoNAC20	282	9.3	32.25563	Nuclear
Dendrobium_GLEAN_10100250	DoNAC21	300	5.37	34.18254	Nuclear
Dendrobium_GLEAN_10098876	DoNAC22	174	9	20.14462	Nuclear
Dendrobium_GLEAN_10096163	DoNAC23	641	8.75	72.99217	Nuclear
Dendrobium_GLEAN_10091520	DoNAC24	284	8.7	31.92515	Chloroplast
Dendrobium_GLEAN_10089166	DoNAC25	316	9.49	35.73608	Nuclear
Dendrobium_GLEAN_10089170	DoNAC26	318	5.75	36.07187	Cytoplasmic
Dendrobium_GLEAN_10088322	DoNAC27	130	4.62	15.0337	Nuclear
Dendrobium_GLEAN_10087230	DoNAC28	162	9.18	18.98751	Mitochondrial
Dendrobium_GLEAN_10083779	DoNAC29	293	8.37	32.75773	Nuclear
Dendrobium_GLEAN_10080940	DoNAC30	262	5.23	30.17884	Nuclear
Dendrobium_GLEAN_10080942	DoNAC31	262	5.32	30.16295	Cytoplasmic
Dendrobium_GLEAN_10079456	DoNAC32	285	5.95	32.94021	Nuclear
Dendrobium_GLEAN_10078554	DoNAC33	101	4.78	11.51279	Cytoplasmic
Dendrobium_GLEAN_10078555	DoNAC34	359	6.05	40.56424	Cytoplasmic
Dendrobium_GLEAN_10077726	DoNAC35	160	9.74	18.50435	Mitochondrial
Dendrobium_GLEAN_10073743	DoNAC36	661	6	75.60385	Nuclear
Dendrobium_GLEAN_10072577	DoNAC37	151	8.76	17.89539	Nuclear
Dendrobium_GLEAN_10069509	DoNAC38	208	8.57	24.32624	Cytoplasmic
Dendrobium_GLEAN_10069510	DoNAC39	267	5.08	31.15486	Nuclear
Dendrobium_GLEAN_10068907	DoNAC40	308	6.49	35.85838	Nuclear
Dendrobium_GLEAN_10066548	DoNAC41	304	7.69	34.92676	Nuclear
Dendrobium_GLEAN_10065340	DoNAC42	221	9.17	25.23475	Nuclear
Dendrobium_GLEAN_10061197	DoNAC43	128	8.28	14.26358	Cytoplasmic
Dendrobium_GLEAN_10061198	DoNAC44	220	8.59	24.31788	Extracellular
Dendrobium_GLEAN_10061199	DoNAC45	112	4.69	12.52021	Extracellular
Dendrobium_GLEAN_10060620	DoNAC46	251	8.78	28.10788	Nuclear
Dendrobium_GLEAN_10060621	DoNAC47	97	4.52	11.14347	Nuclear
Dendrobium_GLEAN_10058619	DoNAC48	278	8.77	31.99242	Cytoplasmic
Dendrobium_GLEAN_10053684	DoNAC49	313	8.92	35.57001	Cytoplasmic
Dendrobium_GLEAN_10053685	DoNAC50	187	8.48	20.87264	Nuclear
Dendrobium_GLEAN_10052939	DoNAC51	251	8.95	28.01581	Nuclear
Dendrobium_GLEAN_10052503	DoNAC52	311	9	34.53148	Nuclear
Dendrobium_GLEAN_10052536	DoNAC53	546	4.51	61.31671	Nuclear
Dendrobium_GLEAN_10051831	DoNAC54	430	6.2	47.6221	Nuclear
Dendrobium_GLEAN_10049252	DoNAC55	319	9.14	35.37876	Nuclear
Dendrobium_GLEAN_10049133	DoNAC56	630	8.22	71.57787	Cytoplasmic
Dendrobium_GLEAN_10048805	DoNAC57	378	8.38	41.8827	Nuclear
Dendrobium_GLEAN_10048260	DoNAC58	321	5.32	36.82507	Nuclear
Dendrobium_GLEAN_10046985	DoNAC59	160	9.32	18.69425	Nuclear
Dendrobium_GLEAN_10046111	DoNAC60	262	6.94	29.87379	Extracellular

Table 1. *Cont.*

Gene ID	Gene Name	Protein Size (AA)	pI	Molecular Weight (kDa)	Subcellular
Dendrobium_GLEAN_10045817	DoNAC61	236	4.94	27.23745	Nuclear
Dendrobium_GLEAN_10043622	DoNAC62	352	6.76	41.10419	Nuclear
Dendrobium_GLEAN_10042843	DoNAC63	345	7.6	39.17742	Nuclear
Dendrobium_GLEAN_10042799	DoNAC64	277	8.24	30.96703	Nuclear
Dendrobium_GLEAN_10042684	DoNAC65	369	6.01	41.49966	Nuclear
Dendrobium_GLEAN_10042333	DoNAC66	334	8.56	38.10982	Nuclear
Dendrobium_GLEAN_10042836	DoNAC67	437	6.26	50.02975	Nuclear
Dendrobium_GLEAN_10042279	DoNAC68	222	8.69	24.99947	Nuclear
Dendrobium_GLEAN_10042421	DoNAC69	157	9.64	18.32222	Cytoplasmic
Dendrobium_GLEAN_10041251	DoNAC70	286	7.63	33.8024	Nuclear
Dendrobium_GLEAN_10040136	DoNAC71	245	5.73	27.80819	Cytoplasmic
Dendrobium_GLEAN_10039845	DoNAC72	283	6.67	32.33944	Nuclear
Dendrobium_GLEAN_10039511	DoNAC73	409	6.26	45.40578	Nuclear
Dendrobium_GLEAN_10037990	DoNAC74	319	5.63	36.12287	Cytoplasmic
Dendrobium_GLEAN_10036377	DoNAC75	116	6.28	13.35206	Nuclear
Dendrobium_GLEAN_10036378	DoNAC76	335	8.37	36.74328	Nuclear
Dendrobium_GLEAN_10036159	DoNAC77	326	8.48	36.90069	Nuclear
Dendrobium_GLEAN_10034487	DoNAC78	146	7.76	17.00228	Cytoplasmic
Dendrobium_GLEAN_10032744	DoNAC79	309	8.62	35.55115	Cytoplasmic
Dendrobium_GLEAN_10032445	DoNAC80	379	9.07	43.11576	Nuclear
Dendrobium_GLEAN_10030109	DoNAC81	291	9.05	32.6348	Nuclear
PEQU_23244-D2	DoNAC82	345	6.09	39.06895	Nuclear
Dendrobium_GLEAN_10030254	DoNAC83	342	5.79	38.68952	Nuclear
Dendrobium_GLEAN_10027918	DoNAC84	534	9.16	60.5709	Nuclear
Dendrobium_GLEAN_10027115	DoNAC85	555	8.97	64.00039	Plasma Membrane
Dendrobium_GLEAN_10026921	DoNAC86	283	6.67	32.33944	Nuclear
Dendrobium_GLEAN_10023733	DoNAC87	593	5.02	66.27489	Nuclear
Dendrobium_GLEAN_10022335	DoNAC88	371	5.27	41.05842	Cytoplasmic
Dendrobium_GLEAN_10018741	DoNAC89	198	6.16	22.51354	Nuclear
Dendrobium_GLEAN_10016232	DoNAC90	330	6.8	37.84653	Nuclear
Dendrobium_GLEAN_10016023	DoNAC91	298	6.08	33.85566	Nuclear
Dendrobium_GLEAN_10015903	DoNAC92	199	5.81	22.27767	Nuclear
Dendrobium_GLEAN_10014831	DoNAC93	130	4.19	14.78027	Nuclear
Dendrobium_GLEAN_10014832	DoNAC94	561	5.74	64.18911	Nuclear
Dendrobium_GLEAN_10014534	DoNAC95	340	5.96	38.01689	Nuclear
Dendrobium_GLEAN_10014535	DoNAC96	183	5.89	20.71126	Cytoplasmic
Dendrobium_GLEAN_10014195	DoNAC97	109	5.35	12.24524	Cytoplasmic
Dendrobium_GLEAN_10013786	DoNAC98	177	9.6	19.70194	Chloroplast
Dendrobium_GLEAN_10010728	DoNAC99	313	7.8	35.8342	Nuclear
Dendrobium_GLEAN_10010634	DoNAC100	148	8.49	17.43163	Nuclear
Dendrobium_GLEAN_10010200	DoNAC101	103	9.94	12.06804	Mitochondrial
Dendrobium_GLEAN_10008910	DoNAC102	321	6.06	36.54962	Nuclear
Dendrobium_GLEAN_10008576	DoNAC103	200	9.05	23.83835	Plasma Membrane
Dendrobium_GLEAN_10007482	DoNAC104	311	5.68	35.97639	Nuclear
Dendrobium_GLEAN_10006697	DoNAC105	178	9.43	21.16647	Cytoplasmic
Dendrobium_GLEAN_10005041	DoNAC106	230	9.61	26.40513	Nuclear
Dendrobium_GLEAN_10003671	DoNAC107	144	9.46	17.19184	Cytoplasmic
Dendrobium_GLEAN_10003465	DoNAC108	275	8.66	31.21719	Nuclear
Dendrobium_GLEAN_10003618	DoNAC109	148	8.39	17.4567	Nuclear
Dendrobium_GLEAN_10000578	DoNAC110	131	7.87	14.95097	Nuclear

2.2. *Phylogenetic and Conserved Motif Analysis of DoNACs*

To understand the phylogenetic relationships of these putative *DoNACs*, the identified 110 DoNACs, together with 105 Arabidopsis and 137 rice NAC proteins, were aligmented to perform phylogenetic analysis. Based on their phylogenetic relationships, these NAC proteins could be divided into 15 groups; homologous pairs were also determined, having a closer phylogenetic relationship to cluster together (Figure 1). According to the classification criteria of Arabidopsis and rice [1], these DoNAC proteins categorized into different subfamilies, providing some clues about the function of these DoNACs. It was shown that the DoNACs were present in all of the 15 subfamilies, suggesting that no obvious NAC gene loss occurred in *D. officinale*. Furthermore, the number of DoNACs in each subfamily ranged from two to 19, of which the NAM subfamily possessed the most members while NAC2, TIPANAC001 and OsNAC8 contained only two DoNACs. Meanwhile, it is obvious that several NAC proteins from the same species generally cluster together, which might be the result of the segmental duplication events of the NAC genes in their genomes [29]. Compared to *D. officinale* and Arabidopsis, duplication events are more frequent in rice.

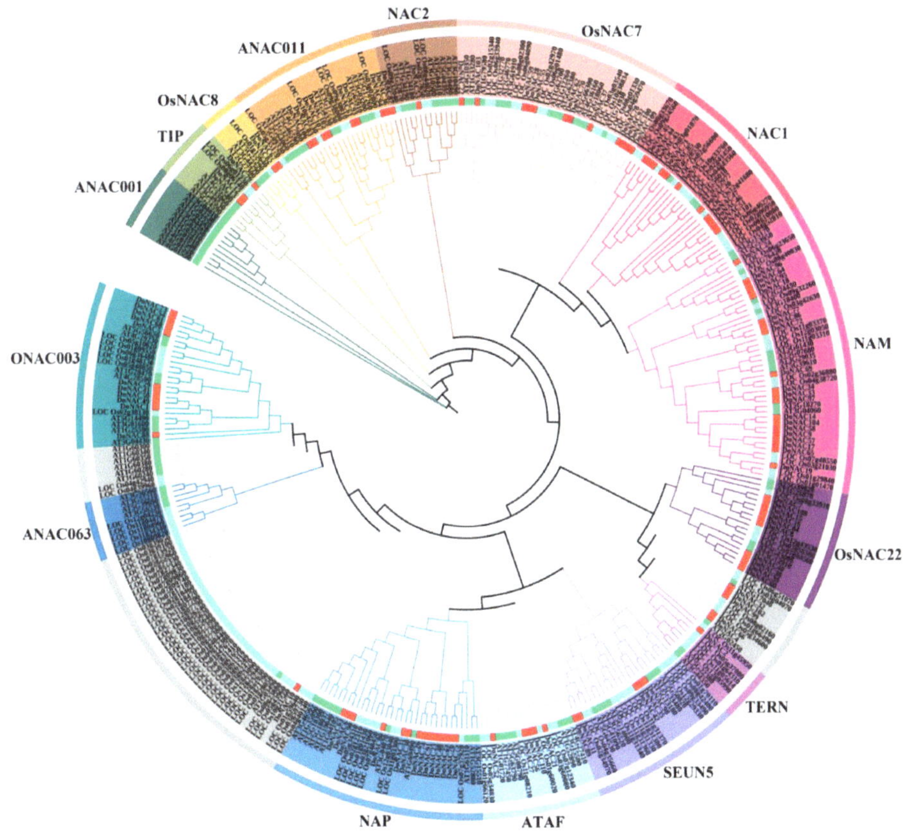

Figure 1. Phylogenetic analysis of the NAC proteins of Arabidopsis, rice and *D. officinale*. The phylogenetic tree was constructed using the neighbor-joining (NJ) method with 1000 bootstrap replications.

Based on the multi-sequence alignment, the conserved region in these 110 DoNAC proteins was further identified (Figure 2). Results showed that the N-terminus seemed rather conservative among them, while the C-terminus was highly variable, which was consistent with previous studies. A total of five conserved regions were found, four of which were located at the N-terminus, and the remaining one was adjacent to the C-terminus. It was found that protein conservation gradually becomes lower from N to C terminus. Furthermore, the conserved functional domain among them were analyzed using the MEME tool to obtain clues about the function that they might be involved in. In total, nine highly conserved functional domains were found and all of them were found adjacent to the N-terminus, with none in found in the C-terminus (Figure 3). Most of the DoNACs possessed Motif 1 to 8, suggesting they are highly conserved in the functional domain composition. It is no accident that all of the DoNAC proteins contained the NAM, ATAF and CUC domains.

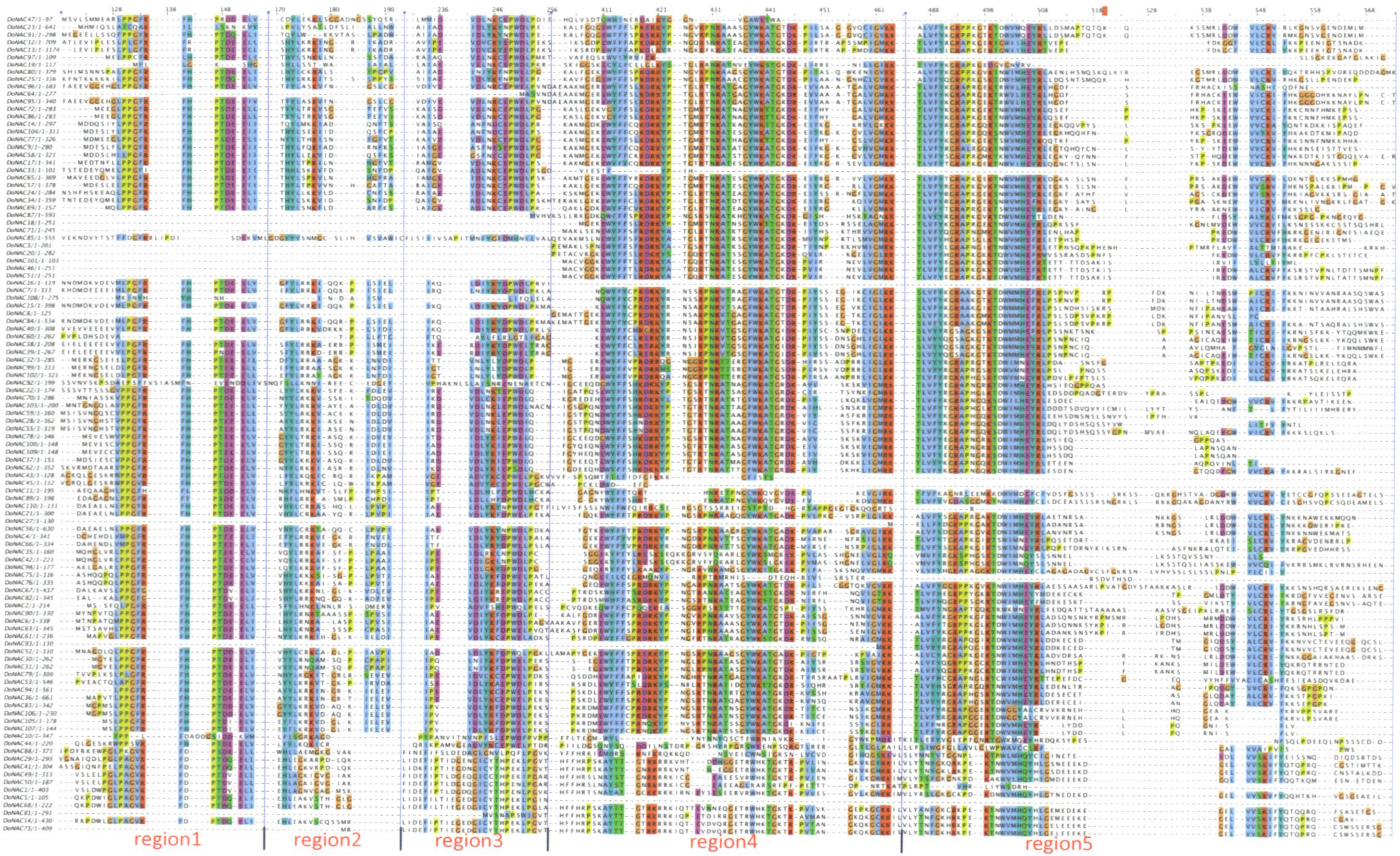

Figure 2. The conserved region identified in the NAC protein in *D. officinale* through multi-sequence alignment.

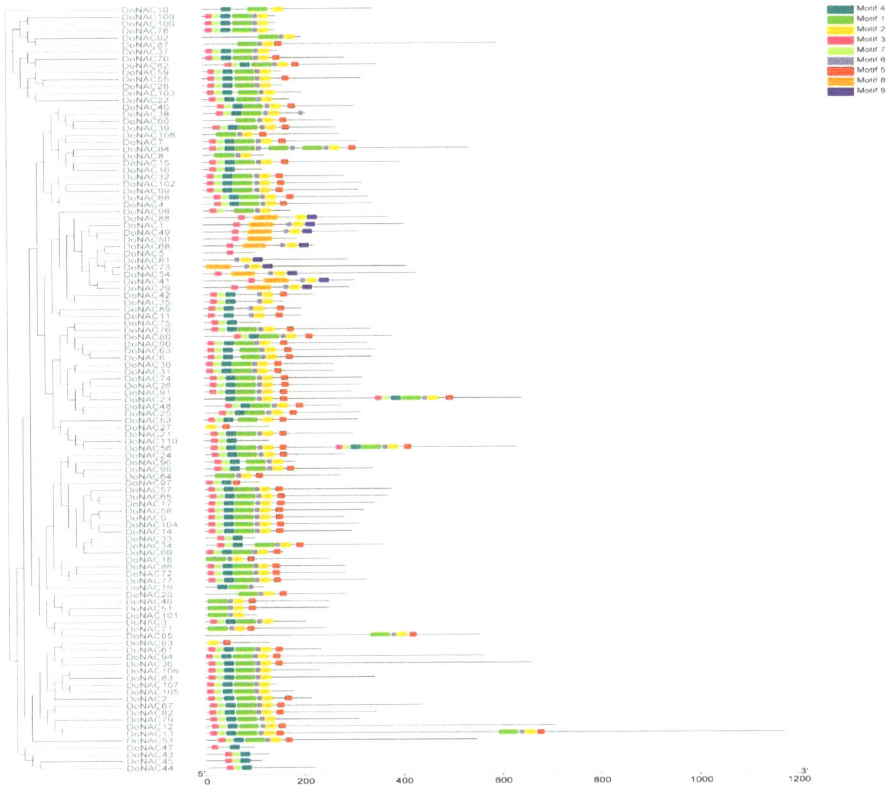

Figure 3. The conserved domain organization in the NAC protein in *D. officinale*.

At the same time, Motif 9 was only found in a few DoNACs, which have relatively high protein sequence divergents in conserved protein regions two, three and four, indicating that Motif 9 might be the novel formed domain with some specific function. Further analysis found that Motif 9 is associated with proto2021_ 04 Eukaryote PS00880, namely the Acyl CoA binding (ACB) domain. These NAC proteins, having Motif 9, may participate in the regulation of energy metabolism. Finally, it is obvious that members in the same subfamily have similar conserved protein regions and motif compositions, suggesting their similar biological function.

2.3. Cis-Element Analysis of DoNACs

The 1.5 kb genomic sequences upstream from the transcription start sites (TSS) of these DoNACs were extracted to predict the cis-acting elements. Totally, 1890 cis-elements were found in them with an average of 17.2 elements per *DoNAC* gene, which are widely associated with growth and development (1004), differentiation and specificity (33) as well as hormone responsiveness (299) and stress responsiveness (562) (Figure 4 and Table S1). A large number of cis-elements associated with growth and development were found in the *DoNACs*, such as CAT-box (related to meristem expression), circadian (related to circadian control), GCN4_motif (related to endosperm expression) and MSA-like (related to cell cycle regulation). Almost each of the *DoNACs* contained the cis-element are involved in light responsiveness, including G-box, Sp1, TCCC-motif and ATCT-motif. HD-Zip I elements that are involved in the differentiation of the palisade mesophyll cells, were found in nine *DoNACs*, and an RY-element that is involved in seed-specific regulation was also found in nine *DoNACs*, suggesting that *DoNAC*s have a function in tissue differentiation and specificity. Particularly, the promoter region of *DoNAC66* possessed three RY-elements, indicating it might play an important role in flower or seed morphogenesis. Furthermore, the cis-elements related to diverse hormones were also widely identified in 93 *DoNACs*, including abscisic acid (ABRE), auxin (AuxRR-core and TGA-box) and gibberellin (P-box and TATC-box), of which DoNAC31 contained 11 ABRE elements, *DoNAC3* contained two TATC-box and *DoNAC101* had two TGA-box. Finally, 105 *DoNACs* (95.5%) were found to have 562 stress-responsive cis-elements, indicating their indispensable role in stress response in *D. officinale*. In detail, 46 *DoNAC*s had MBS elements (involved in drought-inducibility) and 36 *DoNAC*s had LTR elements (involved in low-temperature responsiveness). Among them, *DoNAC90* had three LTR elements and *DoNAC56* and *DoNAC76* had two LTR elements, suggesting they might be involved in cold stress response.

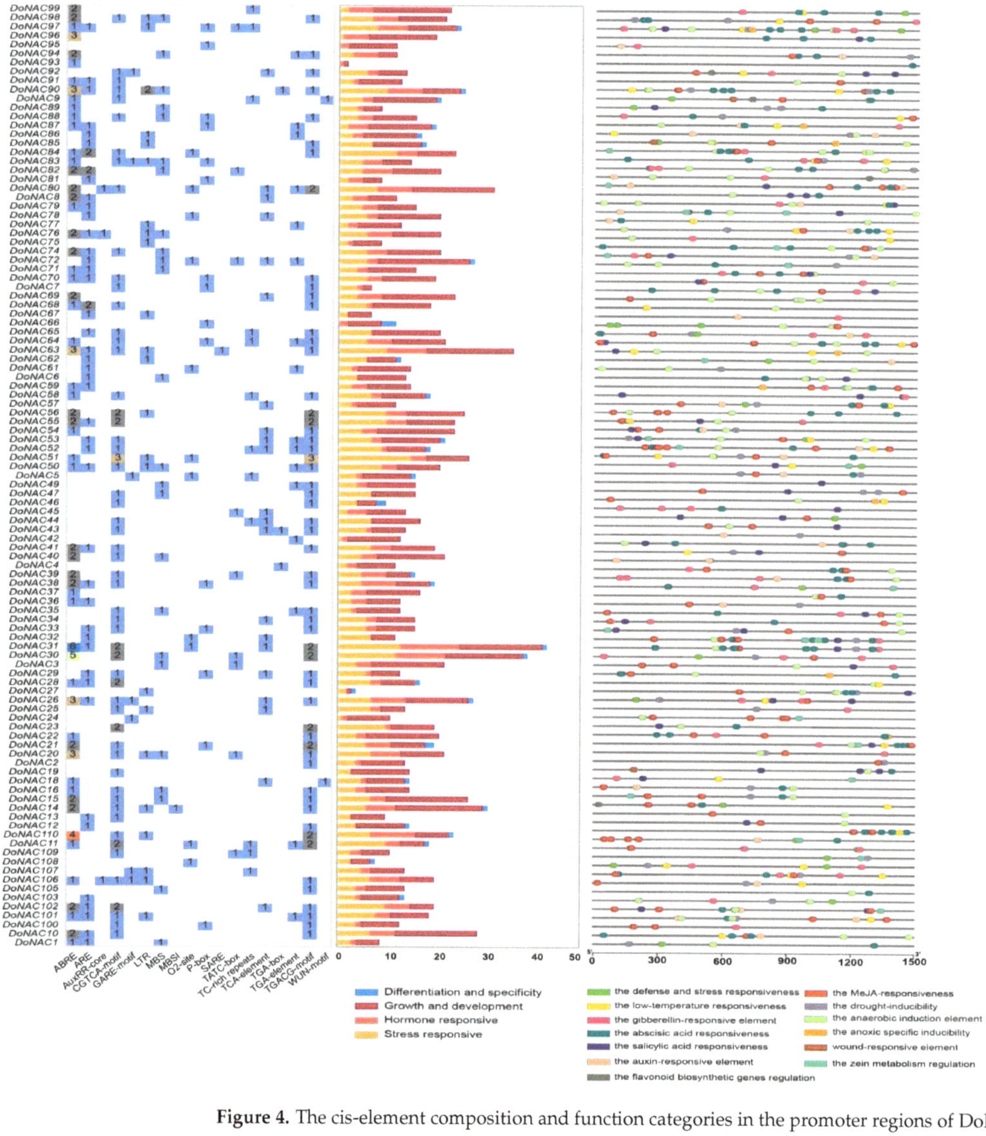

Figure 4. The cis-element composition and function categories in the promoter regions of DoNACs.

2.4. Expression Profile of DoNACs in Different Tissues

The expression profiles of these *DoNACs* were investigated in eight different tissues based on the public RNA-seq data (PRJNA348403) (Figure 5 and Table S2). Results showed that 105 out of 110 *DoNACs* were expressed in at least one of the tissues. Based on their expression profiles, they could be clustered into 10 groups, including eight tissue-specific groups, one group showing expression in all tissues and one group with no obvious expression tendency. We found that 20 DoNACs displayed continuously high expression in all of the tested tissues, including *DoNAC35*, *DoNAC42*, *DoNAC52*, *DoNAC53*, *DoNAC87*, *DoNAC98* and so on, which might function as crucial regulators associated with the growth and development of *D. officinale*. Meanwhile, tissue-specific DoNACs were also identified. *DoNAC66* showed significant high expression in sepal, lip, leaf and flower buds, which was consistent with it having three RY-elements in the promoter region. DoNAC17, DoNAC22 and DoNAC103 displayed highest expression in root tissue compared to other tissues, indicating their important role in root development. DoNAC26 showed high expression in leaf tissue, and *DoNAC28* was found to be highly expressed in stem and root tip tissues and *DoNAC100* was highly expressed in both column tissue and flower buds. *DoNAC108* was found to be lip—specific. These tissue-specific DoNACs provided the potential target for further functional study to reveal their regulation function.

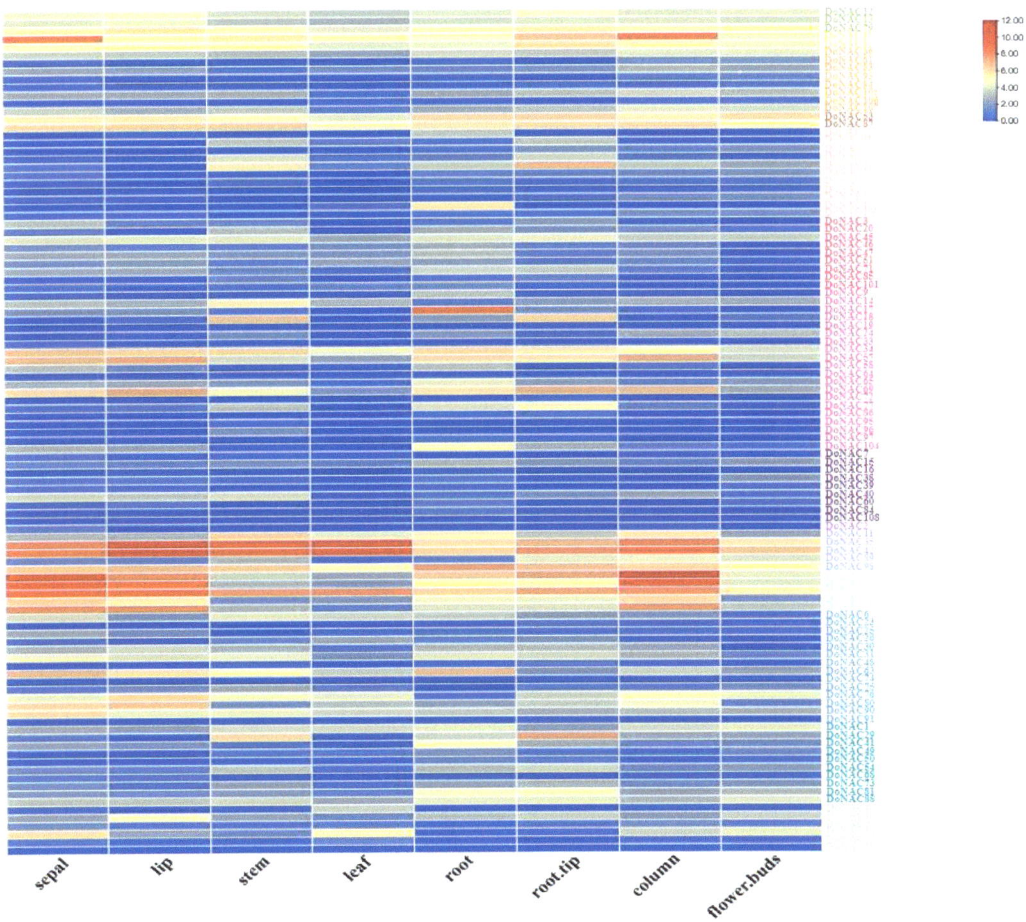

Figure 5. Expression patterns of DoNACs in 8 different tissues.

2.5. Expression Analysis of DoNACs under Cold Stress

To identify the candidate DoNACs involved in response to cold stress, we further investigated their expression patterns under cold treatment (Figure 6 and Table S3). The results showed that a total of 87 out of 110 DoNACs showed expression under cold treatment, proving the crucial role of NAC in cold response. Compared to the control, most of the DoNACs showed lower expression levels under cold stress and the differentially expressed DoNACs between them were detected. In total, 31 DoNACs were found to be differentially expressed between the cold and control treatments, which could be considered cold-responsive DoNACs. Among them, 11 were up-regulated and 20 were down-regulated, respectively. The expression level of *DoNAC102* showed 9.8-fold higher under cold stress than that of control, which is the most significantly up-regulated expression gene, followed by *DoNAC56, DoNAC57* and *DoNAC26*. While *DoNAC58* and *DoNAC39* were the most significantly down-regulated expression genes, which displayed 42 and 37 times lower expression under cold stress compared to CK. These cold-responsive DoNACs could be used as important candidates for revealing their regulatory function in the cold stress response and tolerance in *D. officinale*.

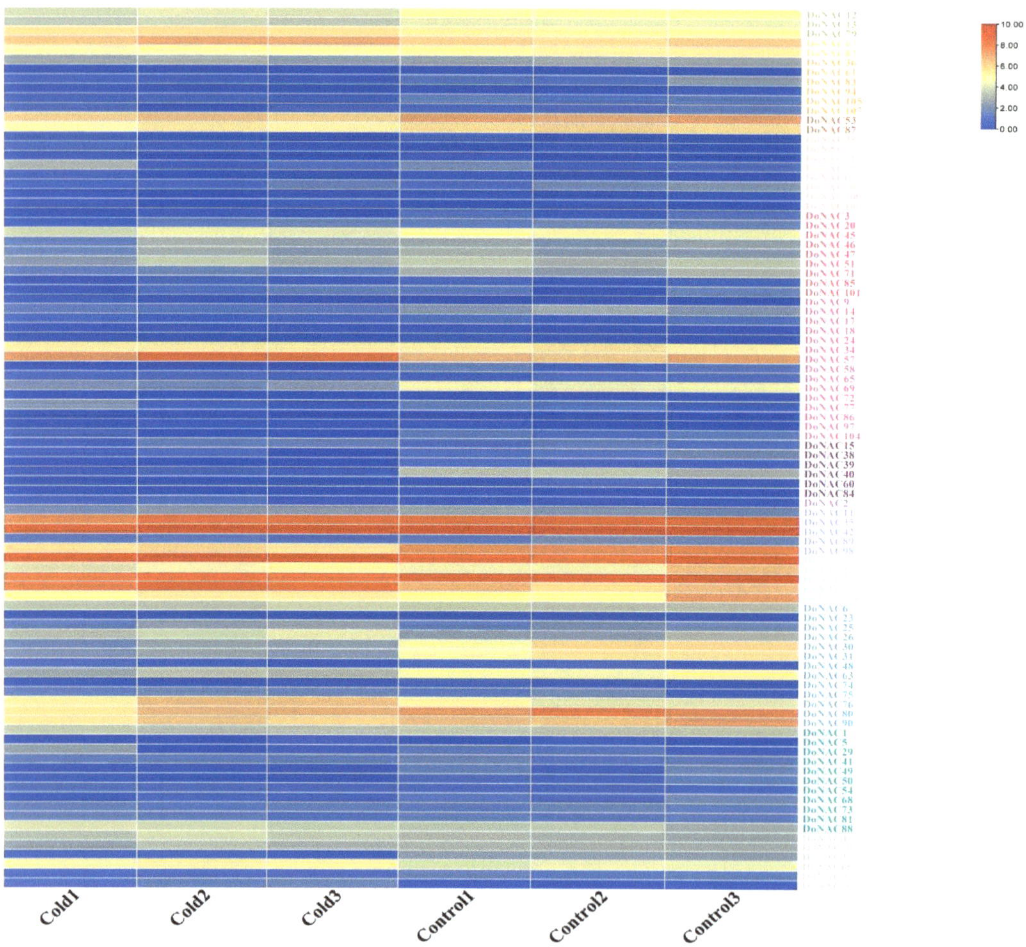

Figure 6. Expression patterns of DoNACs under cold stress. Cold and control represent under cold and control treatment. Three biological replications were used.

2.6. Validation of the Cold-Responsive DoNACs by qRT-PCR Analysis and Subcellular Localization

To explore the key NAC gene underlying cold response, five up-regulated (*DoNAC26, DoNAC56, DoNAC57, DoNAC76* and *DoNAC102*) and five down-regulated (*DoNAC30, DoNAC32, DoNAC39, DoNAC40* and *DoNAC58*) candidates were randomly selected based on the RNA-seq analysis to verify their expression by qRT-PCR analysis (Figure 7). The results showed that the expression based on the qRT-PCR method was completely consistent with that of RNA-seq analysis. The *DoNAC102* was validated to be the most significantly up-regulated gene under cold stress, which was induced by cold to more than 10 times higher compared to CK. *DoNAC26, DoNAC56, DoNAC57* and *DoNAC76* were also induced to up-regulated expression by cold with different levels. It is interesting that all five down-regulated genes were validated to be induced to low expression by cold and their expression levels were rather low or seemed to show no expression. These validated cold-responsive DoNACs could be considered the key candidates underlying cold stress response, which are useful targets for further studies to reveal their function and role in cold adaptation and tolerance in *D. officinale*, especially the identified novel down-regulated candidates, which provided the valuable target to improve cold tolerance through a genome-editing approach [35].

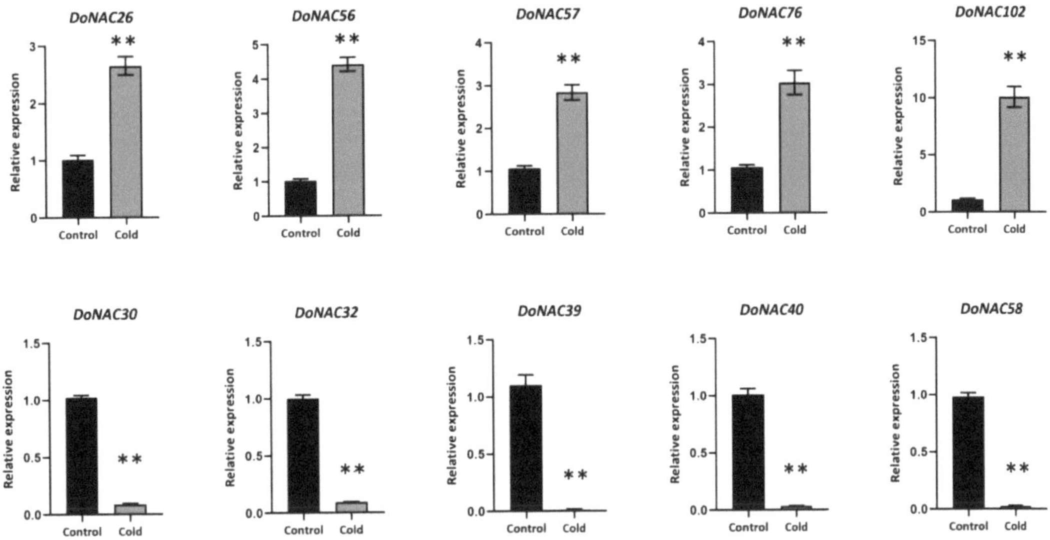

Figure 7. The expression levels of 10 randomly selected cold-responsive DoNACs under control and cold stress by qRT-PCR methods. ** represents significant level at 0.01.

Furthermore, the subcellular localization of two down-regulated candidates (*DoNAC39* and *DoNAC58*) was investigated. OsNAC-GFP fusion protein transient vectors were constructed and then injected into tobacco leaf cells (Figure 8). The results showed that both of them were localized in the nucleus, demonstrating their role as transcription factors to regulate downstream biological processes.

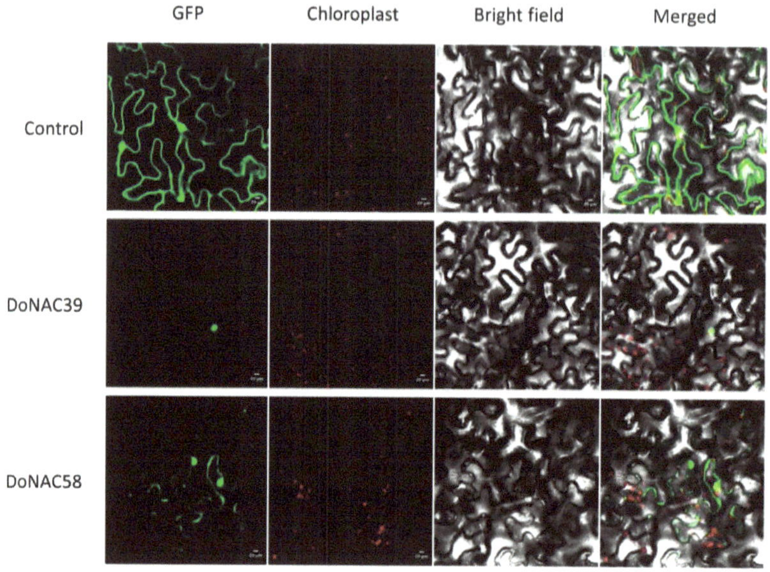

Figure 8. Subcellular localization of 2 recombinant CaMV35S: DoNAC-GFP-fused proteins transiently expressed in tobacco leaf cells.

3. Discussion

Dendrobium officinale is a rare and precious herb with huge ornamental, medicinal and cultural value which has been used as an ingredient in traditional medicine for thousands of years in China and is also widely considered a medicinal material in many other Asian countries [31]. Recently, a large number of studies have reported that *D. officinale* is rich in bio-active compounds, including polysaccharides, flavonoids, alkaloids and multiple amino acids, which have been demonstrated to have anti-oxidation, immune regulation and anti-cancer effects [36,37]. With the shortage of wild *D. officinale*, it is urgently needed to produce more *D. officinale*. As we know, *D. officinale* originates from tropical and subtropical regions, with an optimum growth temperature range between 15 and 28 °C [38]. Thus, cold stress is one of the most destructive environmental factors limiting the planting and production of *D. officinale*, especially in northern China. Improvement of cold tolerance is the most efficient, rapid and economic approach for accelerating production of *D. officinale* [33,39].

The NAC transcription factor family is a plant-specific transcription factor family, which plays key roles in stress response and tolerance by regulating the expression of stress-related genes in plants [1,40]. Previous studies have demonstrated that NAC genes also have a great impact on plant cold tolerance and some key cold-related regulation pathways have been revealed, such as the NAC-DREB and NAC-CBF-COR signaling pathways [41]. However, NAC genes, especially those involved in cold tolerance in *D. officinale*, have not been well understood up to now. In this study, 110 NAC genes were identified in *D. officinale* through a genome-wide search method, which is similar to Arabidopsis (105), rice (137), poplar (163) and potato (110) [29]. Based on phylogenetic analysis and the classification criteria of Arabidopsis and rice, these 110 DoNACs can be further categorized into 15 subfamilies, and the potential functions of them were also pried by the homologs in Arabidopsis and rice. Sequence analysis found that the N-terminus of DoNACs was rather conserved while the C-terminus was highly variable. Furthermore, nine conserved protein motifs were predicted, of which, Motifs 1–8 were widely found in almost all of the DoNACs while Motif 9 was only present in a few DoNACs, adjacent to the C-terminus. It is interesting that DoNACs containing Motif 9 were completely clustered into independent evolutionary branches in the evolutionary tree and displayed high expression in root (such as *DoNAC29*) and stem (such as *DoNAC6*) tissues. Functional prediction found that Motif 9 was related to plant energy metabolism, speculating that it might be involved in regulating nutrient transport during root and stem development.

Based on RNA-seq data, a total of 31 DoNACs showed significantly differential expression between cold and control treatment, of which 11 were up-regulated and 20 were down-regulated. To verify the cold-responsive candidates, the expressions of five up-regulated and five down-regulated genes were further validated by the qRT-PCR method. There was very good consistancy between the results from the RNA-seq and qRT-PCR, demonstrating their role in cold response. The cold-responsive DoNAC102 was significantly induced to be 10 times up-regulated expression by cold stress, which belonged to the NAC22 subfamily and had rich ABRE, TGA-element and CGTCA-motif elements in its promoter [42]. It suggested that *DoNAC102* might mediate hormones to respond to cold stress. The cold-responsive *DoNAC56*, belonging to the ATAF subfamily, was also induced to up-regulated expression by cold stress. Previous studies have demonstrated that the ATAF subfamily generally play a vital role in abiotic stress tolerance. Furthermore, there were two LTR elements as well as many other stress-responsive elements in its promoter regions. These results suggested that *DoNAC56* might be a key candidate involved in cold tolerance in *D. officinale*. Additionally, some DoNACs were induced to down-regulated expression by cold, which might be considered a valuable target for gene editing study.

4. Materials and Methods

4.1. Genome-Wide Identification of NAC Family in D. officinale

The whole-genome reference and annotated protein sequences of the *D. officinale* genome were retrieved from the Herbal Medicine Omics Database (http://202.203.187.112/herbalplant/ (accessed on 15 March 2023)) and then used as the local protein database. NAC proteins in Arabidopsis and rice were downloaded from the TAIR (https://www.arabidopsis.org/ (accessed on 15 March 2023)) and Rice databases (http://rice.plantbiology.msu.edu/ (accessed on 15 March 2023)), respectively, to perform a BLASTP search against the local protein database with the threshold of E-value < 10^{-5}. Furthermore, the Hidden Markov Model (HMM) profile of the NAC (PF01849) and NAM (PF02365) were downloaded from the PFAM database (http://pfam.xfam.org/ (accessed on 15 March 2023)) to search against the local protein database by HMMER 3.0 with the threshold of E-value < 10^{-5}. The results from these two methods were integrated together and the redundant was manually removed to obtain the putative NAC proteins in *D. officinale* (DoNACs). To confirm the accuracy of prediction, these putative DoNACs were further submitted to the PFAM databases (https://pfam.xfam.org (accessed on 25 March 2023)) and SMART databases (https://smart.embl-heidelberg.de/ (accessed on 25 March 2023)) and only those with NAC or NAM domains remained as candidate DoNAC proteins. Finally, the EXPASy online software (https://www.expasy.org/ (accessed on 10 May 2023)) was used to predict the molecular weight (MW), length of amino acid (AA) and isoelectric point (pI) and their subcellular localization was predicted using the Plant-mPLoc subcellular location tool (http://www.csbio.sjtu.edu.cn/bioinf/plant-multi/ (accessed on 10 May 2023)).

4.2. Phylogenetic Relationship, Conserved Motif and Cis-Element Analysis

The protein sequences of the identified DoNACs, together with 105 AtNACs and 137 OsNACs, were used to perform a multiple sequence alignment based on the ClustalW tool and the phylogenetic tree was constructed using RAxML software with a bootstrap value of 1000 replications. The obtained tree was further edited and visualized using the Interactive Tree Of Life online tool (https://itol.embl.de/ (accessed on 10 May 2023)). The conserved regions were determined based on the alignment file. Meanwhile, the conserved protein motif was predicted using the MEME online tool (http://alternate.meme-suite.org/tools/meme (accessed on 10 May 2023)) with the maximum motif set to 10. Additionally, the upstream 1500 bp of the TSS (transcription start site) of the identified DoNACs were extracted as the putative promoter sequences, and the cis-acting elements were predicted using the PlantCARE online tool (https://bioinformatics.psb.ugent.be/webtools/plantcare/html/ (accessed on 10 May 2023)).

4.3. Expression Analysis in Diverse Tissues and under Cold Stress

RNA-seq datasets of 8 tissues, including column, sepal, root, root tip, stem, leaf, lip and flower buds, were downloaded from the NCBI sequence read archive (SRA) database with the accession No. PRJNA348403, and the RNA-seq data under cold stress and control conditions were also downloaded from the SRA database with the accession No. PRJNA314400 (Table S4). After quality control, the clean reads were mapped onto the reference sequences of *D. officinale* using the Hisat2 tool [43], and the transcripts per million (TPM) were calculated by StringTie v2.1.2 [44]. The differential expressed gene was detected by the DESeq2 tool with the cut-off parameter: FDR < 0.05 and $|\log2foldchange| > 1$. The heatmap of the expression patterns was drawn by R software version 4.1.0.

4.4. qPCR Validation of the Cold-Responsive DoNACs

Plant materials were prepared following the method described previously [33] and the leaves were collected after 24 h of treatment under control and cold stress conditions with three to five plants mixed together. Three biological replications were performed. All samples were stored at −80 °C until RNA extraction using the RNA Easy Fast Plant Tissue Kit (Tiangen, Beijing, China)and 2 μg total RNA was used to synthesize the cDNA using Geneseed® II First Strand cDNA Synthesis Kit (Geneseed, Guangzhou, China) according to the manufacturer's protocol. A total of 10 cold-responsive DoNACs were randomly selected for validation by qRT-PCR analysis, including 5 up- and 5 down-regulated candidates, respectively. The qRT-PCR reaction was performed using the ABI 7500 instrument (ABI7500, ABI, Foster City, CA, USA) with Geneseed® qPCR SYBR® Green Master Mix (Geneseed, Guangzhou, China) with the primers listed in Table S5. The reaction mixture was 20 μL in volume, including 10 μL of Geneseed® qPCR SYBR® Green Master Mix, 0.5 μL of each primer (10 μM), 0.4 μL 50× ROX Reference Dye 2, 2 μL of the cDNA template and 7.6 μL of RNase free H_2O. The thermal cycling program was as follows: 95 °C for 5 min, followed by 40 cycles at 95 °C for 10 s and 60 °C for 34 s. DoGAPDH was used as the internal reference gene. The expression levels of the targeted genes were calculated by the $2^{-\Delta\Delta Ct}$ method.

4.5. Subcellular Localization of DoNAC-GFP Fusion Proteins

To get further insights into the biological function of the cold-responsive DoNACs, 2 candidates were selected for validating their subcellular localization, including DoNAC39 and DoNAC58. They were firstly cloned to integrate into the pBI121-GFP vector. Then, the constructed recombined vector was separately transformed into Agrobacterium tumefaciens strain GV3101 and was further injected into 4-week-old tobacco leaves for transient expression of DoNACs. Subcellular localization of them was determined by the imaging of the leaves after 48 h of agro-infiltration using a laser confocal microscope (Olympus, Tokyo, Japan).

5. Conclusions

In this study, a total of 110 NAC genes belonging to 15 subfamilies were identified in *D. officinale* at the genome level. The genomic organization, phylogenetic relationship, conserved domain and cis-element of these DoNACs were systematically investigated. It is obvious that the genes with closer phylogenetic relationships shared more similar protein motifs and cis-element compositions. Furthermore, the tissue-specific and cold-responsive DoNACs were obtained based on RNA-seq data. Then, 10 cold-responsive DoNACs were further validated by qRT-PCR analysis, and the subcellular localization of two down-regulated candidates (DoNAC39 and DoNAC58) was also revealed to obtain some key cold-responsive candidates. This study provided the targets for further functional studies, which will contribute to the genetic improvement of cold tolerance in *D. officinale* and other herbs.

Supplementary Materials: The following supporting information can be downloaded at: https://www.mdpi.com/article/10.3390/plants12203626/s1, Table S1: The cis-elements identified in the promoter regions of the 110 DoNAC genes; Table S2: The FPKM value of each DoNAC gene in different organs; Table S3: The FPKM value of each DoNAC gene in different organs; Table S4: The information of the RNA-seq data used in this study; Table S5: Primers used for qPCR analysis in this study.

Author Contributions: Conceptualization, Q.Y. and H.W.; methodology, Q.Y., X.W. and Z.L.; software, Y.N.; formal analysis, Q.Y., X.F. and G.Z.; resources, Q.Y., F.L. and X.W.; writing—original draft preparation, Q.Y.; writing—review and editing, Q.Y. and L.L.; supervision, H.W. and C.J.; funding acquisition, Q.Y. and H.W. All authors have read and agreed to the published version of the manuscript.

Funding: This study was funded by the National Natural Science Foundation of China (31800522); the National Key R&D Program of China (2022YFF1303004); and the ecosystem service function and value assessment of green land in the Minsheng Taoqing Hepan community.

Data Availability Statement: All of the datasets supporting the results of this article are included within the article and its additional files.

Acknowledgments: We thank the High-Performance Computing Center of Northwest A&F University for providing computational resources.

Conflicts of Interest: The authors declare no conflict of interest.

References

1. Ooka, H.; Satoh, K.; Doi, K.; Nagata, T.; Otomo, Y.; Murakami, K.; Matsubara, K.; Osato, N.; Kawai, J.; Carninci, P.; et al. Comprehensive analysis of NAC family genes in *Oryza sativa* and *Arabidopsis thaliana*. *DNA Res.* **2003**, *10*, 239–247. [CrossRef] [PubMed]
2. Olsen, A.N.; Ernst, H.A.; Leggio, L.L.; Skriver, K. NAC transcription factors: Structurally distinct, functionally diverse. *Trends Plant Sci.* **2005**, *10*, 79–87. [CrossRef] [PubMed]
3. Yang, Q.; Li, B.; Rizwan, H.; Sun, K.; Zeng, J.; Shi, M.; Guo, T.; Chen, F. Genome-wide identification and comprehensive analyses of NAC transcription factor gene family and expression analysis under *Fusarium kyushuense* and drought stress conditions in Passiflora edulis. *Front. Plant Sci.* **2022**, *13*, 972734. [CrossRef] [PubMed]
4. Zhang, H.; Jin, J.P.; Tang, L.; Zhao, Y.; Gu, X.C.; Gao, G.; Luo, J.C. PlantTFDB 2.0: Update and improvement of the comprehensive plant transcription factor database. *Nucleic Acids Res.* **2011**, *39*, D1114–D1117. [CrossRef]
5. Souer, E.; van Houwelingen, A.; Kloos, D.; Mol, J.; Koes, R. The no apical meristem gene of Petunia is required for pattern formation in embryos and flowers and is expressed at meristem and primordia boundaries. *Cell* **1996**, *85*, 159–170. [CrossRef]
6. Aida, M.; Ishida, T.; Fukaki, H.; Fujisawa, H.; Tasaka, M. Genes involved in organ separation in Arabidopsis: An analysis of the cup-shaped cotyledon mutant. *Plant Cell* **1997**, *9*, 841–857. [CrossRef]
7. Dang, X.; Zhang, B.; Li, C.; Nagawa, S. FvNST1b NAC protein induces secondary cell wall formation in strawberry. *Int. J. Mol. Sci.* **2022**, *23*, 13212. [CrossRef]
8. Christianson, J.A.; Dennis, E.S.; Llewellyn, D.J.; Wilson, I.W. ATAF NAC transcription factors: Regulators of plant stress signaling. *Plant Signal. Behav.* **2010**, *5*, 428–432. [CrossRef]
9. Zhu, G.; Chen, G.; Zhu, J.; Zhu, Y.; Lu, X.; Li, X.; Hu, Y.; Yan, Y. Molecular characterization and expression profiling of NAC transcription factors in *Brachypodium distachyon* L. *PLoS ONE* **2015**, *10*, e0139794. [CrossRef]
10. Zhu, T.; Nevo, E.; Sun, D.; Peng, J. Phylogenetic analyses unravel the evolutionary history of NAC proteins in plants. *Evolution* **2012**, *66*, 1833–1848. [CrossRef]
11. Kikuchi, K.; Ueguchi-Tanaka, M.; Yoshida, K.; Nagato, Y.; Matsusoka, M.; Hirano, H.Y. Molecular analysis of the NAC gene family in rice. *Mol. Gen. Genet.* **2000**, *262*, 1047–1051. [CrossRef]
12. Puranik, S.; Sahu, P.P.; Srivastava, P.S.; Prasad, M. NAC proteins: Regulation and role in stress tolerance. *Trends Plant Sci.* **2012**, *17*, 369–381. [CrossRef]
13. Diao, P.; Chen, C.; Zhang, Y.; Meng, Q.; Lv, W.; Ma, N. The role of NAC transcription factor in plant cold response. *Plant Signal. Behav.* **2020**, *15*, 1785668. [CrossRef] [PubMed]
14. Singh, S.; Koyama, H.; Bhati, K.K.; Alok, A. The biotechnological importance of the plant-specific NAC transcription factor family in crop improvement. *J. Plant Res.* **2021**, *134*, 475–495. [CrossRef] [PubMed]
15. Huang, D.; Wang, S.; Zhang, B.; Shang-Guan, K.; Shi, Y.Y.; Zhang, D.; Liu, X.; Wu, K.; Xu, Z.; Fu, X.; et al. A gibberellin-mediated DELLA-NAC signaling cascade regulates cellulose synthesis in rice. *Plant Cell* **2015**, *27*, 1681–1696. [CrossRef]
16. Mao, C.; Ding, W.; Wu, Y.; Yu, J.; He, X.W.; Shou, H.X.; Wu, P. Overexpression of a NAC-domain protein promotes shoot branching in rice. *New Phytol.* **2007**, *176*, 288–298. [CrossRef]

17. Huysmans, M.; Buono, R.A.; Skorzinski, N.; Radio, M.C.; De Winter, F.; Parizot, B.; Mertens, J.; Karimi, M.; Fendrych, M.; Nowack, M.K. NAC transcription factors ANAC087 and ANAC046 control distinct aspects of programmed cell death in the Arabidopsis columella and lateral root cap. *Plant Cell* **2018**, *30*, 2197–2213. [CrossRef] [PubMed]
18. Guo, Y.F.; Gan, S.S. AtNAP, a NAC family transcription factor, has an important role in leaf senescence. *Plant J.* **2006**, *46*, 601–612. [CrossRef] [PubMed]
19. Sablowski, R.W.; Meyerowitz, E.M. A homolog of NO APICAL MERISTEM is an immediate target of the floral homeotic genes APETALA3/PISTILLATA. *Cell* **1998**, *92*, 93–103. [CrossRef]
20. Jensen, M.K.; Skriver, K. NAC transcription factor gene regulatory and protein-protein interaction networks in plant stress responses and senescence. *Iubmb Life* **2014**, *66*, 156–166. [CrossRef]
21. Hu, H.; You, J.; Fang, Y.; Zhu, X.; Qi, Z.; Xiong, L. Characterization of transcription factor gene SNAC2 conferring cold and salt tolerance in rice. *Plant Mol. Biol.* **2008**, *67*, 169–181. [CrossRef] [PubMed]
22. Hong, Y.; Zhang, H.; Huang, L.; Li, D.; Song, F. Overexpression of a stress-responsive NAC transcription factor gene ONAC022 improves drought and salt tolerance in rice. *Front. Plant Sci.* **2016**, *7*, 4. [CrossRef] [PubMed]
23. Thirumalaikumar, V.P.; Devkar, V.; Mehterov, N.; Ali, S.; Ozgur, R.; Turkan, I.; Mueller-Roeber, B.; Balazadeh, S. NAC transcription factor JUNGBRUNNEN 1 enhances drought tolerance in tomato. *Plant Biotechnol. J.* **2018**, *16*, 354–366. [CrossRef]
24. De Clercq, I.; Vermeirssen, V.; Van Aken, O.; Vandepoele, K.; Murcha, M.W.; Law, S.R.; Inzé, A.; Ng, S.; Ivanova, A.; Rombaut, D.; et al. The membrane-bound NAC transcription factor ANAC013 functions in mitochondrial retrograde regulation of the oxidative stress response in Arabidopsis. *Plant Cell* **2013**, *25*, 3472–3490. [CrossRef] [PubMed]
25. Ng, S.; Ivanova, A.; Duncan, O.; Law, S.R.; Van Aken, O.; De Clercq, I.; Wang, I.; Carrie, C.; Xu, L.; Kmiec, B.; et al. A membrane-bound NAC transcription factor, ANAC017, mediates mitochondrial retrograde signaling in Arabidopsis. *Plant Cell* **2013**, *25*, 3450–3471. [CrossRef]
26. Shan, W.; Kuang, J.F.; Lu, W.J.; Chen, J.Y. Banana fruit NAC transcription factor MaNAC1 is a direct target of MaICE1 and involved in cold stress through interacting with MaCBF1. *Plant Cell Environ.* **2014**, *37*, 2116–2127. [CrossRef]
27. Shiriga, K.; Sharma, R.; Kumar, K.; Yadav, S.K.; Hossain, F.; Thirunavukkarasu, N. Genome-wide identification and expression pattern of drought-responsive members of the NAC family in maize. *Meta Gene* **2014**, *2*, 407–417. [CrossRef]
28. Le, D.T.; Nishiyama, R.; Watanabe, Y.; Mochida, K.; Yamaguchi-Shinozaki, K.; Shinozaki, K.; Tran, L.S.P. Genome-wide survey and expression analysis of the plant-specific NAC transcription factor family in soybean during development and dehydration stress. *DNA Res.* **2011**, *18*, 263–276. [CrossRef]
29. Rui, Z.; Pan, W.; Zhao, Q.; Hu, H.; Li, X.; Xing, L.; Jia, H.; She, K.; Nie, X. Genome-wide identification, evolution and expression analysis of NAC gene family under salt stress in wild emmer wheat (*Triticum dicoccoides.* L). *Int. J. Biol. Macromol.* **2023**, *230*, 123376. [CrossRef]
30. Liu, X.H.; Lyu, Y.S.; Yang, W.; Yang, Z.T.; Lu, S.J.; Liu, J.X. A membrane-associated NAC transcription factor OsNTL3 is involved in thermotolerance in rice. *Plant Biotechnol. J.* **2020**, *18*, 1317–1329. [CrossRef]
31. Ng, T.B.; Liu, J.; Wong, J.H.; Ye, X.; Wing, S.S.C.; Tong, Y.; Zhang, K.Y. Review of research on Dendrobium, a prized folk medicine. *Appl. Microbiol. Biotechnol.* **2012**, *93*, 1795–1803. [CrossRef]
32. Wang, Y.H. Traditional uses, chemical constituents, pharmacological activities, and toxicological effects of Dendrobium leaves: A review. *J. Ethnopharmacol.* **2021**, *270*, 113851. [CrossRef]
33. Yang, Q.; Xiang, W.; Li, Z.; Nian, Y.; Fu, X.; Zhou, G.; Li, L.; Zhang, J.; Huang, G.; Han, X.; et al. Genome-wide characterization and expression analysis of HD-ZIP gene family in Dendrobium officinale. *Front. Genet.* **2022**, *13*, 797014. [CrossRef]
34. Mao, X.; Zhang, H.; Qian, X.; Li, A.; Zhao, G.; Jing, R. TaNAC2, a NAC-type wheat transcription factor conferring enhanced multiple abiotic stress tolerances in Arabidopsis. *J. Exp. Bot.* **2012**, *63*, 2933–2946. [CrossRef]
35. Wang, N.; Tang, C.; Fan, X.; He, M.; Gan, P.; Zhang, S.; Hu, Z.; Wang, X.; Yan, T.; Shu, W.; et al. Inactivation of a wheat protein kinase gene confers broad-spectrum resistance to rust fungi. *Cell* **2022**, *185*, 2961–2974. [CrossRef]
36. Guo, L.; Qi, J.; Du, D.; Liu, Y.; Jiang, X. Current advances of Dendrobium officinale polysaccharides in dermatology: A literature review. *Pharm. Biol.* **2020**, *58*, 664–673. [CrossRef]
37. Xu, X.; Zhang, C.; Wang, N.; Xu, Y.; Tang, G.; Xu, L.; Feng, Y. Bioactivities and mechanism of actions of Dendrobium officinale: A comprehensive review. *Oxidative Med. Cell Longev.* **2022**, *2022*, 6293355. [CrossRef]
38. Wang, Y.; Tong, Y.; Adejobi, O.I.; Wang, Y.; Liu, A. Research advances in multi-omics on the traditional Chinese herb Dendrobium officinale. *Front. Plant Sci.* **2022**, *11*, 808228. [CrossRef] [PubMed]
39. Yu, Z.; Dong, W.; Teixeira da Silva, J.A.; He, C.; Si, C.; Duan, J. Ectopic expression of DoFLS1 from Dendrobium officinale enhances flavonol accumulation and abiotic stress tolerance in Arabidopsis thaliana. *Protoplasma* **2021**, *258*, 803–815. [CrossRef] [PubMed]
40. Liu, H.; Chen, S.; Wu, X.; Li, J.; Xu, C.; Huang, M.; Wang, H.; Liu, H.; Zhao, Z. Identification of the NAC transcription factor family during early seed development in *Akebia trifoliata* (Thunb.) Koidz. *Plants* **2023**, *12*, 1518. [CrossRef] [PubMed]
41. Jin, C.; Li, K.Q.; Xu, X.Y.; Zhang, H.P.; Chen, H.X.; Chen, Y.H.; Hao, J.; Wang, Y.; Huang, X.S.; Zhang, S.L. A novel NAC transcription factor, PbeNAC1, of Pyrus betulifolia confers cold and drought tolerance via interacting with PbeDREBs and activating the expression of stress-responsive genes. *Front. Plant Sci.* **2017**, *8*, 1049. [CrossRef]
42. Nakashima, K.; Yamaguchi-Shinozaki, K. ABA signaling in stress-response and seed development. *Plant Cell Rep.* **2013**, *32*, 959–970. [CrossRef] [PubMed]

43. Kim, D.; Langmead, B.; Salzberg, S.L. HISAT: A fast spliced aligner with low memory requirements. *Nat. Methods* **2015**, *12*, 357–360. [CrossRef] [PubMed]
44. Pertea, M.; Pertea, G.M.; Antonescu, C.M.; Chang, T.C.; Mendell, J.T.; Salzberg, S.L. StringTie enables improved reconstruction of a transcriptome from RNA-seq reads. *Nat. Biotechnol.* **2015**, *33*, 290–295. [CrossRef] [PubMed]

plants

MDPI

Article

Biochemical Alterations in Triticale Seedlings Pretreated with Selective Herbicide and Subjected to Drought or Waterlogging Stress

Zornitsa Katerova, Dessislava Todorova, Elena Shopova, Liliana Brankova, Ljudmila Dimitrova, Margarita Petrakova and Iskren Sergiev *

Institute of Plant Physiology and Genetics, Bulgarian Academy of Sciences, Acad G. Bonchev Str., Bl. 21, 1113 Sofia, Bulgaria; zornitsa@bio21.bas.bg (Z.K.); dessita@bio21.bas.bg (D.T.); kostei@abv.bg (E.S.); lbrankova@abv.bg (L.B.); dim.lyudmila@gmail.com (L.D.); margarita.p02@abv.bg (M.P.)

* Correspondence: iskren@bio21.bas.bg

Abstract: Waterlogging and drought disrupt crop development and productivity. Triticale is known to be relatively tolerant to different stress factors. In natural conditions, plants are rather subjected to multiple environmental factors. Serrate® (Syngenta) is a systemic selective herbicide suitable for cereal crops such as triticale and wheat to restrain annual grass and broadleaf weeds. Triticale (×*Triticosecale* Wittm., cv. Rozhen) was grown as soil culture under controlled conditions. Seventeen-day-old plantlets were leaf sprayed with Serrate®. The water stress (drought or waterlogging) was applied after 72 h for 7 days, and then the seedlings were left for recovery. The herbicide does not provoke sharp alterations in the antioxidant state (stress markers level, and antioxidant and xenobiotic-detoxifying enzymes activity). The water stresses and combined treatments enhanced significantly the content of stress markers (malondialdehyde, proline, hydrogen peroxide), non-enzymatic (total phenolics and thiol groups-containing compounds), and enzymatic (activities of superoxide dismutase, catalase, guaiacol peroxidase, glutathione reductase) antioxidants, and xenobiotic-detoxifying enzymes (activities of glutathione S-transferase, NADPH:cytochrome P450 reductase, NADH:cytochrome *b5* reductase). These effects were more severely expressed after the drought stress, suggesting that this cultivar is more tolerant to waterlogging than to drought stress.

Keywords: antioxidants; Serrate®; stress markers; triticale; water stress; xenobiotic-detoxifying enzymes

Citation: Katerova, Z.; Todorova, D.; Shopova, E.; Brankova, L.; Dimitrova, L.; Petrakova, M.; Sergiev, I. Biochemical Alterations in Triticale Seedlings Pretreated with Selective Herbicide and Subjected to Drought or Waterlogging Stress. *Plants* **2023**, *12*, 2803. https://doi.org/10.3390/plants12152803

Academic Editor: Andrzej Bajguz

Received: 26 June 2023
Revised: 21 July 2023
Accepted: 27 July 2023
Published: 28 July 2023

1. Introduction

Triticale (×*Triticosecale* Wittm.) is a crop, artificially obtained after the hybridization of wheat (*Triticum* spp.) with rye (*Secale cereale* L.), which became commercially available approximately a half century ago [1,2]. It is assumed to be relatively tolerant to environmental challenges, due to its rye traits, and it has high genetic diversity for abiotic stress responses [2–4]. Drought and waterlogging are among the devastating abiotic stresses, which influence negatively crop yield [5]. Water stresses (drought and waterlogging) negatively affect photosynthesis, lead to redox imbalance and oxidative stress, alter optimal physiological and biochemical processes and ultimately reduce crop yield [6–8]. These stresses were studied in detail when applied individually in both field and laboratory conditions [9]. In natural conditions, plants are subjected to combinations of different factors, but the information for the effects is insufficient [10,11]. There are few articles reporting the effect of abiotic stress combination (including herbicides) on the biochemical status of crops [12–17]. Generally, the resultant effect of multiple treatments is specific and unpredictable [9,18,19]. The information if the effect of treatment combinations on plants is negative (cross-synergism) or positive (cross-adaptation) could be obtained experimentally.

Among other specific physiological alterations, environmental stresses force plants to intensify accumulation of different reactive oxygen species (ROS) [6,20,21]. Depending on their types and levels in plants, at low concentrations ROS may trigger specific biochemical signals and alterations and could function as signaling molecules, while at high concentrations they cause oxidative stress. Plants respond to rising ROS through well-organized antioxidative defense, integrating enzymatic system (superoxide dismutase (SOD), catalase (CAT), peroxidases, glutathione reductase (GR), glutathione S-transferase (GST), etc.) and non-enzymatic scavengers (glutathione, ascorbate, phenolics, etc.) to keep the optimal cellular redox state. Based on their results Raisat et al. [22] suggested that enzymatic antioxidants might be used to screen for drought tolerant triticale genotypes. The NADPH:cytochrome P450 reductase (CPR) and NADH:cytochrome *b5* reductase (B5R) also take part in defensive mechanisms of plants against abiotic stress [23–25]. CPR is a membrane-bound flavoprotein, localized mainly in the endoplasmic reticulum and its function is to transfer reducing equivalents from NADPH to a variety of P450 monooxygenases, involved in secondary metabolism reactions (biosynthesis of phenylpropanoids, terpenoids, alkaloids, fatty acids, signaling molecules, plant hormones, etc.) [24–26]. B5R is also localized in the endoplasmic reticulum, but transfers electrons from NADH to cytochrome b5 and next on different lipid-modification reactions as fatty-acid and sterol precursors desaturation, fatty acids hydroxylation, and in P450-interposed reactions [27,28]. In addition, CPR and B5R, together with glutathione and GST, are involved in Phase I and Phase II of xenobiotic catabolism, including herbicide detoxification [26].

Spontaneous weed growth is undesirable in farmlands used for crop cultivation. Nowadays, different chemical compounds and their combinations are utilized in arable land to control this process. Serrate® (Syngenta) contains the inhibitor of fatty acid biosynthesis–clodinafop-propargyl (prop-2-ynyl(R)-2-[4-(5-chloro-3-fluoro-2-pyridyloxy) phenoxy]propionate) together with the inhibitor of amino acid biosynthesis–pyroxsulam (([N-(5,7-dimethoxy[1,2,4]triazolo [1,5-a]pyrimidin-2-yl)-2-methoxy-4-(trifluoro methyl)pyridine-3-sulfonamide])), and a herbicide safener, cloquintocet-mexyl ((RS)-1-methylhexyl [(5-chloro-8-quinolyl)oxy]acetate). This herbicide formulation aims to control annual grass and broadleaf weeds due to its efficacy in cereal crop protection [29]. Serrate® is systemic and selective for triticale, wheat and rye, which are tolerant to its application. The reports concerning triticale subjected to multiple abiotic factors containing water stresses does not include herbicides [19,30–33].

Previously, the combined effect of Serrate® and subsequent water stress (soil drought or waterlogging) was studied in wheat plants, [13–15,34]. Serrate® application was documented to provoke a synergistic physiological effect in seedlings subjected to waterlogging, whereas the biochemical response of wheat was only modulated under drought [13,14]. The current study explores the effect of the same treatment combinations in triticale, as it could be expected to be more tolerant to abiotic stressors than wheat. The study is important in the view of the climate change leading to unstable and extreme weather phenomena (including water stress) and the associated global search for resilient cereals [35,36].

This study aimed to assess the effect of pre-treatment with the selective herbicide Serrate® and subsequent water stress (drought or waterlogging) on the physiological response of triticale by the examination of oxidative stress markers and selected enzymatic and non-enzymatic antioxidants.

2. Results

2.1. Stress Marker Contents: Hydrogen Peroxide, Malondialdehyde and Proline

To evaluate the effects of water stress the following stress marker contents were monitored: hydrogen peroxide, malondialdehyde and proline.

The hydrogen peroxide (Figure 1A) concentration was raised (by 120%) at the 7th day of stress and at the 4th day of recovery due to the Serrate® application. Drought increased it substantially by 160% (at the 4th day of stress) and by 2046% (at the 7th day of stress) reaching a value of 42.1 μmol g^{-1} FW, but recovery H_2O_2 levels reverted to 26% above the respective control. The combined stress (HB + D) caused similar alterations, but at recovery, the H_2O_2 level was reverted to 60% above the respective control. Waterlogging slightly decreased (by 20%) the H_2O_2 concentrations at the 4th day of stress but then by an increment of 280% at the 7th day of stress and 60% at the recovery was observed. Similarly, the combined stress (HB + W) increased the concentration of H_2O_2 to 290% (at the 7th day of stress), and at recovery, the level of H_2O_2 reverted to 70% above the respective control.

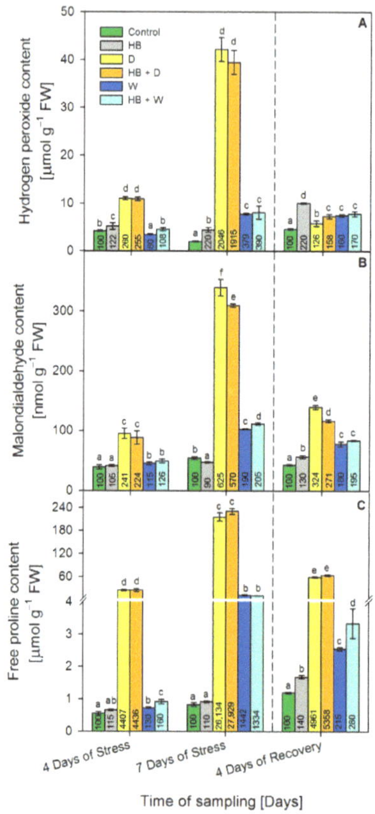

Figure 1. Content of hydrogen peroxide (**A**), malondialdehyde (**B**) and free proline (**C**) in the leaves of triticale treated with selective herbicide (HB) and exposed to waterlogging (W) or drought (D). The numbers in each bar represent % to the respective controls. The statistical significance between treatments at $p \leq 0.05$ is indicated by different small letters within the column groups.

The herbicide induced moderate alterations in MDA concentrations (Figure 1B): a reduction by 10% (at the 7th day of stress) and an increment of 30% at recovery. Drought applied alone or in combination with herbicide (HB + D) increased considerably MDA levels: by 525% (339.3 nmol g^{-1} FW after drought), by 470% (309.4 nmol g^{-1} FW after H + D) at the 7th day of stress, by 224% (139.5 nmol g^{-1} FW after drought) and by 171% (116.6 nmol g^{-1} FW after H + D) at recovery. Waterlogging had a weaker effect than drought: MDA concentrations rose to 90% at the 7th day of stress and by 80% at recovery. Combined stress (HB + W) altered the MDA levels similarly to waterlogging alone.

Herbicide application raised proline levels by 40% only at the recovery stage (Figure 1C). Drought increased severely proline concentrations as follows: by 4307% (24.8 µmol g^{-1} FW at the 4th day of stress), by 26,034% (215.9 µmol g^{-1} FW at the 7th day of stress) by 4861% (59.0 µmol g^{-1} FW at recovery) in drought-treated seedlings, by 4336% (24.9 µmol g^{-1} FW at the 4th day of stress), by 27,829% (230.8 µmol g^{-1} FW at the 7th day of stress) and by 5258% (63.7 µmol g^{-1} FW at recovery) in triticale subjected to combined (HB + D) treatment.

Waterlogging also raised the proline levels by 30% (at the 4th day of stress), by 1342% (11.9 µmol g^{-1} FW at the 7th day of stress) but at recovery it reverted to 2.6 µmol g^{-1} FW

(115% above the control). Similarly, the combined stress (HB + W) increased proline concentrations by 60% (at the 4th day of stress), by 1234% (11.0 µmol g^{-1} FW at the 7th day of stress) and at recovery to 180% above the respective control.

2.2. Non-Enzymatic Antioxidants: Total Phenolics and Thiol Groups Containing Compounds

The herbicide increased the phenolics content (Figure 2A) by 20% at the 4th day of stress and by 50% at recovery. Drought applied alone or in combination with herbicide (HB + D) raised the phenolics gradually to 390% and 360%, respectively, after 7 days of stress, and at recovery, they were reverted to about 20% above the respective control. Phenolics were increased by 60% at the 7th day of stress due to waterlogging, and then were reduced to 40% above the respective control at recovery. These compounds were raised (by 20% and 80%, respectively at the 4th and 7th days) due to the combined treatment (HB + W); then they were reverted to 30% above the respective control at recovery.

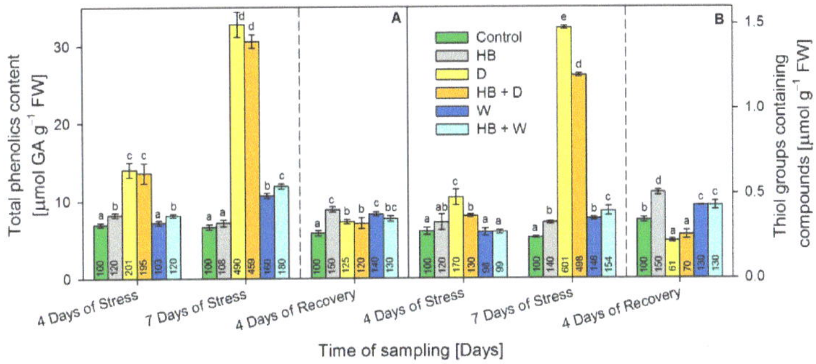

Figure 2. Content of total phenolics (**A**) and thiol groups containing compounds (**B**) in the leaves of triticale treated with selective herbicide (HB) and exposed to waterlogging (W) or drought (D). The numbers in each bar represent % to the respective controls. The statistical significance between treatments at $p \leq 0.05$ is indicated by different small letters within the column groups.

Serrate® increased thiols by 40% at the 7th day of stress and by 50% (at the 4th day of recovery above the respective control (Figure 2B). Thiol levels raised during drought treatment by 70% at the 4th day of stress and by 500% at the 7th day of stress, but at recovery, a decline by 40% below the respective control was observed. A similar tendency was found due to the combined treatment (HB + D): an increase by 30% and 400% at the 4th and 7th days of stress, respectively, and a decrease by 30% at recovery. Waterlogging and combined treatment (HB + W) increased thiol groups containing compounds by approximately 50% at the 7th day of stress and by 30% at recovery above the respective control.

2.3. Activity of Enzymatic Antioxidants

Serrate® slightly induced SOD activity (Figure 3A) up to 40% at the 4th day of stress above the respective control. Drought gradually raised the activity of SOD up to 230% at the 7th day of stress, but at the 4th day of recovery, it was decreased by 50% above the respective control. The combined treatment (HB + D) caused similar alterations in the leaves of triticale. The activity of SOD progressively increased during the waterlogging treatment up to 40% at the 7th day of stress, but at recovery, it reached the respective control level. The combined treatment (HB + W) caused alterations similar to waterlogging, but at recovery, the SOD activity remained increased by 30% above the respective control.

The herbicide did not cause significant alteration in the CAT activity during the monitored period (Figure 3B). Drought and combined (HB + D) treatment increased the CAT activity by 208% and by 207%, respectively, at the 7th day of stress, but at recovery, the CAT activity was close to the respective control. Waterlogging led to slight increment of CAT activity by 20% at the 4th day of stress only in the combined (HB + W) treatment.

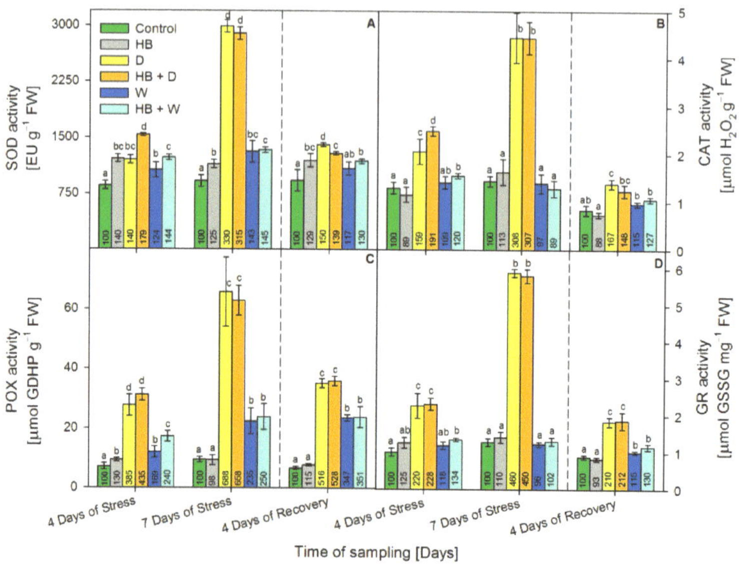

Figure 3. Activity of superoxide dismutase (SOD) (**A**), catalase (CAT) (**B**), guaiacol peroxidase (POX) (**C**), and glutathione reductase (GR) (**D**) in the leaves of triticale treated with selective herbicide (HB) and exposed to waterlogging (W) or drought (D). The numbers in each bar represent % to the respective controls. The statistical significance between treatments at $p \leq 0.05$ is indicated by different small letters within the column groups.

Serrate® increased POX activity (Figure 3C) by 30% only at the 4th day of stress. The activity of POX was increased steadily during drought and combined (HB + D) treatment. The activity was amplified by 588% due to drought and by 558% due to H + D at the 7th day of stress. After 4 days of recovery, it tended to decline but remained increased by 416% and by 428%, respectively, as compared to the control. Although, waterlogging provoked weaker induction of POX activity than drought, the activity rose gradually during the whole monitored period and reached 250% above the respective control at recovery due to both waterlogging and the combined (HB + W) treatment.

The herbicide did not cause significant alteration in GR activity (Figure 3D). Drought increased substantially GR activity at the 4th (by 120%) and 7th (by 360%) days of stress and at recovery (by 110%). A similar increment was found due to the combined (HB + D) treatment. The activity of GR increased slightly (by 20%) at recovery due to waterlogging. The combined treatment (HB + W) raised it by 30% at the 4th day of stress and at recovery.

2.4. Activity of Xenobiotic-Detoxifying Enzymes

The alteration of the GST activity after water stress and herbicide treatment in triticale is presented on Figure 4A. The herbicide raised the activity of GST during treatment (40% and 70% above the control at the 4th day and 7th day of stress, respectively) and reached control values at recovery. Drought increased GST activity by 65% at the 4th day of stress, by 280% at the 7th day of stress and by 70% at recovery above the respective control. The combined treatment (HB + D) raised GST activity even more, by 140% at the 4th day of stress, by 320% at the 7th day of stress and by 120% at recovery above the respective control. Waterlogging provoked a weaker effect on the activity of GST than drought, and it increased the GST activity by 120% only at the 7th day of stress. The combined treatment (HB + W) raised the GST activity by 70% at the 4th day of stress, by 124% at the 7th day of stress and by 21% at recovery above the control.

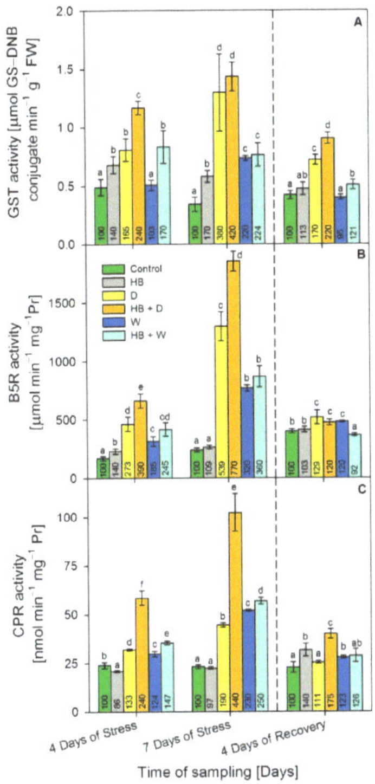

Figure 4. Activity of glutathione *S*-transferase (GST) (GS–DNB: conjugate of GSH and CDNB) (**A**), cytochrome *b5* reductase (B5R) (**B**), and cytochrome P450 reductase (CPR) (**C**) in the leaves of triticale treated with selective herbicide (HB) and exposed to waterlogging (W) or drought (D). The numbers in each bar represent % to the respective controls. The statistical significance between treatments at $p \leq 0.05$ is indicated by different small letters within the column groups.

Serrate® led to a temporary induction (by 40%) of Cytochrome *b5* reductase (B5R) activity only at the 4th day of stress (Figure 4B). Drought increased substantially the activity of B5R by 439% at the 7th day of stress and reverted the activity to 30% over the control at the recovery. The combined treatment (HB + D) further raised B5R activity by 290% at the 4th day of stress, by 670% at the 7th day of stress, but at recovery reverted it to 20% above the control.

Similarly to drought, waterlogging also raised the activity of B5R up to 220% at the 7th day of stress, but at recovery, it was 20% above the control. The combined treatment (HB + W) increased the activity of B5R up to 260% at the 7th day of stress and dropped it slightly (by 8%) below the respective control at the recovery point.

Serrate® caused slight decline (by 10%) of NADPH:cytochrome P450 reductase (CPR) activity at the 4th day of stress but enhanced it by 40% at recovery (Figure 4C). The activity of CPR gradually increased during drought treatment up to 90% at the 7th day of stress, but after 4 days of recovery it reached the respective control level. The combined treatment (HB + D) raised CPR activity much more—by 140% at the 4th day of stress and by 340% at the 7th day of stress, but the activity decreased to 75% over the control during the recovery stage. Both waterlogging and combined (HB + W) treatment showed similar effect on the CPR activity. The increase of CPR enzymatic activity peaked (by 130% and 150% for waterlogging and the combined treatment) at the 7th day of stress and then dropped close to the respective control at the recovery point.

3. Discussion

Earlier we have published studies concerning treatment combination, applied in a consequent manner–herbicide (Serrate®) pretreatment and water stress (drought or waterlogging) in wheat plants [13–15,34]. The current study puts triticale on focus because it is expected to be relatively more tolerant to water stress than wheat due to its rye traits, and is an interesting object in the light of the climate changes and the need for more tolerant crops [2,35,36]. Additionally, in real conditions plants are exposed to a number of environmental factors and their effects are not possible to be predicted but just experimentally tested [9,18,19]. Similarly to other environmental challenges, water stress affects ROS and the antioxidative system (enzymatic and non-enzymatic). Under standard growth conditions, ROS and the antioxidative system are in balance and plants keep their cellular redox homeostasis under control [20,21,37]. ROS are implicated in multiple processes as photosynthesis, primary and secondary metabolism, participation in chain reactions, etc., and affect the whole plant's physiological state. Under abiotic stress conditions, such as water stress, the redox homeostasis is impaired [6].

To assess the oxidative stress effects in our model system, the changes in hydrogen peroxide, malondialdehyde and proline content were monitored as well-known plant stress markers. In general, herbicides were reported as a possible cause of oxidative stress and toxicity in various plants [6]. The herbicide Clodinafop, in particular, was reported to increase concentrations of superoxide anion and MDA as markers of lipid peroxidation in leaf discs and in intact 7-day-old winter wheat and rye [38]. The current study also showed a minor induction of stress markers due to the Serrate® application in triticale, which was found at the recovery period. This finding agrees with the consideration that the selectivity of herbicides is not absolute, and they could induce stress responses in tolerant crops as well [12]. The content of the stress markers was observed to be much higher when water stress or the combination treatment (HB + water stress) was applied in triticale. Proline over-accumulation, due to drought and the combined treatment (HB + D) in triticale, was probably linked with cellular toxicity [39,40] because its concentration remains very high even at the recovery. On the opposite, waterlogging, including HB + W, seems to provoke minor increase of stress markers as compared to drought-treated seedlings during the entire experimental period, which confirmed better triticale tolerance to waterlogging than wheat. The herbicide application in combined stress (HB + D and HB + W) did not demonstrate additive negative effects on the oxidative stress markers accumulation in triticale.

Drought boosted the oxidative stress markers more dramatically than waterlogging and kept that tendency at recovery, pointing to the higher susceptibility of triticale to a water deficit than to waterlogging. Earlier, we reported an opposite effect for wheat plants—waterlogged seedlings were unable to recover at the monitored recovery period, whereas those subjected to drought could recover [13,14,34]. These observations are also well illustrated by the phenotype alterations of triticale as compared to wheat [34] (Figure 5).

Phenolic compounds in plants are part of the secondary metabolites with multiple functions in plants [21,41]. Stress-induced phenolics are predominantly linked with their ROS-scavenging properties, including cell protection from radical chain reactions [42]. Therefore, the slight induction of the total phenolic compounds level due to herbicide application could correspond to the slight induction of lipid peroxidation and hydrogen peroxide. Phenolic compounds in triticale were less altered due to waterlogging as compared to drought treatment. Drought was reported to cause different effects on phenolics content: it does not induce phenylpropanoid biosynthesis in stress-tolerant agave [43]; reduces phenolic compounds in grapevine seedlings [44]; or increases them in amaranth [45]. Hura et al. [46] reported that drought-sensitive triticale cultivars had increased total phenolic content due to water deficit, whereas drought-tolerant cultivars had not. The aforementioned statement is in line with our results showing the highest phenolics' raise being due to drought and HB + D, and corresponding well with the assumption that triticale cv. Rozhen is more sensitive to drought than to waterlogging.

Figure 5. Differences and similarities in phenotype alterations of triticale (left panel) and wheat (right panel—adapted from Todorova et al., 2022 [34]) evoked by herbicide and water stress. C—control; HB—herbicide; D—drought; HB + D—herbicide + drought; W—waterlogging; HB + W—herbicide + waterlogging.

One of the most important low-molecular thiol containing compounds is glutathione, which has a major role in stress tolerance, ROS-scavenging, xenobiotic detoxification, regulation of cellular redox status, redox signaling and cross-talk, etc. [21,47–50]. The slight increase of the low-molecular weight thiols in herbicide-treated plants and in those subjected to waterlogging or to combined treatment (HB + W) might signify the antioxidant and detoxification roles of glutathione. On the contrary, its significant boost due to drought and combined treatment (HB + D) could imply an imbalance of the cellular redox status. The subsequent sharp decrease in low-molecular weight thiols levels at the recovery below the control could also confirm the impaired redox status in triticale plantlets subjected to drought and the combined treatment (HB + D). Such circumstances may hamper triticale plants repairing from drought and the combined treatment (HB + D). The GR enzyme catalyzes the reduction of oxidized (GSSG) to reduced (GSH) glutathione, maintains high the cellular GSH/GSSG ratio and is one of the enzymes of the ascorbate–glutathione cycle, which supports optimal levels of the cellular redox state [21]. Arabidopsis plants overexpressing GR were reported to overcome oxidative stress caused by aluminum via the repression of ROS and the detoxification of lipid peroxide-derived reactive carbonyl species [50]. A fine coordination between non-enzymatic and enzymatic systems is necessary to maintain an optimal redox status in plants [21,47]. For example, although the slight induction of SOD activity due to herbicide application was found, it seems that triticale keeps its redox homeostasis. The increment in H_2O_2 due to the Serrate® application at recovery might be detoxified by other peroxidases (such as ascorbate peroxidase) because of the lack of enhanced enzymatic activities of CAT and POX. The moderate induction of SOD, CAT, POX and GR activities, together with the relevant raise of H_2O_2 levels imply for good adaptability of triticale after waterlogging, including HB + W. On the contrary, the huge raise of these activities due to drought and HB + D treatment along with the increased stress marker levels during the stress period signify severe oxidative stress events. These observations support our findings that the triticale cultivar Rhozen tolerates waterlogging better than the drought.

Beside typical antioxidant enzymes, we also studied the activity of selected xenobiotic-detoxifying enzymes. The presence of an antidote in the herbicide formulation aims to induce specific metabolic enzymes, which assist the herbicide detoxification, and this could probably be a reason for the enhanced activities of GST, CPR and B5R after the herbicide application. The same explanation could be valid at least partially for the increased activities of detoxification enzymes (GST, CPR and B5R) due to the combined treatments (HB + D and HB + W) as compared with the activities obtained after water stress only. In addition, the participation of GST, B5R and CPR activities in the antioxidative defense could not be excluded [23,24,50]. Interestingly, the enzymatic activity of both B5R and CPR obtained after HB + D was much higher than that measured after drought. In general, CPR provides electrons from NAD(P)H to cytochrome P450 monooxigenases and participates in a vast number of reactions related to secondary metabolites biosynthesis, including phenolics [51]. Our results suggest that in triticale subjected to HB + D, other secondary metabolites might also be involved, apart from phenolics, because the CPR activity is much higher than after drought alone, though the phenolic levels are similar. Plants usually have many CPR genes and their enzymes are classified in class I (constitutively expressed) and class II (stress-inducible) [52] but the monocotyledonous crops are considered as more resistant to herbicides, which could be mediated by P450 monooxigenases [53]. Therefore, the substantially higher CPR activity in triticale after HB + D compared with droughted plants might be due to the expression of a higher number of CPR isoforms. Conventionally, the electron transport chains (ETC) in the endoplasmic reticulum are considered to be NADH:B5R with cytochrome *b5* as an electron donor and NADPH:CPR with P450 [54,55]. In addition, CPR activity is rather related with secondary metabolites synthesis [56] whether B5R is rather linked with the alteration of biomembrane fluidity by increasing unsaturated fatty acids [27,28]. Some stresses, such as desiccation, have been reported to affect membrane integrity and membrane lipid composition. It is well known that the fluidity of the lipid membrane is determined by the content of unsaturated fatty acids, and the change in their amount can affect the plant's tolerance to stress. Several studies have shown a relationship between the ability of plants to maintain or increase their unsaturated fatty acid content and their resistance to drought [57–59]. It was suggested that various ETC could be formed such as NADH:B5R:cytochrome *b5* or NADPH:CPR:cytochrome *b5* [51,60,61]. Based on their results, Zhao et al. [51] and Wayne et al. [62] suggested that both reductases (CPR and B5R) could operate overlapped functions. Therefore, in the current model system the functions of CPR and B5R might also be interrelated. The smaller alteration in GST, CPR and B5R activities, due to waterlogging and HB + W, in comparison to these after drought and HB + D, imply smaller changes in the secondary metabolites and biomembrane fluidity.

In conclusion, the selective herbicide Serrate® alone does not harshly alter the physiological status of triticale, evidenced by the changes in antioxidant state (stress markers, oxidative and xenobiotic-detoxification status). As expected, drought and waterlogging boosted the levels of stress markers and non-enzymatic compounds, as well as the activities of the antioxidative and xenobiotic-detoxifying enzymes in triticale. In general, the combined treatments (HB + D and HB + W) caused similar trends in the biochemical parameters as compared to those of the water stresses themselves. The effects were more intense after drought than after waterlogging which could mean that triticale cv. Rozhen is more susceptible to drought than to waterlogging. However, the gene expression analyses of key antioxidant and xenobiotic-detoxifying enzymes will expand our observations.

4. Materials and Methods

4.1. Model System

Triticale (×*Triticosecale* Wittm.) cv. Rozhen is a certified Bulgarian triticale cultivar originating from the south region of Bulgaria. The seeds were obtained from the Institute of Plant Genetic Resources "K. Malkov" (Sadovo, Bulgaria). Seedlings were grown in a growth chamber under controlled conditions: 60% relative humidity, at 22/19 °C and with a 16/8 h photoperiod (day/night). Seventeen-day-old plantlets grown as a soil culture were leaf sprayed with the herbicide Serrate® according to the manufacturer's instructions. After 72 h, part of the triticale seedlings were exposed to drought or waterlogging as described earlier [13–15,34]. Drought was achieved by withholding watering for 7 days, while waterlogging was realized by relocating the pots into an outer container filled with water at a level 2 cm higher than the soil. The stress program lasted for 7 days, then the plants were transferred to a normal irrigation regime (recovery period). The leaf plant material was randomized within each treatment group, collected after 4 days and 7 days of water stress and after 4 days of recovery. The sample material was weighed, fixed in liquid nitrogen and kept at −70 °C until analysis.

4.2. Biochemical Extraction and Analyses

Leaf material (around 250 mg) was ground in 0.1% cold trichloroacetic acid (TCA) and then centrifuged for 30 min at 15,000× *g*, 4 °C. The resultant supernatant was used for the analyses of stress markers (hydrogen peroxide, malondialdehyde (MDA) and free proline) and non-enzymatic antioxidants (phenolic compounds and free thiol-groups-containing compounds). The concentration of hydrogen peroxide was assessed after the dark-incubation (1 h) of 75 µL supernatant with 75 µL 1 M KI [63]. The absorbance was measured at 390 nm, and the concentration of H_2O_2 was quantified by a standard curve. The level of lipid peroxidation was estimated by the concentration of MDA in plant tissues [64]. A half milliliter of supernatant was incubated for 45 min at 100° C with 1 mL 0.5% thiobarbituric acid in 20% TCA. The absorbance of the resulting thiobarbituric reaction products was read at 532 nm and 600 nm. The content of MDA was calculated by using the extinction coefficient (155 mM cm^{-1}). Free proline was determined according to Bates et al. [65] with some adaptations. Reaction mixture consisted of 0.5 mL supernatant, 0.5 mL 0.1% TCA, 1 mL concentrated CH_3COOH, and 1 mL ninhydrin reagent (1.25 g ninhydrin, 30 mL concentrated CH_3COOH, 20 mL 6M H_3PO_4) was incubated on a water bath for 1 h at 100 °C, then the reaction was terminated in ice bath and the absorbance was read at 520 nm.

The total content of the phenolic compounds was assessed using the procedure reported by Swain and Goldstein [66] with a few modifications. Twenty µL supernatant was incubated for 3 min at room temperature with 130 µL distilled water and 50 µL Folin–Ciocalteu reagent. Next, the reaction was supplemented with 50 µL 1M Na_2CO_3 and developed in light for 2 h at room temperature. The absorbance was measured at 725 nm, and the results were evaluated by a standard curve prepared with gallic acid (GA). The content of free thiol-groups-containing compounds was assessed by using the Elman's reagent [67]. The reaction mixture comprised 40 µL supernatant and 150 µL Elman's reagent. The absorbance was measured at 412 nm after 10 min incubation at room temperature.

Roughly 200 mg leaf material was homogenized in cold 100 mM potassium phosphate buffer (pH 7.0 with 1 mM EDTA) plus 1% PVP for the assessment of the activities of antioxidant enzymes and GST. The homogenate was centrifuged at 15,000× *g* for 30 min at 4 °C. Catalase (EC 1.11.1.6) activity was evaluated by the monitoring of H_2O_2 degradation [68]. The reaction mixture consisted of 50 µL supernatant, 2.930 mL reaction 0.05 M potassium phosphate buffer (pH 7.0) and 20 µL 6% hydrogen peroxide. The enzymatic activity was assessed by following the degradation of H_2O_2 for 1 min at 240 nm. Guaiacol peroxidase (EC 1.11.1.7) activity was assessed by using guaiacol as an electron donor. The reaction mixture comprised 20 µL supernatant, 1.1 mL reaction buffer (0.05 M potassium phosphate buffer, pH 7.0), 360 µL 1% guaiacol and 20 µL 15% H_2O_2. The absorbance change was monitored at 470 nm [69]. To measure the activity of superoxide dismutase (EC 1.15.1.1) the inhibition of the photochemical reduction of nitroblue tetrazolium was used. The amount of the enzyme required to inflict a 50% inhibition was defined as one unit of SOD [70]. The activity of glutathione reductase was assessed by following the method described by Smith et al. [71]. The reaction mixture contained 100 μL supernatant, 1.180 mL reaction buffer (0.05M potassium phosphate buffer, pH 7.5, 1 mM EDTA), 20 μL 50 mM DTNB, 0.1 mL 7.5 mM oxidized glutathione and 0.1 mL 1.5 mM NADPH. The reaction was assessed at 412 nm for 60 s. The GST activity with 1-chloro-2,4-dinitrobenzene (CDNB, extinction coefficient 9.6 mM cm^{-1} at 340 nm) as a substrate was determined according to Gronwald et al. [72].

Cold 100 mM potassium phosphate buffer (pH 7.5), containing 1 mM EDTA, 0.2 mM PMSF, 1% PVP and 0.3 M sucrose, was used as an extraction solution during the homogenization of the plant material for the assessment of the enzymatic activities of B5R and CPRs. The homogenate was centrifuged at 15,000× *g* for 30 min at 4 °C. Potassium ferricyanide was used as an artificial electron acceptor for the in vitro B5R (EC 1.6.2.2) activity assay [73]. The rate of reduction was evaluated by an extinction coefficient of 1.02 mM cm^{-1} at 420 nm. Cytochrome *c* was used as an artificial electron acceptor for the in vitro CPR (EC 1.6.2.4) activity assay [73]. The rate of reduction was evaluated by an extinction coefficient of 21.1 mM cm^{-1} at 550 nm. The protein content was determined according to Bradford [74]. The herbicide Serrate® was purchased from a local distributor of Syngenta (Basel, Switzerland). All chemical compounds used for the biochemical analyses were obtained from Sigma-Aldrich, (Saint Louis, MO, USA). The measurements of the stress markers, SOD activity and non-enzymatic antioxidants were performed on a Multiskan Spectrum spectrophotometer with a microplate reader (Thermo Electron Corporation, Vantaa, Finland). The enzymatic activities were measured on a Shimadzu UV-1601 spectrophotometer (Shimadzu, Kyoto, Japan). A refrigerated Sigma 2-16K centrifuge (SciQuip, Wem, UK) was used to obtain the respective supernatants.

4.3. Statistical Analysis

The results presented are obtained from three independent biological experiments with three internal replicates. The statistical significance between the treatments were evaluated using one-way ANOVA with a post-hoc Duncan's multiple range test ($p \leq 0.05$). The data in the figures represent the average values ± standard error (SE).

Author Contributions: I.S. and D.T., conceptualized and coordinated the research; L.D., Z.K., M.P. and D.T., grew and treated the plants; I.S., D.T., E.S., L.B., M.P. and Z.K., performed the laboratory analyses, collected and interpreted the data; I.S., prepared figures and photos; D.T. and Z.K., prepared original draft of manuscript; I.S. and D.T., reviewed and edited the manuscript. All authors have read and agreed to the published version of the manuscript.

Funding: This research was funded by the grant KP-06-N36/3 (30.09.2020) of National Science Fund, Ministry of Education and Science, Republic of Bulgaria.

Data Availability Statement: Not applicable.

Conflicts of Interest: The authors declare no conflict of interest.

References

1. Mergoum, M.; Singh, P.K.; Peña, R.J.; Lozano-del Río, A.J.; Cooper, K.V.; Salmon, D.F.; Gómez Macpherson, H. Triticale: A "New" Crop with Old Challenges. In *Cereals*; Carena, M.J., Ed.; Springer: New York, NY, USA, 2009; pp. 267–287. [CrossRef]
2. Blum, A. The Abiotic Stress Response and Adaptation of Triticale—A Review. *Cereal Res. Commun.* **2014**, *42*, 359–375. [CrossRef]
3. Arsenuik, E. Triticale Abiotic Stresses—An Overview. In *Triticale*; Eudes, F., Ed.; Springer International Publishing: Cham, Switzerland, 2015; pp. 69–81. [CrossRef]
4. Din, A.M.U.; Mao, H.T.; Khan, A.; Raza, M.A.; Ahmed, M.; Yuan, M.; Zhang, Z.W.; Yuan, S.; Zhang, H.Y.; Liu, Z.H.; et al. Photosystems and antioxidative system of rye, wheat and triticale under Pb stress. *Ecotoxicol. Environ. Safety* **2023**, *249*, 114356.
5. Kaur, G.; Singh, G.; Motavalli, P.P.; Nelson, K.A.; Orlowski, J.M.; Golden, B.R. Impacts and management strategies for crop production in Waterlogged/Flooded soils: A review. *Agron. J.* **2020**, *112*, 1475–1501. [CrossRef]

6. Hasanuzzaman, M.; Bhuyan, M.B.; Parvin, K.; Bhuiyan, T.F.; Anee, T.I.; Nahar, K.; Hossen, M.S.; Zulfiqar, F.; Alam, M.M.; Fujita, M. Regulation of ROS metabolism in plants under environmental stress: A review of recent experimental evidence. *Int. J. Mol. Sci.* **2020**, *21*, 8695. [CrossRef]
7. Zhou, W.; Chen, F.; Meng, Y.; Chandrasekaran, U.; Luo, X.; Yang, W.; Shu, K. Plant waterlogging/flooding stress responses: From seed germination to maturation. *Plant Physiol. Biochem.* **2020**, *148*, 228–236. [CrossRef] [PubMed]
8. Pan, J.; Sharif, R.; Xu, X.; Chen, X. Mechanisms of Waterlogging Tolerance in Plants: Research Progress and Prospects. *Front. Plant Sci.* **2021**, *11*, 627331. [CrossRef]
9. Suzuki, N.; Rivero, R.M.; Shulaev, V.; Blumwald, E.; Mittler, R. Abiotic and biotic stress combinations. *New Phytol.* **2014**, *203*, 32–43. [CrossRef] [PubMed]
10. Zandalinas, S.I.; Fritschi, F.B.; Mittler, R. Signal transduction networks during stress combination. *J. Exp. Bot.* **2020**, *71*, 1734–1741. [CrossRef] [PubMed]
11. Cohen, I.; Rapaport, T.; Chalifa-Caspi, V.; Rachmilevitch, S. Synergistic effects of abiotic stresses in plants: A case study of nitrogen limitation and saturating light intensity in *Arabidopsis thaliana*. *Physiol. Plant.* **2019**, *165*, 755–767. [CrossRef]
12. Radchenko, M.; Ponomareva, I.; Pozynych, I.; Morderer, Y. Stress and use of herbicides in field crops. *Agric. Sci. Pract.* **2021**, *8*, 50–70. [CrossRef]
13. Todorova, D.; Sergiev, I.; Katerova, Z.; Shopova, E.; Dimitrova, L.; Brankova, L. Assessment of the Biochemical Responses of Wheat Seedlings to Soil Drought after Application of Selective Herbicide. *Plants* **2021**, *10*, 733. [CrossRef]
14. Katerova, Z.; Sergiev, I.; Todorova, D.; Shopova, E.; Dimitrova, L.; Brankova, L. Physiological responses of wheat seedlings to soil waterlogging applied after treatment with selective herbicide. *Plants* **2021**, *10*, 1195. [CrossRef]
15. Brankova, L.; Dimitrova, L.; Shopova, E.; Katerova, Z.; Sergiev, I.; Todorova, D. Microsomal P450-related electron transfer components, glutathione and glutathione S-transferase contribution in stress response of herbicide-treated wheat to drought and waterlogging. *C. R. Acad. Bulg. Sci.* **2022**, *75*, 1089–1096. [CrossRef]
16. Stefanovic, L.; Zaric, L. Effect of herbicides and low temperatures on certain maize genotypes. *Plant Prot.* **1991**, *42*, 345–356.
17. Lastochkina, O.; Yakupova, A.; Avtushenko, I.; Lastochkin, A.; Yuldashev, R. Effect of Seed Priming with Endophytic *Bacillus subtilis* on Some Physio-Biochemical Parameters of Two Wheat Varieties Exposed to Drought after Selective Herbicide Application. *Plants* **2023**, *12*, 1724. [CrossRef] [PubMed]
18. Rivero, R.M.; Mittler, R.; Blumwald, E.; Zandalinas, S.I. Developing climate-resilient crops: Improving plant tolerance to stress combination. *Plant J.* **2022**, *109*, 373–389. [CrossRef]
19. Grzesiak, M.T.; Hura, K.; Jurczyk, B.; Hura, T.; Rut, G.; Szczyrek, P.; Grzesiak, S. Physiological markers of stress susceptibility in maize and triticale under different soil compactions and/or soil water contents. *J. Plant Interact.* **2017**, *12*, 355–372. [CrossRef]
20. Kerchev, P.I.; Van Breusegem, F. Improving oxidative stress resilience in plants. *Plant J.* **2022**, *109*, 359–372. [CrossRef]
21. Sharma, P.; Jha, A.B.; Dubey, R.S.; Pessarakli, M. Reactive oxygen species, oxidative damage, and antioxidative defense mechanism in plants under stressful conditions. *J. Bot.* **2012**, *2012*, 217037. [CrossRef]
22. Riasat, M.; Kiani, S.; Saed-Mouchehsi, A.; Pessarakli, M. Oxidant related biochemical traits are significant indices in triticale grain yield under drought stress condition. *J. Plant Nutr.* **2019**, *42*, 111–126. [CrossRef]
23. Brankova, L.; Ivanov, S.; Alexieva, V. The induction of microsomal NADPH:cytochrome P450 and NADH: Cytochrome b5 reductases by long-term salt treatment of cotton (*Gossypium hirsutum* L.) and bean (*Phaseolus vulgaris* L.) plants. *Plant Physiol. Biochem.* **2007**, *45*, 691–695. [CrossRef] [PubMed]
24. Gao, X.; Zhang, F.; Hu, J.; Cai, W.; Shan, G.; Dai, D.; Huang, K.; Wang, G. MicroRNAs modulate adaption to multiple abiotic stresses in *Chlamydomonas reinhardtii*. *Sci. Rep.* **2016**, *6*, 38228. [CrossRef] [PubMed]
25. Pandian, B.A.; Sathishraj, R.; Djanaguiraman, M.; Prasad, P.V.V.; Jugulam, M. Role of cytochrome P450 enzymes in plant stress response. *Antioxidants* **2020**, *9*, 454. [CrossRef] [PubMed]
26. Siminszky, B. Plant cytochrome P450-mediated herbicide metabolism. *Phytochem. Rev.* **2006**, *5*, 445–458. [CrossRef]
27. Kumar, R.; Tran, L.-S.P.; Neelandan, A.K.; Nguen, H.T. Higher plant cytochrome b5 polypeptydes modulate fatty acids desaturation. *PLoS ONE* **2012**, *7*, e31370.
28. Oh, Y.J.; Kim, H.; Seo, S.H.; Hwang, B.G.; Chang, Y.S.; Lee, J.; Lee, D.W.; Sohn, E.J.; Lee, S.J.; Lee, Y.; et al. Cytochrome b5 reductase 1 triggers serial reactions that lead to iron uptake in plants. *Mol. Plant.* **2016**, *9*, 501–513. [CrossRef] [PubMed]
29. Chhokar, R.S.; Sharma, R.K.; Sharma, I. Weed management strategies in wheat—A review. *J. Wheat Res.* **2012**, *4*, 1–21.
30. Mashhady, A.S.; Sayed, H.I.; Heakal, M.S. Effect of soil salinity and water stresses on growth and content of nittrogent, chloride and phosphate of wheat and triticale. *Plant Soil* **1982**, *68*, 207–216. [CrossRef]
31. Bijanzadeh, E.; Emam, Y.; Pessarakli, M. Biochemical responses of water-stressed triticale (×*Triticosecale wittmack*) to humic acid and jasmonic acid. *J. Plant Nutr.* **2020**, *44*, 252–269. [CrossRef]
32. Alagoz, S.M.; Hadi, H.; Toorchi, M.; Pawłowski, T.A.; Shishavan, M.T. Effects of Water Deficiency at Different Phenological Stages on Oxidative Defense, Ionic Content, and Yield of Triticale (×*Triticosecale wittmack*) Irrigated with Saline Water. *J. Soil Sci. Plant Nutr.* **2022**, *22*, 99–111. [CrossRef]
33. Abu-Altemen, M. Effect of Excess Nitrogen Fertilization on Triticale Production under Rainfed and Supplemental Irrigation Conditions. Ph.D. Thesis, American Uiversity of Beirut, Beirut, Lebanon, April 2022.
34. Todorova, D.; Aleksandrov, V.; Anev, S.; Sergiev, I. Photosynthesis Alterations in Wheat Plants Induced by Herbicide, Soil Drought or Flooding. *Agronomy* **2022**, *12*, 390. [CrossRef]

35. Toulotte, J.M.; Pantazopoulou, C.K.; Sanclemente, M.A.; Voesenek, L.A.C.J.; Sasidharan, R. Water stress resilient cereal crops: Lessons from wild relatives. *J. Integr. Plant Biol.* **2022**, *64*, 412–430. [CrossRef] [PubMed]
36. Verslues, P.E.; Bailey-Serres, J.; Brodersen, C.; Buckley, T.N.; Conti, L.; Christmann, A.; Dinneny, J.R.; Grill, E.; Hayes, S.; Heckman, R.W.; et al. Burning questions for a warming and changing world: 15 unknowns in plant abiotic stress. *Plant Cell* **2023**, *35*, 67–108. [CrossRef]
37. Gill, S.S.; Tuteja, N. Reactive oxygen species and antioxidant machinery in abiotic stress tolerance in crop plants. *Plant Physiol. Biochem.* **2010**, *48*, 909–930. [CrossRef] [PubMed]
38. Lukatkin, A.S.; Gar'kova, A.N.; Bochkarjova, A.S.; Nushtaeva, O.V.; da Silva, J.A.T. Treatment with the herbicide TOPIK induces oxidative stress in cereal leaves. *Pesticide Biochem. Physiol.* **2013**, *105*, 44–49. [CrossRef]
39. Rizhsky, L.; Liang, H.; Shuman, J.; Shulaev, V.; Davletova, S.; Mittler, R. When defense pathways collide. The response of *Arabidopsis* to a combination of drought and heat stress. *Plant Physiol.* **2004**, *134*, 1683–1696. [CrossRef]
40. Kavi Kishor, P.B.; Sreenivasulu, N. Is proline accumulation per se correlated with stress tolerance or is proline homeostasis a more critical issue? *Plant Cell Environ.* **2014**, *37*, 300–311. [CrossRef]
41. Marchiosi, R.; dos Santos, W.D.; Constantin, R.P.; de Lima, R.B.; Soares, A.R.; Finger-Teixeira, A.; Mota, T.R.; de Oliveira, D.M.; Foletto-Felipe, M.D.P.; Abrahão, J.; et al. Biosynthesis and metabolic actions of simple phenolic acids in plants. *Phytochem. Rev.* **2020**, *19*, 865–906. [CrossRef]
42. Kumar, M.; Tak, Y.; Potkule, J.; Choyal, P.; Tomar, M.; Meena, N.L.; Kaur, C. Phenolics as Plant Protective Companion Against Abiotic Stress. In *Plant Phenolics in Sustainable Agriculture*; Lone, R., Shuab, R., Kamili, A., Eds.; Springer: Singapore, 2020; pp. 277–308. [CrossRef]
43. Puente-Garza, C.A.; Meza-Miranda, C.; Ochoa-Martínez, D.; García-Lara, S. Effect of in vitro drought stress on phenolic acids, flavonols, saponins, and antioxidant activity in *Agave salmiana*. *Plant Physiol. Biochem.* **2017**, *115*, 400–407. [CrossRef] [PubMed]
44. Król, A.; Amarowicz, R.; Weidner, S. Changes in the composition of phenolic compounds and antioxidant properties of grapevine roots and leaves (*Vitis vinifera* L.) under continuous of long-term drought stress. *Acta Physiol. Plant.* **2014**, *36*, 1491–1499. [CrossRef]
45. Sarker, U.; Oba, S. Response of nutrients, minerals, antioxidant leaf pigments, vitamins, polyphenol, flavonoid and antioxidant activity in selected vegetable amaranth under four soil water content. *Food Chem.* **2018**, *252*, 72–83. [CrossRef]
46. Hura, T.; Grzesiak, S.; Hura, K.; Thiemt, E.; Tokarz, K.; Wędzony, M. Physiological and biochemical tools useful in drought-tolerance detection in genotypes of winter triticale: Accumulation of ferulic acid correlates with drought tolerance. *Ann. Bot.* **2007**, *100*, 767–775. [CrossRef]
47. Zagorchev, L.; Seal, C.E.; Kranner, I.; Odjakova, M.A. Central role for thiols in plant tolerance to abiotic stress. *Int. J. Mol. Sci.* **2013**, *14*, 7405–7432. [CrossRef] [PubMed]
48. Szalai, G.; Kell, T.; Kocsy, G. Glutathione as an antioxidant and regulatory molecule in plants under abiotic stress conditiond. *Plant Growth Regul.* **2009**, *28*, 66–80. [CrossRef]
49. Khunpon, B.; Cha-Um, S.; Faiyue, B.; Uthaibutra, J.; Saengnil, K. Regulation on antioxidant defense system in rice seedlings (*Oryza sativa* L. ssp. indica cv. 'Pathumthani 1') under salt stress by paclobutrazol foliar application. *Not. Bot. Hortic. Agrobot. Clju Napoca* **2019**, *47*, 368–377. [CrossRef]
50. Yin, L.; Mano, J.I.; Tanaka, K.; Wang, S.; Zhang, M.; Deng, X.; Zhang, S. High level of reduced glutathione contributes to detoxification of lipid peroxide-derived reactive carbonyl species in transgenic *Arabidopsis* overexpressing glutathione reductase under aluminum stress. *Physiol. Plant.* **2017**, *161*, 211–223. [CrossRef]
51. Zhao, X.; Zhao, Y.; Gou, M.; Liu, C.J. Tissue-preferential recruitment of electron transfer chains for cytochrome P450-catalyzed phenolic biosynthesis. *Sci. Adv.* **2023**, *9*, eade4389. [CrossRef]
52. Ro, D.K.; Ehlting, J.; Douglas, C.J. Cloning, functional expression, and subcellular localization of multiple NADPH-cytochrome P450 reductase from hybrid poplar. *Plant Physiol.* **2002**, *130*, 1837–1851. [CrossRef]
53. Batard, Y.; Hehn, A.; Nedelkina, S.; Schalk, M.; Pallett, K.; Schaller, H.; Werck-Reichhart, D. Increasing Expression of P450 and P450-Reductase Proteins from Monocots in Heterologous Systems. *Arch. Biochem. Biophys.* **2000**, *379*, 161–169. [CrossRef]
54. Fukuchi-Mizutani, M.; Mizutani, M.; Tanaka, Y.; Kusumi, T.; Ohta, D. Microsomal electron transfer in higher plants: Cloning and heterologous expression of NADH-cytochrome b5 reductase from Arabidopsis. *Plant Physiol.* **1999**, *119*, 353–362. [CrossRef]
55. Pollak, N.; Dölle, C.; Ziegler, M. The power to reduce: Pyridine nucleotides—Small molecules with a multitude of functions. *Biochem. J.* **2007**, *402*, 205–218. [CrossRef]
56. Parage, C.; Foureau, E.; Kellner, F.; Burlat, V.; Mahroug, S.; Lanoue, A.; de Bernonville, T.D.; Londono, M.A.; Carqueijeiro, I.; Oudin, A.; et al. Class II cytochrome P450 reductase governs the biosynthesis of alkaloids. *Plant Physiol.* **2016**, *172*, 1563–1577. [CrossRef] [PubMed]
57. Repellin, A.; Thi, A.P.; Tashakorie, A.; Sahsah, Y.; Daniel, C.; Zuily-Fodil, Y. Leaf membrane lipids and drought tolerance in young coconut palms (*Cocos nucifera* L.). *Europ. J. Agron.* **1997**, *6*, 25–33. [CrossRef]
58. Gigon, A.; Matos, A.R.; Laffray, D.; Zuily-Fodil, Y.; Pham-Thi, A.T. Effect of drought stress on lipid metabolism in the leaves of *Arabidopsis thaliana* (ecotype Columbia). *Ann Bot.* **2004**, *94*, 345–351. [CrossRef] [PubMed]
59. Wang, Y.; Zhang, X.; Huang, G.; Feng, F.; Liu, X.; Guo, R.; Gu, F.; Zhong, X.; Mei, X. Dynamic changes in membrane lipid composition of leaves of winter wheat seedlings in response to PEG-induced water stress. *BMC Plant Biol.* **2020**, *20*, 84. [CrossRef] [PubMed]
60. Porter, T.D. The roles of cytochrome b5 in cytochrome P450 reactions. *J. Biochem. Mol. Toxicol.* **2002**, *16*, 311–316. [CrossRef]

J. Cytochrome b5: A versatile electron carrier and regulator for plant metabolism. *Front. Plant Sci.* **2022**, *13*, 984174. [CrossRef]

Wayne, L.L.; Wallis, J.G.; Kumar, R.; Markham, J.E.; Browse, J. Cytochrome b5 reductase encoded by CBR1 is essential for a functional male gametophyte in Arabidopsis. *Plant Cell* **2013**, *25*, 3052–3066. [CrossRef]

63. Alexieva, V.; Sergiev, I.; Mapelli, S.; Karanov, E. The effect of drought and ultraviolet radiation on growth and stress markers in pea and wheat. *Plant Cell Environ.* **2001**, *24*, 1337–1344. [CrossRef]
64. Kramer, G.; Norman, H.; Krizek, D.; Mirecki, R. Influence of UV-B radiation on polyamines, lipid peroxidation and membrane lipids in cucumber. *Phytochemistry* **1991**, *30*, 2101–2108. [CrossRef]
65. Bates, L.; Waldren, R.; Teare, I. Rapid determination of free proline for water-stress studies. *Plant Soil* **1973**, *39*, 205–207. [CrossRef]
66. Swain, T.; Goldstein, L. *Methods in Polyphenol Chemistry*; Pridham, J.B., Ed.; Pergamon Press: Oxford, UK, 1964; pp. 131–146.
67. Ellman, G.L. Tissue sulfhydryl groups. *Arch. Biochem. Biophys.* **1959**, *82*, 70–75. [CrossRef]
68. Aebi, M. Catalase in vitro. *Methods Enzymol.* **1984**, *105*, 121–126. [PubMed]
69. Dias, I.; Costa, M. Effect of low salt concentration on nitrate reductase and peroxidase of sugar beet leaves. *J. Exp. Bot.* **1983**, *34*, 537–543. [CrossRef]
70. Beauchamp, C.; Fridovich, I. Superoxide dismutase. Improved assay and an assay applicable to acrylamide gels. *Anal. Biochem.* **1971**, *44*, 276–287. [CrossRef] [PubMed]
71. Smith, I.K.; Vierheller, T.L.; Thorne, C.A. Assay of glutathione reductase in crude tissue homogenates using 5,50-dithiobis(2-nitrobenzoic acid). *Anal. Biochem.* **1988**, *175*, 408–413. [CrossRef]
72. Gronwald, J.; Fuerst, P.; Eberlein, C.; Egli, M. Effect of herbicide antidotes on glutathione content, and glutathione S-transferase activity of sorgum shoots. *Pestic. Biochem. Phys.* **1987**, *29*, 66–76. [CrossRef]
73. Tanigaki, F.; Ishihara, A.; Yoshida, K.; Hara, T.; Shinozaki, M.; Iwamura, H. Interaction of microsomal cytochrome P-450s and N-phenylcarbamates that induce flowering in Asparagus seedlings. *Z. Naturforsch* **1993**, *48*, 879–885. [CrossRef]
74. Bradford, M. A rapid and sensitive method for the quantification of microgram quantities of protein utilizing the principle of protein-dye binding. *Analyt. Biochem.* **1976**, *72*, 248–254. [CrossRef]

MDPI AG
Grosspeteranlage 5
4052 Basel
Switzerland
Tel.: +41 61 683 77 34

Plants Editorial Office
E-mail: plants@mdpi.com
www.mdpi.com/journal/plants

www.ingramcontent.com/pod-product-compliance
Lightning Source LLC
Chambersburg PA
CBHW041358210926
51850CB00013B/3119
9783725831685